AF684377

To Don:

With bottomless thanks for all you do to keep this work going!

Lee

Ubiquity, Mobility, Security

Ubiquity, Mobility, Security

The Future of the Internet III

Janna Quitney Anderson
and
Lee Rainie

Cambria Press
Amherst, New York

Printed in the United States of America

Requests for permission should be directed to:
permissions@cambriapress.com, or mailed to:
Cambria Press
20 Northpointe Parkway, Suite 188
Amherst, NY 14228

Library of Congress Cataloging-in-Publication Data

Rainie, Harrison.
Ubiquity, mobility, security / Janna Quitney Anderson and Lee Rainie.
p. cm. — (The future of the Internet ; v. 3)
Includes bibliographical references and index.
ISBN 978-1-60497-615-1 (alk. paper)
1. Internet—Social aspects. 2. Internet—Forecasting. 3. Information technology—Social aspects. I. Anderson, Janna Quitney, 1955- II. Title. III. Series.

HM851.R357 2009
303.48'33—dc22

2009008448

For Jonathan B. Postel

A member of the creative computer science community that innovated the foundations of the Internet in the 1960s through the '90s, Jon Postel was a modest man whose devotion to quality helped set high goals for Internet standards and practices. He was the first member of the Internet Society, and he served as the Request for Comments (RFCs) editor for nearly 30 years, demanding excellence as drafts were moved through the open-consensus processes of the Internet Engineering Task Force (IETF) and other contributors. He wrote or coauthored more than 200 RFCs. His technical contributions were key to the foundation of simple mail transfer protocol (SMTP), the e-mail protocol, and the Domain Name System (DNS), the basis for all search and identity online. He served as the administrator for the Internet Assigned Numbers Authority until his death in 1998. He is seen as the archetype of an Internet pioneer, and his personal integrity still inspires many. One of his RFCs, 793, is known as Postel's law, or the "robustness principle"; it includes the line "be conservative in what you do; be liberal in what you accept from others."

A survey of experts shows that they expect major tech advances by 2020, as the phone becomes a primary device for online access, voice recognition improves, and the structure of the Internet itself improves. They disagree about whether this will lead to more social tolerance, more forgiving human relations, or better home lives.

Table of Contents

Foreword

I am writing this on the day of Barack Obama's inauguration, and there is a palpable sense in the technology community that a new kind of patron sits in the White House. The new president is interested in the use of the Internet and cell phones in a way that his predecessors were not. Indeed, one of the interesting minor themes of the previous year's campaign was the difference between Obama and his Republican opponent John McCain on their personal relationship to technology, as well as their policy prescriptions. The future of technology seems mapped differently at the dawn of a period when a technophile sits at the head of the table.

This makes life even more interesting for those of us who are trying to think about the future of technology. We are already having a challenging time just keeping up with the present. Between the time our last survey about the future was fielded in early 2006 and the time this survey was fielded in early 2008, the world's Internet-using population surged well past a billion and the mobile phone population moved toward 3 billion, the proportion of Americans with broadband connections at home surged past 50%, social networking services such as MySpace and Facebook garnered tens of millions of users, most Internet users had some connection to cloud computing, and the Internet environment itself became a video, as well as a text-based, experience. Just 2 years before this survey was administered, iPhones were still being designed at Apple, iPods did not store and display videos, the iTunes service was just getting off of the ground as a music service, and netbooks (mini-laptops) were not yet conceived as market makers. None of these things figured much in the imagination of respondents to our second survey. In this current effort, those "far-out" technologies of 2006 became touchstones.

It also was surprising to me to see how some of the things we had probed in our first survey in 2004 had already become realities by 2008. The 2004 survey was meant to explore the world in 2014 and beyond, not the world 4 years hence. As I look back at some of the questions we asked in 2004, I see that the initial survey was overtaken by events a lot sooner than I imagined. I write these words now with an inescapable sense that the future is speeding up.

Yet even as change occurs, it is striking how certain metathemes recur in experts' responses to our questions no matter how we frame the queries and what topics we ask the experts to address. These metathemes include the following:

- **Improvements in technology will continue unabated:** There is no sense among our expert respondents that upgrades in computing capacity, digital-storage capacity, and communications speeds (bandwidth capacity) have hit any kind of natural limit. There is a powerful view that devices will get smaller, cheaper, and more powerful. As über blogger and digital strategist Jeff Jarvis sees it, the majority of people in the most developed areas of the world in coming years will avail themselves to many connected devices that plug into the communications grid: "This will lead to a world that is not only connected but also live and immediate. Witnesses will share news as they witness it. We can get answers to any question, anytime. We can stay in constant touch with the people we know, following their lives as we [now] follow RSS and Twitter feeds."

We asked about two new kinds of technology improvements in this survey and got the same positive response to them. The first was that mobile phones (or the device they "grow up" to be) would be the primary means of connection for most people by 2020. The only challenge we received on that assertion was from those who argued—probably rightly—that the current upgrades in cell phones have already produced a world where mobile connectivity is becoming common.

The second new thrust we queried related to user interfaces. Some believe that 2020 user interfaces will offer advanced talk, touch, and typing options, and some respondents added that they think advances in understanding brain functions will facilitate a thought-based technology interface.

- **Challenges to privacy will multiply:** In the world of ubiquitous computing and always-available connectivity, people's digital footprints will become bigger, and that will make their identities more difficult to monitor and manage. Indeed, a sense of fatalism pervades many experts' answers. They feel privacy will be sacrificed and probably replaced by different expectations and norms. Technology analyst and consultant Jerry Michalski foresees the day when "the old 'never trust anyone over 30' will turn into 'never trust anyone who doesn't have embarrassing stuff online.'"

  One reasonably plausible scenario was penned by communications scholar danah boyd: "We are going to see a lot of chaos around privacy in the next 13 years, yet I don't think that we will have equilibrium by then. Realistically, the only comfort we will reach will be over embarrassing material. I think that we'll be far less embarrassed by our pasts once everyone's are out there in some form or another. My prediction is that we will find ways of using content to talk at different levels, just as writers have in the past and just as Chinese activists do now. Much of the 'private' content will be produced in a way that is publicly palatable and can be read at multiple levels by those who are closer to the individual."

- **Intellectual property wars will persist:** It is not easy to characterize the battle between copyright and trademark holders and those who want to gain access to that material. Some experts see it as close to a hot war; others would describe it as a cold war. Most see it as a long war that will produce, at least, some recalibration of the relationship between creators and the people who used to be their audience. Perhaps, some speculate, the economics of intellectual property will change and attention will become the resource over which firms battle.

  Those who fight over the creators' ownership of their work product are fighting a 20th-century battle, argues Fred Baker, the former chairman of the Internet Society. He believes a "new capitalism" will arise and be built around new kinds of added value, new licensing agreements, advertising, and metrics of attention.

- **Technology will not overcome human foibles:** Hopeful notes are consistently sounded by some of our respondents. They are convinced that technology affords access to new and different people and that is the way some people will use the Internet and cell phones. Contrarians argue that human nature is more or less immutable, and this means people will use new tools to meet primal needs. Some of those needs are noble; some are base. This was well captured by William Winton, project manager for digital media at the 1105 Government Information Group. He wrote, "The Internet is a two-edged sword. Its openness and ease of communication have also fostered the rise in online Jihadists, resurrected a flagging neo-Nazi movement, and enabled all sorts of intolerant movements, ideas, and people to flourish online. The jury will probably still be sequestered in 2020 as to whether the Internet has fostered 'tolerance' or merely 'siloed' hate." This summarizes a view we have consistently heard through all three of our surveys.

- **Enabled by technology, the demands of work and work colleagues will continue to invade leisure sanctums:** One scenario that many respondents thought was too much a reality in the present to qualify as a prediction about life in 2020 involved the relationship of work and leisure: *In 2020 well-connected knowledge workers in more developed nations have willingly eliminated the industrial-age boundaries between work hours and personal time. Outside of formally scheduled activities, work and play are seamlessly integrated in most of these workers' lives.* This drew a 56–29% verdict (and 15% did not respond) and more than a small number of polite complaints that it was hardly a prediction. "This is already happening in my life and the lives of those around me," the protesters protested. They also pushed back against the prediction's assertion that the net impact would be beneficial and satisfying. "There are too many ways the electronic leash can be exploited by those who control workplace realities," respondents proclaimed. Power arrangements might be rebalanced, but they will not vanish entirely.

  One of my favorite of the sober responses came from Roberto Gaetano, ICANN board member: "The 9-to-5 approach will disappear completely, with few exceptions. The current separation between 'work time' and 'free time' is a by-product of the industrial revolution and is bound to disappear with it. Whether this is positive or negative, I don't know, because the pressure of being 'always at work' just because you have the ability to be 'always connected' will be high. But we would need to build a new way of life that has to cope with this."

Beyond the recurring metathemes, this survey produced new material for us to assess about the future. The most surprising to me was the respondents' overwhelming view that improvements to the Internet would be built into the existing network structure rather than a wholly new reengineered architecture. The respondents cite two major changes that are already in process: The first is the transition to IPv6, a new protocol that has a much larger address space than the current IPv4 protocol. This expansion provides flexibility in allocating addresses and routing traffic. The IPv6 protocol also makes Internet-address assignments simpler.

The second is implementation of elements of the Semantic Web, which will make it easier to find and link related information. This challenges the view of some of the most prominent Internet creators who believe that a "do-over" of the Internet is necessary because the current architecture is too vulnerable to exploitation by bad actors and too inefficient. At the time we posed this question, I guessed that the reengineer solution would have more support than it turned out to have. Whatever the outcome in 2020, though, there is no mistaking the tone of the written comments of survey participants: The future is speeding up.

This emotion is a by-product of, at least, two forces that are at work in these experts' lives. One is the fact that most of them are busy inventing the future or, at least, adjusting to the forces that seem inevitable. They cannot help but be focused on life over the horizon because they know they will reach the horizon soon enough. The other is that the business of futurism or futurology seems to be thriving these days. Each month yields invitations to a substantial number of conferences and meetings titled "The Future of [fill in the blank]." All of them have technological change at the center of their deliberations. Clearly, the number of people whose livelihoods are centered on predicting the future also seems to be growing. We are surrounded by a flow of information that is cast in the future tense.

That alone is reason enough to inspire us to stay at the work of monitoring the futurism-industrial-complex and soliciting the views of those who are hoping to see over the horizon. We also know that the wisdom of crowds can frequently be correct in futures markets. That is why we hope this book inspires your own contribution to our work. We invite you to add your own insights to the collective intelligence amassed at http://www.imaginingtheinternet.org. Respond to the scenarios that anchored our predictions in this survey, or tell us what scenario ought to inspire our next survey. The future looks a bit different at the dawn of the Obama presidency—in its particulars and in its velocity.

Lee Rainie, Director
Pew Internet & American Life Project

Acknowledgments

Pew Internet & American Life Project

The Pew Internet Project is a nonprofit, nonpartisan "fact tank" that explores the impact of the Internet on children, families, communities, the workplace, schools, health care, and civic and political life. The Project takes no positions on policy questions. Support for the Project is provided by The Pew Charitable Trusts and is an initiative of the Pew Research Center. The Project's Web site is http://www.pewinternet.org.

At The Pew Charitable Trusts, the support of two people in particular has sustained the Project: President and CEO Rebecca Rimel and the Managing Director for Information and Civic Initiatives Donald Kimelman. A third person at the Charitable Trusts has come into the picture in recent years and has become a vital enabler of the Project's work, including the "future" series: Tim Durkin. Inside the Pew Research Center, the Project profits handsomely from the expertise of President Andrew Kohut and Executive Vice President Paul Taylor.

Lee Rainie is especially thankful for the help and support of his colleagues at the Pew Internet Project: Susannah Fox, John Horrigan,

Amanda Lenhart, Mary Madden, Sydney Jones, Jessica Vitak, and Cornelia Carter. He also appreciates the help on the *Future of the Internet* surveys from Professor Steve Jones of the University of Illinois–Chicago, a longtime advisor to the Project.

Princeton Survey Research Associates

PSRA conducted the survey that is covered in this report. It is an independent research company specializing in social and policy work. The firm designs, conducts, and analyzes surveys worldwide. Its expertise also includes qualitative research and content analysis. The firm can be reached at 911 Commons Way, Princeton, NJ, 08540, by telephone at 609-924-9204, by fax at 609-924-7499, or by e-mail at ResearchNJ@PSRA.com.

Elon University School of Communications

The Imagining the Internet Center, an initiative of the School of Communications at Elon University, has teamed with the Pew Internet Project to complete a number of significant research studies, including the building of the Early '90s Predictions Database and the series of *Future of the Internet* surveys (http://www.imaginingtheinternet.org), and a 2001 ethnographic study of families' use of the Internet in a small town, "One Neighborhood, One Week on the Internet" (http://org.elon.edu/ pew/oneweek/), both under the direction of Janna Quitney Anderson. For contact regarding the Predictions Database, send e-mail to predictions@elon.edu. The university's Web site is http://www.elon.edu/.

Many people at the university made this work possible. We owe special thanks to President Leo Lambert; Executive Vice President Gerry Francis; Provost Steven House; School of Communications leaders Paul Parsons and Connie Ledoux Book; Dan Anderson, assistant vice president for university relations; and Christopher Eyl, assistant director of design.

Thanks also to editorial assistant Kacie Anderson.

Ithiel de Sola Pool Inspired the Work

Pool, leading communications researcher of the 20th century, died in the mid-1980s, but he was a significant influence when vital decisions were being made regarding freedom on the Internet in the decades to follow. He inspired many Internet stakeholders with his book *Technologies of Freedom*. In it, he projected that interconnected computing devices would be joined to form an open-ended, all-encompassing structure. He described it as "the largest machine that man has ever constructed—the global telecommunications network; the full map of it no one knows; it changes every day" (1983, p. 56), and he projected that it would be questioned by regulators fearing the challenge to the economic and political status quo. He warned that a positive future would be delayed if regulators chose to interfere. His research was an inspiration for the initiation of the Imagining the Internet project and surveys.

Survey Respondents

We are thankful for the thoughtful and revealing contributions of the thousands of international participants in the *Future of the Internet* surveys. Their insights are helping to create significant knowledge about the past, present, and future of information technologies.

Summary of Findings

Technology stakeholders and critics were asked to assess scenarios about the future social, political, and economic impact of the Internet, and they predicted the following:

- The mobile device will be the primary connection tool to the Internet for most people in the world in 2020.
- The transparency of people and organizations will increase, but this will not necessarily yield more personal integrity, social tolerance, or forgiveness.
- Talk and touch user interfaces with the Internet will be more prevalent and accepted by 2020.
- Those working to perfect intellectual property law and copyright protection will remain in a continuing "arms race" with the "crackers," who will find ways to copy and share content without payment.
- The divisions between personal and work time and between physical and virtual reality will be further erased for everyone who is connected, and the results will be mixed in terms of social relations.

- "Next-generation" engineering of the network to improve the current Internet architecture is more likely than an effort to rebuild the architecture from scratch.

There is less clear-cut sentiment on several other possibilities about the future of the Internet. For instance, there was an even split when people were asked if the new transparency of humans and organizations will increase personal integrity and forgiveness—roughly half believe human relations will soften; half think that it will not happen. Similarly, when asked to assess whether many humans in 2020 will regularly spend time in enhanced or augmented worlds, there was some agreement, but it was not overwhelming.

ABOUT THE METHODOLOGY AND INTERPRETING THE FINDINGS

This is the third canvassing of Internet specialists and analysts by the Pew Internet & American Life Project.[1] While a wide range of opinion from experts, organizations, and interested institutions was sought, this survey should not be taken as a representative canvassing of Internet experts. By design, this survey was an "opt-in," self-selecting effort. That process does not yield a random, representative sample.

Some 578 leading Internet activists, builders, and commentators responded in this survey to scenarios about the effect of the Internet on social, political, and economic life in the year 2020. An additional 618 stakeholders also participated in the study, for a total of 1,196 participants who shared their views.

Experts were located in two ways. First, nearly a thousand were identified in an extensive canvassing of scholarly, government, and business documents from 1990 to 1995 to see who had ventured predictions about the future impact of the Internet. Several hundred of them participated in the first two surveys conducted by Pew Internet and Elon University, and they were recontacted for this survey.

Second, expert participants were handpicked due to their positions as stakeholders in the development of the Internet, or they were reached through the leadership listservs of top technology organizations including the Internet Society (ISOC), Association for Computing Machinery, the World Wide Web Consortium (W3C), the United Nations' Multistakeholder Group on Internet Governance, Internet2, Institute of Electrical and Electronics Engineers (IEEE), Internet Corporation for Assigned Names and Numbers (ICANN), International Telecommunication Union, Computer Professionals for Social Responsibility, Association of Internet Researchers, and the American Sociological Association's Information Technology Research section. For the first time, some respondents were invited to participate through personal messages sent using a social network, Facebook. In all, 578 experts identified through these channels responded to the survey.

While many respondents are at the pinnacle of Internet leadership, some of the survey respondents are "working in the trenches" of building the Web. Most of the people in this latter segment of responders came to the survey by invitation because they are on the e-mail list of the Pew Internet & American Life Project or are otherwise known to the Project. They are not necessarily opinion leaders for their industries or well-known futurists, but it is striking how much their views were distributed in ways that paralleled those who are celebrated in the technology field. In all, 618 additional respondents participated in this survey from these quarters. Thus, the expert results are reported as the product of 578 responses, and the lines listing "all responses" include these additional 618 participants.

In the survey, participants were asked if they mostly agreed or mostly disagreed with eight scenarios about the potential evolution of the Internet we might see by 2020.

They were given the opportunity to elaborate on their answers. The scenarios—woven from material collected in recent industry and research reports and from predictive statements by leaders in science, technology, business, and politics—were layered with overlapping elements to spur discussion and elicit nuanced views of the future. They were constructed in a way in which to provoke responses and conversation. They were not written to reflect the views of the Pew Internet & American Life Project or Elon University about the most likely or desirable future.

Neither Pew Internet nor Elon takes positions on the policy matters or forecasts the likely impact of technological change. More responses to each of the scenarios are cited at http://imaginingtheinternet.org on specific Web pages devoted to each scenario. These URLs are given in the chapters devoted to the scenarios.

How Respondents Assessed *Future of the Internet III* Scenarios for 2020

Scenarios Presented in Order They Were Posed in the Survey	*Mostly Agree*	*Mostly Disagree*	*Did Not Respond*
The mobile phone is the primary connection tool for most people in the world. In 2020, while One Laptop Per Child (OLPC) and other initiatives to bring networked digital communications to everyone are successful on many levels, the mobile phone—now with significant computing power—is the primary Internet connection and the only one for a majority of the people across the world, providing information in a portable, well-connected form at a relatively low price. Telephony is offered under a set of universal standards and protocols accepted by most operators internationally, making for reasonably effortless movement from one part of the world to another. At this point, the "bottom" three-quarters of the world's population account for at least 50% of all people with Internet access—up from 30% in 2005.	**578 Experts 77%** **1,196 Total Respondents 81%**	**Experts 22%** **Total Respondents 19%**	**Experts *%** **Total Respondents *%**
Social tolerance has advanced significantly due in great part to the Internet. In 2020 people are more tolerant than they are today thanks to wider exposure to others and their views that has been brought about by the Internet and other information and communication technologies. The greater tolerance shows up in several metrics, including declining levels of violence, lower levels of sectarian strife, and reduced incidence of overt acts of bigotry and hate crimes.	**578 Experts 32%** **1,196 Total Respondents 33%**	**Experts 56%** **Total Respondents 55%**	**Experts 13%** **Total Respondents 11%**

*indicates that the number of respondents was less than 1% of the total sample

(*continued on next page*)

How Respondents Assessed *Future of the Internet III* Scenarios for 2020 (*continued*)

Scenarios Presented in Order They Were Posed in the Survey	*Mostly Agree*	*Mostly Disagree*	*Did Not Respond*
Content control through copyright protection technology dominates. In 2020 strict content controls are in place thanks to the efforts of legislatures, courts, the technology industry, and media companies. Those who use copyrighted materials are automatically billed by content owners, and Internet service providers automatically notify authorities when they identify clients who try to subvert this system. Protesters rarely prevail when they make claims that this interferes with free speech and stifles innovation.	**578 Experts 31%** **1,196 Total Respondents 31%**	**Experts 60%** **Total Respondents 61%**	**Experts 9%** **Total Respondents 8%**
Transparency heightens individual integrity and forgiveness. In 2020 people are even more open to sharing their personal information, opinions, and emotions than they are now. The public's notion of privacy has changed. People are generally comfortable exchanging the benefits of anonymity for the benefits they perceive in data being shared by other people and organizations. As people's lives have become more transparent, they have become more responsible for their own actions and more forgiving of the sometimes unethical pasts of others. Being "outed" for some past indiscretion in a YouTube video or other pervasive media form no longer does as much damage to a person's reputation as it did back in the first decade of the 21st century. Carefully investigated reputation corrections and clarifications are a popular daily feature of major media outlets' online sites.	**578 Experts 45%** **1,196 Total Respondents 44%**	**Experts 44%** **Total Respondents 45%**	**Experts 11%** **Total Respondents 10%**

Scenario			
Many lives are touched by the use of augmented reality or spent interacting in artificial spaces. In 2020 virtual worlds, mirror worlds, and augmented reality are popular network formats thanks to the rapid evolution of natural, intuitive technology interfaces and personalized information overlays. To be fully connected, advanced organizations and individuals must have a presence in the "metaverse" and/or the "geoWeb." Most well-equipped Internet users will spend some part of their waking hours—at work and at play—at least partially linked to augmentations of the real world or in alternate worlds. This lifestyle involves seamless transitions between artificial reality, virtual reality, and the status formerly known as "real life."	**578 Experts 55%** **1,196 Total Respondents 56%**	**Experts 30%** **Total Respondents 31%**	**Experts 15%** **Total Respondents 13%**
Talk and touch are common technology interfaces. In 2020 the most commonly used communications appliances prominently feature built-in voice recognition. People have adjusted to hearing individuals dictating information in public to their computing devices. In addition, haptic technologies based on touch feedback have been fully developed, so, for instance, a small, handheld Internet appliance allows you to display and use a full-size virtual keyboard on any flat surface for those moments when you prefer not to talk aloud to your networked computer. It is common to see people air typing as they interface with the projection of a networked keyboard visible only to them.	**578 Experts 64%** **1,196 Total Respondents 67%**	**Experts 21%** **Total Respondents 19%**	**Experts 15%** **Total Respondents 14%**

(*continued on next page*)

How Respondents Assessed *Future of the Internet III* Scenarios for 2020 (*continued*)

Scenarios Presented in Order They Were Posed in the Survey	***Mostly Agree***	***Mostly Disagree***	***Did Not Respond***
Next-generation research will be used to improve the current Internet; it will not replace it. In 2020 the original Internet architecture is in the continuing process of refinement—it has not been replaced by a completely new system. Research into network innovation, with help from the continued acceleration of technologies used to build, maintain, enhance, and enlarge the system, has yielded many improvements. Search, security, and reliability on the Internet are easier and more refined, but those who want to commit crimes and mischief are still able to cause trouble.	**578 Experts 78%** **1,196 Total Respondents 80%**	**Experts 6%** **Total Respondents 6%**	**Experts 16%** **Total Respondents 14%**
Few lines divide professional from personal time, and that is OK. In 2020 well-connected knowledge workers in more developed nations have willingly eliminated the industrial-age boundaries between work hours and personal time. Outside of formally scheduled activities, work and play are seamlessly integrated in most of these workers' lives. This is a net positive for people. They blend personal/professional duties wherever they happen to be when they are called upon to perform them—from their homes, the gym, the mall, a library, and possibly even their company's communal meeting space, which may exist in a new virtual-reality format.	**578 Experts 56%** **1,196 Total Respondents 57%**	**Experts 29%** **Total Respondents 29%**	**Experts 15%** **Total Respondents 14%**

Source. Pew Internet & American Life Project Survey, December 26, 2007–March 3, 2008. This was a nonrandom, Web-based survey sample of Internet users recruited via e-mail and social networks. Data are based on a nonrandom sample; a margin of error cannot be computed.

Thinking Ahead to 2020: Themes Many Respondents Struck in Their Answers

The following are some of the major themes that run through respondents' answers:

- **The mobile phone will be the dominant connection tool:** More than three-quarters of the expert respondents (77%) agreed with a scenario that posited that the mobile computing device—with more significant computing power in 2020—will be the primary Internet communications platform for a majority of people across the world. They agreed that connection will generally be offered under a set of universal standards internationally, though many respondents registered doubts about corporations' and regulators' willingness to make it happen. Most agreed that the benefits of the Internet will be extended further to less developed regions. Some said a new "killer app" will replace the smartphone.

- **Heightened social tolerance will not be a Web 2.0 result:** Respondents were asked if people will be more tolerant in 2020 than they are today. Of the expert respondents, 56% disagreed with a scenario positing that social tolerance will advance significantly by then, saying communication networks also expand the potential for hate, bigotry, and terrorism. Of them, 32% predicted tolerance will grow. A number of the survey participants indicated that the divide between the tolerant and intolerant could possibly be deepened by the options for collaboration offered by the Internet.

- **Air typing, touch interfaces, and talking to devices will become common:** A notable majority of the respondents—64%—favored the idea that by 2020, user interfaces will offer advanced talk, touch, and typing options, and some added a fourth "t"—think. Those who chose to elaborate in extended responses disagreed on which of the four options will make the most progress by 2020. There was a fairly even yes-no split on the likely success of voice recognition or significant wireless keyboard advances and mostly positive support for the advance of interfaces involving touch and gestures—this was highly influenced by the introduction of the iPhone and various multitouch surface computing platforms in 2007 and 2008. A number of respondents projected the possibility of a thought-based interface—neural networks offering mind-controlled human-computer interaction. Many expressed concerns over rude, overt public displays by people using information and communication technologies (ICTs) ("yakking away on their phones about their latest foot fungus") and emphasized the desire for people to keep private communications private in future digital interfaces.

- **IP law and copyright will remain unsettled:** Three out of five respondents (60%) disagreed with the idea that legislatures, courts, the technology industry, and media companies will exercise effective content control by 2020. They said "cracking" technology will stay ahead of technology to control intellectual property (IP) or policy regulating IP. And they predicted that regulators will not be able to come to a global agreement about intellectual property. Many respondents suggested that new economic models will have to be implemented, assuming that what was once classified as paid content will have to be offered free or in exchange for attention or some other unit of value. Nearly one-third of the survey respondents (31%) agreed that IP regulation will be successful by 2020; they said more content will be privatized, some adding that this control might be exercised at the hardware level through Internet-access devices such as smartphones.

- **The division between personal and professional time will disappear:** A majority of expert respondents (56%) agreed with the statement that in 2020 "few lines (will) divide professional from personal time, and that is OK." While some people were hopeful about a hyperconnected future with more freedom, flexibility, and life enhancements, others expressed fears that mobility and ubiquity of networked computing devices will be harmful for most people by adding to stress and challenging family and social life.

- **Network engineering research will build on the status quo—there is not likely to be a "next-gen" Internet:** Nearly four out of five respondents (78%) said they think the original Internet architecture will still be in place in 2020, even as it is continually being refined. They stated that they do not believe the current Internet will be replaced by a completely new next-generation system between now and 2020. Those who wrote extended elaborations to their answers projected the expectation that IPv6 and the Semantic Web will be vital elements in the continuing development of the Internet over the next decade. Among other predictions: There will be more "walled gardens," separated Internet spaces, created by governments and corporations to maintain network control; governments and corporations will leverage security fears to retain power over individuals; and crime, piracy, terror, and other negatives will always be common elements in an open system.
- **Transparency may or may not make the world a better place:** Respondents were split evenly on whether the world will be a better place in 2020 due to the greater transparency of people and institutions afforded by the Internet. Forty-five percent of expert respondents agreed that transparency of organizations and individuals will heighten individual integrity and forgiveness, and 44% disagreed. The comments about this prediction were varied. Some respondents argued that transparency is an unstoppable force that has positives and

negatives; it might, somehow, influence people to live lives in which integrity and forgiveness are more likely. Others posited that transparency will not have any positive influence; in fact, it will make everyone vulnerable, and bad things will happen because of it. Still others argued that the concept of "privacy" is changing, it is becoming scarce, and it will be protected and threatened by emerging innovations; tracking and databasing will be ubiquitous; reputation maintenance and repair will be required; some people will have multiple digital identities; and some people will withdraw.

- **Augmented reality and interactive virtual spaces might see more action:** More than half of the respondents (55%) agreed with the notion that many lives will be touched in 2020 by virtual worlds, mirror worlds, and augmented reality. Yet, 45% either disagreed with or did not answer this question, so the sentiment was not overwhelming. People's definitions for the terms *augmented reality* (AR) and *virtual reality* (VR) were quite varied. Smartphones and global positioning systems (GPSs) help people augment reality to a certain extent today and are expected to do more soon; many respondents said they think today's social networks qualify as a form of virtual reality while others defined it in terms of Second Life or something even more immersive. Some noted that by 2020, augmented reality and virtual reality will have reached the point of blurring with reality. Many respondents indicated that this will enhance the world, providing new opportunities for conferencing, teaching, and 3-D modeling, and some added that breakthroughs to come may bring significant change, including fusion with other developments, such as genetic engineering. Some respondents expressed fear of the negatives of AR and VR, including new extensions of the digital divide, an increase in violence and obesity, and the potential for addiction or overload. There was agreement that user interfaces have to be much more intuitive for AR and VR to become more universally adopted.

A Sample of Revealing Quotations and Predictions Selected From the Thousands Submitted

The Evolution of the Device for Connection

"People in Africa turned paid telephone minutes into an ad hoc, grassroots e-currency…There are already reasons why people at the bottom of the economic system need and can use cheap telecommunication. Once they are connected, they will think of their own ways to use connectivity plus computation to relieve suffering or increase wealth." —**Howard Rheingold**, Internet sociologist and author of *Virtual Community* and *Smart Mobs*

"By 2020, the network providers of 'telephony' will have been disintermediated. We'll have standard network connections around the world…Billions of people will have joined the Internet who don't speak English. They won't think of these things as 'phones' either—these devices will be simply lenses on the online world." —**Susan Crawford**, founder of OneWebDay and an Internet Corporation for Assigned Names and Numbers (ICANN) board member

"Traditional carriers have little incentive to include poor populations, and the next 5 years will be rife with battles between carriers, municipal, and federal governments, handset makers, and content creators. I don't know who will win." —**danah boyd**, Harvard University's Berkman Center for Internet and Society

"Telephones in 2020 will be archaic, relics of a bygone era—like transistor radios are today. Telephony, which will be entirely IP-based by then, will be a standard communications chip on many devices. We'll probably carry some kind of screen-based reading device that will perform this function, though I assume when we want to communicate verbally, we'll do so through a tiny, earplug-based device." —**Josh Quittner**, executive editor of *Fortune* magazine; longtime technology journalist and editor

"I agree, but I don't see this as entirely positive, as it perpetuates 'soundbite' dissemination and thinking and the continuing move toward shorter attention spans and dumbing down of content." —**Anonymous respondent**

The Evolution of Social Tolerance

"Not in mankind's nature. The first global satellite linkup was [in] 1967, BBC's 'Our World: The Beatles "All You Need Is Love,"' and we still have war, genocide, and assassination (Lennon's poignantly)." —**Adam Peake**, policy analyst for the Center for Global Communications and participant in the World Summit on the Information Society

"Polarization will continue and the people on the extremes will be less tolerant of those opposite them. At the same time, within homogenous groups (religious, political, social, financial, etc.), greater tolerance will likely occur." —**Don Heath**, Internet pioneer and former president and CEO of the Internet Society

"Tribes will be defined by social enclaves on the Internet, rather than by geography or kinship, but the world will be more fragmented and less tolerant, since one's real-world surroundings will not have the homogeneity of one's online clan." —**Jim Horning**, chief scientist for information security at SPARTA, Inc. and a founder of InterTrust's Strategic Technologies and Architectural Research Laboratory

The Evolution of Intellectual Property Law and Copyright

"Many people want IP protection, but everyone wants to steal. Regardless of the legal mechanisms so far—e.g., automatic damages, compulsory copyrights—many people would prefer the illegal route, perhaps, because it runs up their adrenaline." —**Michael Botein**, founding director of the Media Law Center at New York University Law School

"Copying data is the natural state of computers; we would have to try to compromise them too much to support this regime." —**Brad Templeton**, chairman of the Electronic Frontier Foundation

"While I applaud the efforts of DRM [digital rights management] opponents, I am discouraged by the progress DRM seems to continue to make in hardware as much as in software. Having purchased an iPhone, I was delighted when Apple updated its software to allow custom ringtones, only to discover that I needed to pay for a ringtone via the iTunes Music Store even though the ringtone I wanted to use was one in which I own the copyright!" —**Steve Jones**, cofounder of the Association of Internet Researchers and editor of *New Media & Society*

"Most people still don't understand the question and will only wake up when it's too late. I foresee a time where NOTHING is free and things like potatoes, chickens, counseling, clichés, and Scrabble will not be able to be accessed without paying a toll at every point." —**Anonymous respondent**

"There will be cross-linking of content provider giants and Internet service provider giants and that they will find ways to milk every last 'currency unit' out of the unwitting and defenseless consumer. Governments will be strongly influenced by the business conglomerates and will not do much to protect consumers. (Just think of the outrageous rates charged by cable and phone company TV providers and wireless phone providers today—it will only get worse.)" —**Steve Goldstein**, ICANN board member, formerly of the U.S. National Science Foundation (NSF)

"Copyright is a dead duck in a digital world. The old regime based its power on high distribution costs. Those costs are going to zero. Bye-bye DRM." —**Dan Lynch**, founder of CyberCash and Interop Company, now a board member of the Santa Fe Institute

"You cannot stop a tide with a spoon. Cracking technology will always be several steps ahead of DRM, and content will be redistributed on anonymous networks." —**Giulio Prisco**, chief executive of Metafuturing Second Life, formerly of CERN

The Evolution of Privacy, Identity, and Forgiveness

"We will enter a time of mutually assured humiliation; we all live in glass houses. That will be positive for tolerance and understanding, but—even more important—I believe that young people will not lose touch with their friends as my generation did, and that realization of permanence in relationships could—or should—lead to more care in those relationships." —**Jeff Jarvis**, top blogger at Buzzmachine.com and professor at City University of New York Graduate School of Journalism

"Gen Y has a new notion of privacy. The old 'never trust anyone over 30' will turn into 'never trust anyone who doesn't have embarrassing stuff online.'" —**Jerry Michalski**, founder and president of Sociate

"Viciousness will prevail over civility, fraternity, and tolerance as a general rule, despite the buildup of pockets or groups ruled by these virtues. Software will be unable to stop deeper and more hard-hitting intrusions into intimacy and privacy, and these will continue to happen." —**Alejandro Pisanty**, ICANN and Internet Society leader and director of computer services at Universidad Nacional Autónoma de México

"By 2020, the Internet will have enabled the monitoring and manipulation of people by businesses and governments on a scale never before imaginable. Most people will have happily traded their privacy—consciously or unconsciously—for consumer benefits such as increased convenience and lower prices. As a result, the line between marketing and manipulation will have largely disappeared." —**Nicholas Carr**, author of the Rough Type blog and *The Big Switch*

"The volume and ubiquity of personal information, clicktrails, personal media, etc., will desensitize us. A super abundance of transparency will lose its ability to shock. Maybe there will be software-driven, real-time reputation insurance service, offering monitoring and repair to dinged reputations. This could be as ordinary as auto insurance or mortgage insurance is today and as automated as the nightly backups performed by most online businesses. I don't agree that this will make us any kinder." —**Havi Hoffman**, senior editor for product development at Yahoo and blogger

The Evolution of Augmented and Virtual Reality

"Mirror worlds are multidimensional experiences with profound implications for education, medicine, and social interaction. 'Real life' as we know it is over. Soon, when anyone mentions reality, the first question we will ask is, 'Which reality are you referring to?' We will choose our realities, and in each reality, there will be truths germane to that reality, and so we will choose our truth as well." —**Barry Chudakov**, principal with the Chudakov Company

"We, in the present, don't think of ourselves as living in 'cyberspace,' even though people of a decade previous would have termed it such. Of the various forms of the metaverse, however, the majority of activity will take place in blended- or augmented-reality spaces, not in distinct virtual/alternative world spaces." —**Jamais Cascio**, a coauthor of the *Metaverse Roadmap Overview*, a report on the potential futures of VR, AR, and the geoWeb

"Augmented reality will become nearly the de facto interface standard by 2020, with 2-D and 3-D overlays over real-world objects providing rich information, context, entertainment, and (yes) promotions and offers. At the same time, a metaverse (especially when presented in an augmented-reality overlay environment) provides compelling ways to facilitate teamwork and collaboration while reducing overall travel budgets." —**Jason Stoddard**, managing partner at Centric/Agency of Change

"The virtual world removes all barriers of human limitation; you can be anyone you want to be instead of being bound by physical and material limitations. That allows people to be who they naturally are, freed of any perception they may have of themselves based on their 'real life'—it is the power of removing the barriers of your own perception of yourself." —**Tze-Meng Tan**, Multimedia Development Corporation in Malaysia, a director at OpenSOS

"We are in the last generation of human fighter pilots. Already, drones in Iraq are piloted in San Diego. What will improve is the ability of the artificial spaces to control physical reality, to expand our reach more effectively in many aspects of the physical universe." —**Dick Davies**, partner at Project Management and Control, Inc. and former president of the Association of Information Technology Professionals

"In a reaction to the virtual world, entrepreneurs will establish 'virt-free' zones where reality is not augmented. In various heavily connected areas, there will be sanctuaries (hotels, restaurants, bars, summer camps, vehicles) which people may visit to separate themselves from adhesion or other realities." —**C. R. Roberts**, Vancouver-based technology reporter

"For some reason, I've never been able to comprehend [how] certain pundits can seriously propose that the wave of the future is chatting using electronic hand puppets. Flight Simulator is not an aircraft, and typing at a screen is not an augmentation of the real world." —**Seth Finkelstein**, author of the Infothought blog, writer and programmer

"A map is not the territory, and a letter is not the person. We have always had multiple facades, for most, most common, work, home, and play. The extension into more immersive 'unreal' worlds is going to happen." —**Hamish MacEwen**, consultant at Open ICT in New Zealand

The Evolution of User Interfaces

"There will be 'subvocal' inputs that detect 'almost speech' that you will, but do not actually voice. Small sensors on teeth will also let you tap commands. Your eyeballs will track desires, sensed by your eyeglasses. And so on." —**David Brin**, futurist and author of *The Transparent Society*

"Wi-Fi– and WiMAX-enabled badges with voice recognition will act as personal assistants—allowing you to talk with someone by saying their name, to post a voice blog, or access directions from the Internet for the task at hand." —**Jim Kohlenberger**, director of Voice on the Net Coalition; senior fellow at the Benton Foundation

"I could see a whole physical way of communicating with our technology tools that could be part of our health and exercise. A day answering e-mails could be a full-on physical workout ;)." —**Tiffany Shlain**, founder of the Webby Awards

"We will see the display interface device separated from the input device over the next 12 years. Display devices will be everywhere, and you will be able to use them with your input device. The input device might be virtual, as in the case of the iPhone or a holographic keyboard, or they might resemble the keyboards and touch pads that people are using today." —**Ross Rader**, a director with Tucows who is active in the ICANN Registrars constituency

"While air typing and haptic gestures are widespread and ubiquitous, the arrival of embedded optical displays, thought transcription, eye-movement tracking, and predictive-behavior modeling will fundamentally alter the human-computer interaction model." —**Sean Steele**, CEO and senior security consultant for infoLock Technologies

The Evolution of Network Architecture

"The control-oriented telco (ITU) next-generation network will not fully evolve, the importance of openness and enabling innovation from the edges will prevail; i.e., Internet will essentially retain the key characteristics we enjoy today, mainly because there's more money to be made." —**Adam Peake**, executive research fellow and telecommunications policy analyst at the Center for Global Communications

"Some parts of the Internet may fragment, as nations pursue their own technology trajectories. The Internet is so vastly complex; incremental upgrades seem to be the only way to get anything done…Places like China may make big leaps and bounds because there is less legacy." —**Anthony Townsend**, research director, the Institute for the Future

"Current Internet-standards bodies and core Internet protocols are ossifying to such an extent that security and performance requirements for next-generation applications will require a totally new base platform. If current Internet base protocols survive, it will be as a substrata paved over by new-generation smarter ways of connecting." —**Ian Peter**, Ian Peter and Associates and the Internet Mark 2 Project

"The Web must still be a messy, fabulous, exciting, dangerous, poetic, depressing, elating place…akin to life; which is not a bad thing." —**Luis Santos**, Universidade do Minho–Braga, Portugal

"When have we ever stopped crime? If it is a choice between having some criminals around and having a repressive government, I will take the former; they are much easier to deal with." —**Leonard Witt**, associate professor at Kennesaw State University in Georgia and author of the weblog PJNet.org

"The Internet is not magical; it will be utterly overmanaged by commercial concerns, hobbled with 'security' micromanagement, and turned into money-shaped traffic for business—the rest, 90% paid-for content download and the rest of the bandwidth used for market feedback." —**Tom Jennings**, University of California–Irvine, creator of FidoNet and builder of *Wired* magazine's first online site

The Evolving Concept of Time for Work, Leisure

> "Corporate control of workers' time—in the guise of work/family balance—now extends to detailed monitoring of when people are on and off work. The company town is replaced by 'company time management,' and it is work time that drives all other time uses. This dystopia challenges the concept of white-collar work, and unionism is increasingly an issue."—**Steve Sawyer**, associate professor in the College of Information Sciences and Technology, Penn State University

> "The real-world interaction of the Internet and the 'long tail' of the Semantic Web will enable everyone to find the perfect job for them, the right opportunity, so that people in general are happier, healthier, and more productive." —**Anonymous respondent**

> "The result may be longer, less-efficient working hours and more stressful home life." —**Victoria Nash**, director of graduate studies and policy and research officer, the Oxford Internet Institute

This Report Builds Upon the Online Resource Imagining the Internet: A History and Forecast

At the invitation of Lee Rainie, director of the Pew Internet & American Life Project, Elon University associate professor Janna Quitney Anderson began a research initiative in the spring semester of 2003 to search for comments and predictions about the future impact of the Internet during the time when the World Wide Web and browsers emerged—between 1990 and 1995.

The idea was to replicate the fascinating work of Ithiel de Sola Pool in his 1983 book, *Forecasting the Telephone: A Retrospective Technology Assessment*. Elon students, faculty, and staff studied government documents, technology newsletters, conference proceedings, trade newsletters, and the business press and gathered predictions about the future of the Internet.

Eventually, more than 4,000 early '90s predictions from about 1,000 people were amassed.

The early 1990s predictions are available in a searchable database online at the site Imagining the Internet: A History and Forecast, and they are also the basis for a book by Anderson titled *Imagining the Internet: Personalities, Predictions, Perspectives* (2005, Rowman & Littlefield).

The fruits of that work inspired additional research into the past and future of the Internet.

The Imagining the Internet: A History and Forecast Web site (http://www.imaginingtheInternet.org/)—now numbering about 6,300 pages—includes results from the entire series of the *Future of the Internet* surveys, video and audio interviews showcasing experts' predictions about the next 10 to 50 years, a children's section, tips for teachers, a Voices of the People section on which anyone can post his or her prediction, and information about the recent history of communications technology.

We expect the site will continue to serve as a valuable resource for researchers, policy makers, students, and the general public for decades to come.

Further, we encourage readers of this report to enter their own predictions at the site.

The series of the *Future of the Internet* surveys are also published in book form by Cambria Press.

Acknowledgments

About the Pew Internet & American Life Project

The Pew Internet Project is an initiative of the Pew Research Center, a nonprofit "fact tank" that provides information on the issues, attitudes, and trends shaping America and the world. Pew Internet explores the impact of the Internet on children, families, communities, the workplace, schools, health care, and civic and political life. The Project is nonpartisan and takes no position on policy issues. Support for the project is provided by The Pew Charitable Trusts. The Project's Web site is http://www.pewInternet.org.

Princeton Survey Research Associates International
PSRAI conducted the survey that is covered in this report. It is an independent research company specializing in social and policy work. The firm designs, conducts, and analyzes surveys worldwide. Its expertise also includes qualitative research and content analysis. With offices in Princeton, New Jersey, and Washington, DC, PSRAI serves the needs of clients around the nation and the world. The firm can be reached at 911 Commons Way, Princeton, NJ, 08540, by telephone at 609-924-9204, by fax at 609-924-7499, or by e-mail at ResearchNJ@PSRA.com.

The Imagining the Internet Center at Elon University's School of Communications
The Imagining the Internet Center at Elon University holds a mirror to humanity's use of communications technologies, informs policy development, exposes potential futures, and provides a historic record. It has teamed with the Pew Internet Project to complete a number of research studies, including the building of Imagining the Internet (a foresight and history Web site), a survey of stakeholders at the UN-administrated Internet Governance Forum in Rio de Janeiro in 2007, and an ethnographic study of a small town, "One Neighborhood, One Week on the Internet," all under the direction of Janna Quitney Anderson. For contact regarding Imagining the Internet, send e-mail to predictions@elon.edu. The university site is http://www.elon.edu/.

BACKGROUND

Predictions inspire lively discussion about the future and help stakeholders prepare to make adjustments to meet the needs associated with technological change. Those who think about the future are best poised to influence it.

Many futurists, scientists, and long-term thinkers today argue that the acceleration of technological change over the past decade has greatly increased the importance of strategic vision. Technology innovations will continue to impact us. The question is whether this process will reflect thoughtful planning or wash over us like an unstoppable wave. This survey is aimed at gathering a collection of opinions regarding the possibilities we all face.

How The Surveys Originated and Have Been Conducted

This research project got its start in mid-2001, when Lee Rainie, the director of the Pew Internet & American Life Project, approached officials at Elon University with an idea that the Project and the university might replicate the work of Ithiel de Sola Pool in his 1983 book *Forecasting the Telephone: A Retrospective Technology Assessment*. Pool and his students had looked at primary official documents, technology community publications, speeches given by government and business leaders, and marketing literature at the turn of the 20th century to examine the kind of impacts experts thought the telephone would have on Americans' social and economic lives.

The idea was to apply Pool's research method to the Internet, particularly focused on the period between 1990 and 1995 when the World Wide Web and Web browsers emerged. In the spring semester of 2003, Janna Quitney Anderson, professor of journalism and communications at Elon, led a research initiative that set out to accomplish this goal. More than 4,200 predictive statements made in the early 1990s by 1,000 people were logged and categorized. The result is available on the site Imagining the Internet: A History and Forecast (http://www.imaginingtheInternet.org/).

We reasoned that if experts and technologists had been so thoughtful in the early 1990s about what was going to happen, they would likely be equally as insightful looking ahead from this moment. In 2004 we asked most of those whose predictions were in the 1990–1995 database and additional experts to assess a number of predictions about the coming decade, and their answers were codified in an initial futures survey, the Future of the Internet (http://www.pewInternet.org/pdfs/PIP_Future_of_Internet.pdf).

Several years later, we repeated the process, presenting a new set of potential future scenarios to an expanded base of experts. In late 2005 and the first quarter of 2006, the Pew Internet Project issued an e-mail invitation to a select group of technology thinkers, stakeholders, and social analysts, asking them to complete the second scenario-based quantitative and qualitative survey, the Future of the Internet II. The results of both the Future I and Future II surveys are already available online and are being released in book form.

And, here, we report the results of a third survey that was conducted between December 26, 2007, and March 3, 2008. Some 1,196 people were generous enough to take the time to respond to this Future of the Internet III online survey.

Nearly half of the Future III respondents are Internet pioneers who were online before 1993. Roughly one-fifth of the respondents said they live and work in a nation outside of North America.

The respondents' answers represent their personal views and in no way reflect the perspectives of their employers. Many survey participants were handpicked due to their positions as stakeholders in the development of the Internet, or they were reached through the leadership listservs of top technology organizations, including the Internet Society, Association for Computing Machinery, the World Wide Web Consortium, the United Nations' Multistakeholder Group on Internet Governance, Internet2, Institute of Electrical and Electronics Engineers, Internet Corporation for Assigned Names and Numbers, International Telecommunication Union, Computer Professionals for Social Responsibility, Association of Internet Researchers, and the American Sociological Association's Information Technology Research section.

The Scenarios Were Built to Elicit Deeply Felt Opinions

The Pew Internet & American Life Project and Elon University do not advocate policy outcomes related to the Internet. The predictive scenarios included in the survey were structured to provoke reaction, not because we think any of them will necessarily come to fruition.

The scenarios for this survey and survey analysis were crafted after a study of the responses from our previous surveys and of the predictions made in reports by the Organisation for Economic Cooperation and Development (OECD), the United Nations Multistakeholder Group on Internet Governance, the Metaverse Roadmap, The Institute for the Future, Global Business Network, and other foresight organizations and individual foresight leaders. Many previous reports and other materials were consulted in researching the questions and content of this study.

The 2020 scenarios were constructed to elicit engaged responses to many-layered issues, so it was sometimes the case that survey participants would agree with most or part of a scenario but not all of it. In addition to trying to pack several ideas into each scenario, we tried to balance them with "good," "bad," and "neutral" outcomes. The history of technology is full of evidence that technology adoption brings *both* positive and negative results.

After each portion of the survey, we invited participants to write narrative responses providing an explanation for their answers. Not surprisingly, the most interesting product of the survey is the ensuing collection of open-ended discussion, predictions, and analyses written by the participants in response to our material. We have included many of these responses in this report. A great number of additional responses are included on the Imagining the Internet site, available at http://www.imaginingtheinternet.org.

Since participants' answers evolved in both tone and content as they went through the questionnaire, the findings in this report are presented in the same order as the original survey. The respondents were asked to "sign" each written response that they were willing to have credited to them in the Elon-Pew database and in this report. The quotations in the report are attributed to those who agreed to have their words quoted. When a quote is not attributed to someone, it is because that person chose not to sign his or her written answer.

To make this report more readable and include many voices, some of the lengthier written elaborations have been edited.

ENDNOTE

1. The results of the first survey can be found at http://www.pewinternet.org/pdfs/PIP_Future_of_Internet.pdf. The results of the second survey are available at http://www.pewinternet.org/pdfs/PIP_Future_of_Internet_2006.pdf. A more extensive review of all the predictions and comments in that survey can be found at the Web site for Imagining the Internet: A History and Forecast at http://www.elon.edu/predictions/default.html.

Ubiquity, Mobility, Security

Scenario 1

The Evolution of Mobile Internet Communications

Prediction: *The mobile phone is the primary connection tool for most people in the world. In 2020, while One Laptop Per Child (OLPC) and other initiatives to bring networked digital communications to everyone are successful on many levels, the mobile phone—now with significant computing power—is the primary Internet connection and the only one for a majority of the people across the world, providing information in a portable, well-connected form at a relatively low price. Telephony is offered under a set of universal standards and protocols accepted by most operators internationally, making for reasonably effortless movement from one part of the world to another. At this point, the "bottom" three-quarters of the world's population account for at least 50% of all people with Internet access—up from 30% in 2005.*

Expert Respondents' Reactions	
Mostly Agree	77%
Mostly Disagree	22%
Did Not Respond	*%

*indicates that the number of respondents was less than 1% of the total sample.

All Respondents' Reactions	
Mostly Agree	81%
Mostly Disagree	19%
Did Not Respond	*%

Note. Since results are based on a nonrandom sample, a margin of error cannot be computed. The "prediction" was composed to elicit responses and is not a formal forecast.

Respondents were presented with a brief set of information outlining the status quo of the issue 2007 that prefaced this scenario. It read:

> According to the UN/ITU *World Information Society Report 2007*, there has been some progress in improving digital inclusion: In 1997 the nearly three-quarters of the world's population who lived in low-income and lower middle-income economies accounted for just 5% of the world's population with Internet access. By 2005, they accounted for just over 30%.[1] A number of commercial and nonprofit agencies are combining forces to bring inexpensive laptop computers to remote regions of the world to connect underserved populations. In addition, by the end of 2008 more than half the world's population is expected to have access to a mobile phone.

OVERVIEW OF RESPONDENTS' REACTIONS: A significant majority agreed with the proposed future. The consensus is that mobile devices will continue to grow in impact because people need to be connected wherever they are; cost-effectiveness and access are motivating factors; the devices of the future will have significant computing power; and there is fear that limits set by governments and/or corporations seeking control might impede positive benefits—the expected "effortless" connectivity is dependent on their willingness to serve the public good.

The overwhelming majority agreeing with this scenario took note of the current boom in cell phone and smartphone use and imagined its extension. "By 2020, we should see several billion cell phones shipping per year, most of which will be Internet capable; this will probably dwarf the volumes of other Internet-capable devices, such as PCs," wrote one anonymous participant.

There are 6.6 billion people in the world, and the United Nations estimated that 1.4 billion have access to and use the Internet (2008 figures). Wireless Intelligence, a market database, reported that it took 20 years for the first billion mobile phones to sell, just 4 years for the second billion, and 2 years for the third billion. It is expected that there will be 4 billion cell phones in the world by the end of 2008 or in early 2009; about 11% were Internet enabled in 2007, and it is expected that this could rise to 15% by the end of 2008. (It is important to remember that some people own more than one mobile phone—in 2007 it was estimated that 700 million people owned more than one—so 3 billion phones do not equate to 3 billion people who have and use mobile phones.)[2]

Several survey participants noted in their written elaborations to the survey question that connectedness serves humanity in so many ways that even people who are struggling to make a dollar a day in the world's least developed nations find the economics of mobile telephony to be manageable and, sometimes, even vital to their lives.

"Communication is a basic human need," responded **Howard Rheingold**, Internet sociologist and author of *Virtual Community* and *Smart Mobs*. "People who are trying to scrape by have immediate need for connection to information about local labor and commodities markets. Public health and disaster relief information can be an SMS [short message service—or "text"] message away. People in Africa turned paid telephone minutes into an ad hoc, grassroots e-currency, because they had the need to transfer small amounts of money. Billions of squatters might live in slums but still ingeniously and often illegally deliver the construction and utilities services they need. There are already reasons why people at the bottom of the economic system need and can use cheap telecommunication. Once they are connected, they will think of their own ways to use connectivity plus computation to relieve suffering or increase wealth."

Geert Lovink, Amsterdam-based media theorist, Net critic, and activist professor, agreed, writing, "To have a communication break will be the most looked-after 'human right' of the 21st century. That's luxury…We in the West still look at modern communication from a moral perspective, as something that's an extra which is not really necessary in order to live a fulfilling life. But mobile phones are not a lifestyle choice. They are vital in order to stay in touch with family and clients in dense metropolitan environments where people are on the road for most of the day."

Several anonymous respondents wrote that small, simple mobile devices are the only affordable choice for connection for the billions of less advantaged people in the world. "There won't be any major shift in economic situations of today's poor by 2020. The cheaper the technology, the better chance it has of being used," wrote an anonymous survey participant. Another wrote, "The mobile phone helps create connectivity in regions that don't have extensive communications infrastructure. So it's a logical entry point for Internet access in the developing world."

Lutfor Rahman, of the Association for Advancement of Information Technology in Bangladesh, said mobile communication is world changing. "Before introducing the mobile phone in remote areas of Bangladesh, the exchange of information was through physically meeting," he wrote. "That wasted much time, and sometimes it became impossible in short time because of lack of communication facilities."

Gbenga Sesan, Nigerian and consultant on the use of the Internet for development for Paradigm Initiative, has written extensively about the use of mobile communications. "With the rise in the number of mobile phone users across the continent, it is only wise to start planning that the future will be driven through mobile phones—governance, businesses, networking, leisure, and more," he commented. "The story will be the same across the world. Regardless of technology choice (GSM, CDMA, etc.), mobile telephones will form the core of human interaction and livelihood. And when you consider the fact that some mobile phones were competing with computers in 2007, you can only wonder if owning a PC will matter by December 31, 2019."

One anonymous respondent gushed, "Mobile technology will allow everyone in the world to access every piece of information out there—from the wealth of the world's academic research to live burlesque shows starting every 10 minutes." Another wrote, "By 2020, the mobile phone will have become the primary 'lifeline' for most of the world's population—and the recent growth in mobile adoption in the third world indicates that cheap handsets and reliable mobile service can bring first-class access to information and communications tools to populations that were previously inaccessible."

Another respondent worried, "I agree, but don't see this as entirely positive, as it perpetuates 'soundbite' dissemination and thinking, and the continuing move toward shorter attention spans and dumbing down of content."

And another wrote, "Information and content are not the same. Push as much content as you want to a mobile platform, but I like my information in a 'book.'"

It Will Be More Computer Than Phone

Many who responded with a further elaboration on this scenario said while the device we will be using will be small and possibly resemble today's wireless phones in its shape, it will actually be a multitasking computer, used less for voice communication than for other tasks. "The computing power that will be able to fit into a phone-size device in 13 years will be incredible," wrote an anonymous respondent.

Another respondent wrote, "It may be that mobile technologies will expand such that a computing phone will be the Internet appliance of 2020." And another wrote, "Thin clients will allow most users to access databases and applications 'in the cloud,' as well as be connected through telephony. Mobile devices will be ubiquitous—somewhere between an ultra-portable PC and a QWERTY phone—but they will be neither phones nor PCs."

"By 2020, a device that more closely resembles today's mobile phone rather than today's computer will certainly be the primary connection tool," said **Paul Miller**, a technology evangelist for Talis, a

U.K.-based Web company, and blogger for ZDNet. "Whether it is at all 'phone'-like, or even used very often for voice-only communication, is more open to question, though."

Susan Crawford, the founder of OneWebDay and an ICANN board member, agreed. "By 2020, we'll stop talking about 'phones,' with any luck," she wrote. "Nor will we be talking about 'telephony.' Those terms, I hope, will be dead. These devices will just be handsets of which we'll be very fond. They'll have screens that are just large enough for us to feel immersed in the visuals provided. What will we be doing? Using the Internet. Interacting, doing work, talking, participating, uploading to the cloud. By 2020, the network providers of 'telephony' will have been (with any luck) disintermediated. We'll have standard network connections around the world, but they won't be optimized on billing (as telephone and wireless connections are now). Billions of people will have joined the Internet who don't speak English. They won't think of these things as 'phones' either—these devices will be simply lenses on the online world."

Rich Miller, CEO for Replicate Technologies and an Internet pioneer with ARPANET, wrote, "The 'phone' as such is more likely to be a personal media server/media gateway. This same personal media server—size not much different than today's mobile phone—permits varieties of 'terminal' devices, including display, voice input/output, etc. Audio and video interfaces are more likely to be separate devices (like today's Bluetooth headset, but with more user-interface controls).

Steve Jones, cofounder of the Association of Internet Researchers and associate dean at the University of Illinois–Chicago, projected, "By 2020, I don't think it will be so easy to distinguish between a mobile phone and a laptop. These will blend into a general 'mobile computing' category of device (for which we probably don't yet have a name)."

Roberto Gaetano, an ICANN board member with 30 years of experience in telecommunications and information technology, wrote, "In 2020 it would be difficult to make the distinction between a mobile telephone and a computer, as we most probably would have an object that would have high computing capability and the possibility to transmit/receive voice and data. I also believe that this object will be an IPv6 mobile node (in the sense that it will have an IPv6 address) and that communication will mostly use the IP channel, although it will have other connection capabilities."

Jim Kohlenberger, executive director of Voice on the Net Coalition, a senior fellow for the Benton Foundation and former White House policy advisor, commented, "The mobile 'phone' will largely be eclipsed and replaced by the open network device—an open mobile computing device also capable of voice. But the assumption is correct that these mobile devices will be more significant and ubiquitous than wired devices. In terms of inclusion, there are already developing countries that have set up open and competitive wireless markets to foster these innovations and reap their benefits. But other developing countries that still have government-run telecom sectors or that haven't enabled wireless competition could be further left behind."

Jeff Jarvis, top blogger at Buzzmachine.com and professor at City University of New York Graduate School of Journalism, and many other respondents said we should not concentrate on the appliance but, rather, the connectivity. "We will have many devices that are constantly connected; in that sense, it's connectivity that will be mobile and the devices will merely plug in," Jarvis explained. "This will lead to a world that is not only connected but also live and immediate. Witnesses will share news as they witness it. We can get answers to any question anytime. We can stay in constant touch with the people we know, following their lives as we follow RSS and Twitter feeds."

Respondents Say Mobility Is Key to Sharing Information Everywhere in the World

In 2007 the bottom three-quarters of the world's population included about 30% of the people who have Internet access. The 2020 scenario proposed to survey respondents that this number will rise to 50%. Participants agreed that mobile communications devices—most of the not-yet Internet connected—have made an amazing impact already and will continue to bridge the digital divide and promote digital inclusion. **Geert Lovink** wrote, “We now still look at the world from a ‘digital divide’ perspective, but that will soon be of little use. The massive use by the ‘emerging’ underclasses of the ‘Global South’ of mobile phones should be interpreted as a necessity of the labour force to gain mobility in order to increase their output.”

Charles Kenny, senior economist for the World Bank, the international aid agency, commented, “The mobile phone will be used for an increasing range of services such as m-banking in developing countries, but it will also remain key as a tool for voice communication. For around a quarter of the world’s population still officially illiterate (and many more functionally illiterate), voice telephony will remain the primary means of communicating over distance.” An anonymous survey participant added, “Voice communication is the most common method used by humans to communicate, and devices with voice capabilities will be key.”

Jonne Soininen, Internet Engineering Task Force and Internet Society leader and manager of Internet affairs for Nokia Siemens Network, added, “In many places, having fixed infrastructure is not possible either physically or economically, thus, making mobile systems the viable option for Internet access.”

Active Internet Society and ICANN participant **Cheryl Langdon-Orr** said she takes issue with the figure of 50% of the world being connected, and she hopes for more. “Mobile-device connectivity to the Internet is, indeed, a cost-effective e-future vision for many,” she wrote, “but in my utopia where the Internet Society states, ‘The Internet is for Everyone,’ we would be looking at much more than 50% of people being online by 2020.”

And **Sudip Aryal**, president of the Nepal Rural Information Technology Development Society, wrote, “To meet this target of 50% or even more than that, each and every country should make ICT as a national priority issue. Just like the awareness of HIV/AIDS and use of condoms, the national and international bodies must launch a program to [make others] aware about the ‘importance of Internet in one’s life’ to the grassroot communities.”

Alejandro Pisanty, a leader in the Internet Society and ICANN and director of computer services at the Universidad Nacional Autónoma de México, wrote, "The fact that poverty will not have diminished significantly by 2020 is a painful guarantee that this statement will be true. A large majority of people will only be able to purchase, own, and operate a single device for all their telecommunications tasks, and that will be a small-form-factor, constantly held device—an evolution of our present mobile phones at least in that sense. It will be better integrated than now to the most pervasive device, the television screen, for use at home."

One anonymous respondent wrote that current economic issues in the world will stop the growth in connectedness: "With the trend in the world today with the middle class disappearing, I see the population at the 'bottom' growing more rapidly than the technological advancement, so that growth would stay relatively the same, 30%." Another respondent provided a similar response: "The price will be high; the standards diverse; the connections for broadband wireless sporadic and different technologies. The bottom of the pyramid will not count for a significant part of the Internet but, rather, remain marginalized."

Michael Botein, a telecommunications law expert at New York University and consultant to the Federal Communications Commission (FCC), said improved, affordable mobile technology could help pave the way to a friendlier world. "It is difficult to foresee a future short of a technological breakthrough in which mobile technology will have enough bandwidth to provide data services, real-time video, and the like," he wrote. "On a positive note, however, cellular will allow the beginnings of universal service in most parts of the world—as already in Latin America and Africa—and thus may help break down long-held hostilities."

Several respondents, including **Neil McIntosh**, director of editorial development for the top news site, http://www.guardian.co.uk, based in London, said, "A greater and more fundamental problem, however, may be poor literacy and continued widespread poverty, which technology by itself can't solve."

One anonymous survey participant suggested that texting by phone is also a likely way communication will continue to improve in developing countries: "Tools which take SMS/texting to another level will create additional ways to connect, communicate, and collaborate cheaply before the Internet in the developing world."

Some Experts Express Doubts About Interoperability and Open Networks

Some of those who chose to mostly agree with this scenario did so while expressing reservations about parts of it. A number of them suggested that governments and/or corporations concerned with retaining or gaining more control over use of the Internet might limit some types of connection in certain parts of the world, and others projected a potential lack of universal standards and protocols in a world of changing technology.

An anonymous respondent noted, "The pace at which affordable access will reach the poorest majority of the world's population is overestimated. See what happens with basic commodities such as tap water. Industry will continue targeting the richest segment with fancy 'innovations.'" Another respondent wrote, "If the Internet doesn't evolve by then, the Internet on cell phones will remain limited to the wealthiest nations." And another wrote, "It's difficult to imagine mobile telephony standards evolving to the point where international roaming is effortless. The business and regulatory drivers would seem to prevent this."

Another anonymous respondent wrote that interoperability and portability as proposed by the scenario will only take place outside of the United States due to "vested interests of competing telecom/Net firms...This will put the U.S. as a whole at risk of lagging seriously behind the rest of the world in technology innovation."

Michael Zimmer, resident fellow at the Information Society Project at Yale Law School, wrote, "I agree almost entirely with this prediction...My only hesitation is whether there will be universal standards and protocols accepted by most operators internationally, since U.S. mobile providers have shown little interest in providing full interoperability and open devices to take full advantage of new mobile services."

Social media research expert **danah boyd** of Harvard University's Berkman Center for Internet and Society, wrote, "Traditional carriers have little incentive to include poor populations, and the next 5 years will be rife with battles between carriers, municipal, and federal governments, handset makers, and content creators. I don't know who will win. If the carriers continue to own the market, network access through mass adoption of the mobile will be far slower than if governments would begin blanketing their land with Wi-Fi (or network access on other spectrum channels) as a public-good infrastructure project and handset makers would begin making cheap accessible handsets for such access. The latter dynamic would introduce network access (and telephony) to many more people, much to the chagrin of carriers."

An anonymous respondent expressed concern about repressive regimes: "Mostly agree with this potential scenario—one caveat may be the way third-world governments act to control Internet access. This sort of empowerment could be construed as a threat to their centralized power and could severely hamper the progress."

Ross Rader, a member of the ICANN Registrars Constituency and executive for Tucows Inc., wrote, "This scenario may likely happen over the next few years, not the next 12. The only real obstacle to this level of adoption and social integration lies with the willingness of the telecommunications industry to resist the temptation to segregate and verticalize its offerings. In other words, the communications network market must be made much more competitive than it is today. Handsets need to be freed from applications, and applications need to be freed from networks. Only truly open networks will drive the sort of adoption envisaged in this scenario. We are starting to see the first glimpses of this today with Google's Android, Verizon's Open Network Initiative, the power of the iPhone; but much work in all of these, and other, areas remains to be done before the networks, applications, and handsets markets are fully competitive."

A few respondents said they believe corporate leaders are interested in the positive diffusion of affordable technology tools to less developed areas of the world. **Peter Kim**, a senior analyst for Forrester Research, commented, "Handset manufacturers have already started to focus on countries with lower GDP [gross domestic product]. Continued efficiency in production and increase in computing power, along with the natural desire of humans to connect, will help make this scenario a reality."

Scott Smith, analyst, consultant, and futurist with Changeist, formerly with Jupiter and the Yankee Group, wrote, "The attractiveness of emerging markets and the overall positive halo technology companies perceive as a benefit of being involved in serving these markets will drive major technology developers to create inexpensive, durable, flexible devices for the developing world. This development will coincide with creation of mobile phones more adapted to the needs of the developing world—cheaper, but more powerful. Telephony by this time will most likely have shifted heavily toward IP-based, software-driven VoIP [voice over Internet protocol] solutions, rather than using today's networks."

Many survey participants expressed concerns about pricing. One anonymous respondent wrote, "The success of the mobile phone as a universal-access device is contingent on adoption of flat-rate style charges, as is normal for Internet applications, rather than high per-minute charges which currently dominate mobile pricing structures."

Another respondent wrote, "Corporate interests will drive up access prices and limit cell phones to mostly a voice-communication tool for the general population." Another responded, "Considering that I can't get an affordable mobile phone that works in various parts of just the United States today, I'm not holding my breath for this." And another pointed out that the United States is lagging due to its marketplace and regulatory structure, "This scenario cannot go forward unless or until the United States joins the world's telephony standards, which it so far has declined to do. Or, perhaps the U.S. will just be left behind, as it already is in so much technology."

Some hope in regard to affordability was expressed by another anonymous respondent: "I live in Brazil, and mobile phone sales have transformed the society here. A telephone was very expensive in the '80s, yet mobile gave voice for all in a short time. I guess we still lack initiatives that involve users and mobile [phones] in a more citizen and informational way, but this will end up happening."

And one anonymous contributor to the survey saw some positives in varied systems. "Mobile phones will be a primary access device," he wrote, "however, I think that there will not be universal protocols. We will find that information diversity will be just as important as biodiversity."

Bandwidth, Screen Size, and Poor User Interface Are Among the Other Potential Limits Cited

Some respondents who mostly disagreed with the scenario wrote that delivery will continue to be more efficient through earth-based connections. "Wireless doesn't ever provide as much bandwidth as wired connections; wireless will always be slower, thus second best," wrote one anonymous respondent. "Primary 'work' will still be done over wired connections, with wireless filling in the gaps and supporting mobile applications." Another wrote, "Will there be enough wireless infrastructure for truly complex Internet applications on a phone?"

Another more multilayered response in regard to limitations of the scenario came from an anonymous survey participant, "Wireless technologies have a number of inherent problems including, but not limited to, interference and capacity. The simple log trend of traffic and data patterns precludes wireless. While some form of ubiquitous wireless access will be available most places, fibre will be more important than ever. Phones also have UI [user interface] restrictions, any conception of phones without other peripheral interfacing technologies such as HUDs [head-up displays that allow a user to see information on a surface at eye level—usually on a transparent surface] and eye movement/brain interfaces simply will not meet the needs."

In the most developed areas of the world today, people are accessing the Internet on 15- to 20-inch flat displays, and 40- to 50-inch plasma or liquid crystal display (LCD) screens are commonly found in people's homes. Even larger innovations such as the multitouch wall are expected to be diffused soon, so there was some pushback from respondents on the idea of a small device being the "primary" connection for most people in the world.

"Mobile phone[s] will be used by more people to connect but not be used as often as laptops and desktops, as some Net functions need the bigger device," wrote one anonymous respondent.

"Unless the phone—which will really be seen as the one device that we carry around that includes voice, text, still/video camera, GPS, AV player, computer, voice-to-digital-information interface, Internet, television, bank account, etc.—has the capacity to project at least a 15" display, it will be too small to use as the primary connection tool for the majority of worldwide users," wrote **Peter Eckart**, director of health information technology for the Illinois Public Health Institute. "The majority of us will carry our digital presence indicator with us from place to place on that device, but the bandwidth and interface will be provided by our home or work or coffee shop, with the device there to maintain digital identity. I do agree that the mobile device will be the primary or only connection for poorer folks. People's wealth or income will be reflected in the size of their display, the number of Ds (2 or 3), their connection speed, amount of digital storage, and most importantly, their level of access to information stores."

Adrian Schofield, a leader in the World Information Technology and Services Alliance and manager of applied research at the Johannesburg Center for Software Engineering in South Africa, wrote that people will use multiple devices. "The phone will be the instrument that enables the less economically empowered people to communicate by voice and text and to perform basic financial and government transactions," he commented. "The PDA will offer the full range of communications and computing facilities, including TV, GPS, and video camera. Using improved solar technology, battery life will be significantly extended, and offices, hotels, and other venues will provide free plasma screens for those who wish to access a larger image than the one offered on the device."

John C. Abell, the new-media project director for the Committee of Concerned Journalists, said it is most likely that people will vary their use of devices, as they do now, writing, "There are fundamental limits to the range of activities possible with a small input and small screen. It is more likely that connectivity will support a device-of-choice world."

Well-known economist and technology expert **Hal Varian**, of Google and the University of California–Berkeley, responded, "The big problem with the cell phone is the UI, particularly on the data side. We are waiting for a breakthrough."

Fabrice Florin, the executive director of NewsTrust.net, a nonprofit social news network, wrote, "Other computing platforms and connectivity options will become widely available by then, such as cheap computers (or wall-based computing environments) with landline or comparable broadband connections. I predict that these faster connections and larger screen platforms will be more affordable and effective from a productivity standpoint than small and slow mobile platforms."

Louis Houle, president of the Internet Society's Quebec chapter, wrote about the dangers of being distracted by mobile computing and communication: "Governments are concerned by the lunatic behaviour of mobile phone users. The first step will be to regulate the use of your phone in your car."

Fredric Litto, a consultant for Pearson Education Global E-Learning and president of the Brazil Distance Learning Association, wrote about tech plans in the pipeline to improve the computer/phone interface. "In early 2007 I saw and handled a normal-sized mobile phone which had the capacity to project, on a horizontal surface before it, a virtual keyboard and to project, on a vertical surface behind it serving as a 'screen,' all the information that would normally be displayed on a laptop or desktop monitor. The apparatus is currently on the market. Surely, these characteristics ensure the universality of use of the mobile phone, bringing together telephony, television, and computation. The future looks good!"

ONE LAPTOP PER CHILD IS SEEN AS LIMITED

One Laptop Per Child is a large-scale, U.S.-based project to provide affordable, practical computing and Internet capabilities to people in underserved communities around the world. The effort has brought together people from the technology industry, nongovernmental organizations, and governments in the process of designing, manufacturing, and distributing these tools.

The *Future of the Internet III* survey was distributed at about the same time that the OLPC computers became available; they have come under some criticism in the popular media, and they met some criticism from survey participants. **Scott Smith** wrote, "OLPC-style efforts are already beginning to fragment at the start of 2008 even before the actual OLPC initiative gains any real ground." **Seth Finkelstein** wrote, "One Laptop Per Child is a classic 'ugly American'-style project."

An anonymous respondent wrote, "One Laptop Per Child and similar schemes will be irrelevant. Technology will continue to become more pervasive due to affordability, rather than through subsidised technology."

Charles Ess, an online culture and ethics researcher from Drury University and a leader of the Association of Internet Researchers, commented, "The One Laptop Per Child initiative is foundering not so much on issues of economics but more on issues of culture. Most of the non-Western 'targets' for the initiative use languages that are not easily captured through the use of the standard Roman keyboard. More broadly, the literacy required to manipulate most computer-based communications technologies and venues is not to be taken for granted among all populations and demographic groups—certainly not within the U.S. and Western Europe, much less through other cultures in which orality still predominates (e.g., indigenous peoples). For that, mobile phones present a relatively straightforward interface—and talking, for most people at least, is easy! In short, talking via a phone is far more universally realizable than presuming everyone will be able and willing to communicate via a Roman keyboard and an expensive computer."

Fernando Barrio, senior lecturer in business law at London Metropolitan University, went so far as to say OLPC may slow digital inclusion. "As soon as the content providers and governments start to produce phone-friendly pages, the phone will be the tool that will lead to an explosion in the number of Internet users and activities carried out by [the] Internet," he wrote. "Initiatives like One Laptop Per Child, while extending the market of the current content providers without major investment from their part, have as a main result the delay in bridging the digital divide through the use of mobile phones."

Some Say 2020 Will Offer a New Paradigm

Some survey participants said this scenario as written is shortsighted, and we will have moved into a different communications environment. "A new technology will blow all of this away," wrote one anonymous respondent, and another wrote, "Another 'killer app' will emerge before 2020 that will change everything; communication will not achieve stability in the 21st century."

Josh Quittner, executive editor of *Fortune* magazine and long-time technology journalist and editor, wrote, "The notion of a 'mobile telephone' in 2020 is quaint. Telephones in 2020 will be archaic, relics of a bygone era—like transistor radios are today. Telephony, which will be entirely IP based by then, will be a standard communications chip on many devices. We'll probably carry some kind of screen-based reading device that will perform this function, though I assume when we want to communicate verbally, we'll do so through a tiny, earplug-based device."

Several anonymous respondents wrote that we need to be inclusive of all ties, human and nonhuman, when we look to the future of connection. One wrote, "In 2020 the network will be a network of things (sensors, etc.), and people will interface with information from the network of things in many ways—including personal devices which may or may not look like cell phones." Another wrote, "Just because a cell phone has the computing power of a PC does not make it a usable PC. A more believable scenario is that a person's phone is one of the 'keys' that establishes an identity, and what we know as 'Internet access' today becomes something...you just walk up to, part of the surroundings. You'd walk up 'close' to a display surface, and it would know who you are by pinging your phone/wallet/[whatever]."

Mike Treder, executive director of the Center for Responsible Nanotechnology and an expert on the social implications of emerging technologies, responded, "It shows a lack of imagination to assume that mobile phones as we know them today will still exist in 2020. While I agree that desktop computers will no longer be the standard interface for Internet connection by then, it seems far more probable to me that some form of ubiquitous wireless communication that goes beyond today's mobile phones will have taken over."

An anonymous respondent wrote, "We'll all have our own personal information communicators. Mobile phones may well have changed to be small voice-activated devices which sit behind your ear to which you give voice commands, e.g., 'Phone Tom,' 'Google Search: how to salsa,' 'weather for London,' 'train times to Liverpool on Wednesday,' etc."

Another respondent wrote, "I don't know if we will only be using mobile phones in 2020. I am hoping for some sort of paper-thin, flat mobile device that is like the eBook or Kindle but can do much, much more than just display a book, but also get the daily paper and act as a touchscreen computer. Or, ideally, we will have come up with materials that can change shape according to whatever we need the mobile device to do. It can be long and thin for phone calls but wide enough with a keyboard for typing. It wouldn't just be mobile phone origami—which we sort of have already—but something much more comprehensive and flexible and durable."

Hamish MacEwan, a consultant at Open ICT in New Zealand, enthusiastically saw an edges-oriented future. "The mobile Internet will dominate usage, but the device will be very different in 13 years from our concept of a 'mobile phone,'" he explained. "So will the providers of connectivity, and another group will provide the services and content. Universal standards will not control access, already WiMAX and other nonproprietary standards are being deployed in competition and combination with the legacy-integrated solution required in the cellular environment...Does your scenario imagine or imply that the legacy dominance of vertically integrated telecommunications services will return? If so, you are very wrong. Operators no longer define the service or the future; the edge, the customer, is now in charge. While we may temporarily embrace or endure the closed proprietary model, with an operator elite, the trend is towards decentralisation, towards control by the edge, with devices that will utilise whatever connectivity is available in a transparent and open mode. As Feynman and Rangaswami and others have explained, there is plenty of room at the bottom."

And **Jonathan Dube**, president of the Online News Association, director of digital media at CBC News, and publisher of CyberJournalist.net, wrote, "It's highly unlikely that telephony will be offered under a set of universal standards and protocols accepted by most operators internationally. More likely, telephony will merge with Internet technology and the two will fuse, so that everyone who is using a mobile phone will always be online and everyone who is online can easily make connections via voice and video. Who knows, maybe by then we'll be too busy running from our robot overlords to spend much time on our mobile phones."

The Current Outlook in the Development of Mobile Devices

Media, government, and industry reports in 2007 and 2008 indicated that the cell phone is currently seen as the top tool for leveraging social and economic opportunity. Sales figures confirm this. In 2007, 1.12 billion mobile phones were sold compared with just 271.2 million PCs and laptops. A 2007 survey on use of mobile phones in the United States by the Pew Internet & American Life Project found that 19% of respondents were already reporting they use their phone to send e-mail; 58% of American adults said they had used a cell phone or PDA to perform at least 1 of 10 mobile, nonvoice data activities, including text messaging, Web surfing, searching for maps or directions, taking a picture, or sending e-mail.[3]

Eighty percent of the people in the world now live within range of a cellular network. The International Telecommunications Union reported that more than 68% of the world's mobile subscribers live in developing nations—this is the first telecommunications technology to have more users in the developing world than in the world's most developed nations.

In Africa, usage is growing at close to 50% annually, and more than 30 African nations have more cell phones networked than landline phones. Cell phones do not demand the high level of literacy required of computer users. Wireless networks are not as expensive to build and maintain as traditional fixed-line networks. Cell phone users don't even have to have a permanent address—they can buy a phone and a prepaid airtime card and communicate anywhere.[4]

In "The Next 4 Billion," a recent report about the 4 billion low- or no-income people living at the base of the world's population pyramid (BOP), it was found that as a family's income grows—from $1 a day to $4, for instance—additional spending on telecommunications is the first consideration in finding a way to leverage even more success.[5] At the BOP, people are buying phones and airtime.

In 2005 a London Business School study reported that for every additional 10 cell phones per 100 people, a country's gross domestic product raises by 0.5%.[6]

Ownership of a cell phone in the developing world gives a person a "fixed location"—his or her telephone number—even if displaced by war, natural disaster, or life circumstances. Public health workers can use text messaging to connect doctors and nurses with homeless or itinerant patients. Fishermen in India have begun selling their catch to the highest bidder before they even bring it ashore. Village phone-sharing programs are making a significant difference in rural areas of Bangladesh, Cameroon, Indonesia, Uganda, and other places; phone sharing is so common that Nokia has begun to produce phones that include individual address book spaces for as many as seven people per phone.

The popularity of mobile phones for people at the base of the pyramid has led to a global movement toward "mobile banking," also known as m-banking, which is allowing people in remote areas to make purchases, forward payments, and withdraw cash over the phone.

Some governments see cell phones as a threat. They are illegal in North Korea, and their use is restricted in some countries, including Myanmar.

The convergence of fully functional computers and phones has been well under way for some time, but breakthroughs in battery capabilities, user interface, and costs are allowing for revolutionary change at a rapid pace. The Organisation for Economic Cooperation and Development's 2008 economic and social regulatory objectives for voice, video, and data services include universal service requirements; promotion of competition; consumer protection; consumer access and choice; access to emergency call services, media plurality, and cultural diversity; data security and consumer privacy; competitive markets offering essential services at cost-based prices; and continued innovation and investment in new technologies, services, and applications.[7]

When Apple's Internet-capable iPhone was introduced in 2007, it became a global phenomenon. Other companies are preparing their computer-phone crossovers. Nokia's entry is expected to be released late in 2008; it has been code-named "Tube" and supports Java and the ability to upload photos. The Phillips X800 is a phone with a touchscreen similar to the iPhone's. Many phones offer slide-out QWERTY keyboards. HTC's X7510 Advantage smartphone has a 5" display. Intel, the world's leading PC chip manufacturer, has started producing processors for what it is calling the mobile Internet device (MID), which is considered to be an emerging product class. Some nanotechnology researchers say it is possible that palm-sized devices with supercomputing capabilities could be developed in the next 10 to 15 years.

Members of today's younger generation are growing up in a world in which they expect and utilize constant network connections. In the most developed nations, today's teens are already maintaining nearly constant connections through Instant Messenger (IM), Twitter, texting, social networking, and voice calls. There is an expectation that consumer-oriented devices and software will continue to offer expanding capabilities. At the Usenix technical conference in Boston in June 2008, Adrian Cockroft, a former Sun Microsystems distinguished engineer now with Netflix, said smartphones will have eight times more CPU power by 2010. This could enable a mobile, continuous-computing device that could also connect wirelessly with your car when you are driving, with a desktop monitor or entertainment-sized flat panel at the office and at home, or with a 3-D portable display wherever you are. Cockroft said the processing capabilities of handhelds are already doubling memory capacity annually.

Google developed Android, a mostly open-source software package and operating system for mobile devices that was first demonstrated in late May 2008 and expected to be available and adopted by users in early 2009. Samsung, LG, HTC, Dell, and other device producers will offer it; Google joined 33 other companies including these to form the Open Handset Alliance in 2007 to bring industry leaders together to develop open standards for mobile devices. Many major companies are not members of this alliance, including Nokia, AT&T, Apple, and Verizon.

The complexities of the wireless industry in the United States are numerous. The marketplace has been dominated by a few companies—AT&T, Verizon, Sprint Nextel, T-Mobile, and Alltel. Verizon is purchasing Alltel, which narrows the field of leading wireless companies in the United States to four.

The number of mobile broadband subscribers using 3G high-speed packet access technology (speeds of 1.8Mbit/second to 7.2Mbit/second) increased by 850% in the past year. The first 3G systems were capable of data transmission at speeds up to 384Kbit/second. NTT DoCoMo is testing Super 3G, which has achieved a download speed of 250Mbit/second in trials. This service probably will not go commercial until 2010 at the earliest. Sprint Nextel is now testing WiMAX, with expected typical speeds in the 2Mbit/second to 4Mbit/second range, and Verizon is developing LTE, a competing system with similar speeds.

A $14.5 billion venture to create a national WiMAX carrier—Clearwire—is being undertaken by Sprint Nextel, Intel, Google, Comcast, Time Warner Cable, and Bright House Networks to provide service for up to 140 million U.S. customers by 2010.

Today's phones generally only connect to a single type of cellular network. Future smartphones will have software capable of sensing various active networks within range and automatically reconfiguring themselves to connect efficiently with the most appropriate one, working equally well on a WiMAX network in Korea, a GSM system in Europe, or an evolution-data optimized or evolution-data only (EV-DO) network in the United States. Chip vendors are continuing to find ways to make smaller, more powerful chips that use less energy—always a concern with mobile devices. Other issues tied to mobile computing include the privacy and security challenges presented when working on public networks and health implications, since it is still unknown whether human health can be damaged by cellular radio-frequency emission.

Major corporations including Cisco and Microsoft are developing plans for a future in which all communications converge; it is known by various names, including "superpresence." This would allow people to communicate through the most efficient and personalized method that is available from moment to moment.

Most consumers are hoping that device manufacturers, industry consortiums, and network operators will cooperate, avoiding walled gardens and building an open system in which many devices can interact economically and globally on different platforms.

The London School of Economics operates a research network centered on what it refers to as "Mobile Interaction & Pervasive Social Devices," offering a compilation of research and articles on the topics at http://mobility.lse.ac.uk/. The Society for the Social Study of Mobile Communications, sponsored by the Institute for the Future, based in the United States, has a mobile society online group with 339 members at http://groups.google.com/group/mobile-society/about.

Additional Responses

Many other participants in the survey shared valuable input on the mobile communications question. Among them:

> "The large-scale models for connectivity, terminal instruments, and use are likely to be set by China." —**Anthony M. Rutkowski**, cofounder of the Internet Society and a founding trustee; longtime leader

in International Telecommunication Union; vice president for regulatory affairs, VeriSign

"'Telephony' is an odd word in this context. That would be like calling e-mail 'textlephony.' The 'phone' by 2020 will be as much computer as voice appliance, and the universal standards and protocols will be data centric rather than voice centric." —**Clay Shirky**, consultant and professor in the Interactive Telecommunications Program at New York University; an expert on the social and economic effects of Internet technologies

"Untethered, 24/7 Internet access will fundamentally redefine how we use information, especially when combined with low-cost, heads-up display technologies ('eyePods,' perhaps?) Augmented reality will automate recognition of real-world objects and 2-D/3-D overlays and interactions should be well within the processing power of an inexpensive handheld device by 2020." —**Jason Stoddard**, managing partner/strategy at Centric/Agency of Change, an interactive strategies company; he is also a popular speaker on social media and virtual worlds

"Three strong drivers—first, Moore's law still drives the diffusion of powerful information and communication technologies to wider and wider markets at lower and lower prices. Cheap chips, cheap phones—manufacturers will continue to deliver less expensive, more powerful mobile devices as they seek to expand into new markets and the price of processors drops. Second, the possibility of openness. Although operators remain significant bottlenecks, fighting to prevent the flowering of innovation from anyone that the Internet's open platform enabled, early indicators like VoIP, the Google open-telephone effort, growth of Wi-Fi, experiments by Nokia might signal that operators might not be able to maintain the degree of control they have exerted so far. Openness is not guaranteed in the world of wireless broadband networks—but it looks possible." —**Howard Rheingold**, Internet sociologist and author; one of the first to illuminate virtual communities; author of *Virtual Reality*, *Smart Mobs*, and *Virtual Community*

"Mobile phones, PCs, TVs, and other devices will all play important roles." —**David Moschella**, global research director for the Computer Sciences Corporation's Leading Edge Forum; *Computerworld*

"The mobile phone is just beginning to touch our digital lives. As these relatively inexpensive devices continue to improve in performance and connectivity, they will serve more as a 'remote control' to many of our electronic touch points, such as purchasing retail items through scanning, serving as an electronic passport, turning on our cars/GPS systems, translating text to talk, and video conferencing with our friends through our online accounts. As we improve visual projection and 'plugging in' to larger display systems, mobile phones can serve as the old laptop we once knew and lugged." —**Drew Diskin**, director of e-strategy, Johns Hopkins Medicine

"Basic AI [artificial intelligence] and speech recognition should improve to the point that many tasks can be done with devices that lack good screens and keyboards, though more intense tasks will demand those things." —**Brad Templeton**, chairman of the Electronic Frontier Foundation, also of the Foresight Institute, is an Internet pioneer who has been active in the computer-networking community since 1979

"Mobile Internet penetration will need a very strong combination of wireless and satellite media in order to achieve ubiquity." —**D. K. Sachdev**, founder and president of SpaceTel Consultancy LLC; early developer of XM Radio

"I agree only if the 'telephone' has really migrated into a computer/telephone, as seems likely. If phone only, or phone with limited capability of today, then it is not robust enough and the $100 laptop with Wi-Fi will predominate." —**Ed Lyell**, professor of business and economics, Adams State College, Regis University, San Luis Valley Board of Educational Services; pioneer in issues regarding Internet and education

"This is more or less a baseline scenario, with the exception of the international interoperability. That seems a bit optimistic." —**Jamais Cascio**, originator of Open the Future, also works with the Institute for the Future, Institute for Ethics and Emerging Technologies, and Worldchanging

"However, the Internet access won't be the way we typically think of it—Web surfing, e-mail, social networking. It will be used in very limited fashion—as limited communication mechanisms for simple business transactions and reaching family members." —**Dian Schaffhauser**, writer and editor for CampusTechnology.com, THEJournal.com, *Redmond Magazine*, *Computerworld*, and Web Worker Daily; founder of Sourcingmag.com

"The mobile phone systems have become the first means of telecommunication. It is only natural that the same system and paradigm is used also for Internet." —**Jonne Soininen**, manager of Internet affairs for Nokia Siemens Networks, and formerly system engineering manager for Nokia; leader in Internet Society and the Internet Engineering Task Force

"Mobile technology will take over in providing access not only to Internet but other multimedia technologies, namely, TV. From a business scope, telecom operators and device manufacturers will target to engage more people usage by implementing the 'economy of scales' notion." —**Hanane Boujemi**, ICT researcher for DiploFoundation, working on educating people about Internet policy and Internet governance, Malta

"Cheap, small, easy-to-maintain phones with handheld computer capabilities have become essential for people in unstable climatic and political situations." —**Cliff Figallo**, social innovator and original member of the first online community—The WELL, now of AdaptLocal.org; expert in fitting and implementing social Web applications to groups

"The mobile phone is actually a killer app. It might be replaced by a new communication tool in a near future for your social reasons: accessibility, usability are among the threats that will chase the actual phone from the front scene. Users will maintain some interest for the phone, but video (now available by phone) demands a screen that mobile will not entirely satisfy for obvious reasons." —**Louis Houle**, president of the Internet Society's Quebec, Canada, chapter

"Assuming the term 'mobile phone' means a device capable of multiple access methods (WiMAX, 4G, 5G, etc.) to deliver TCP-/IP-based services to the user—i.e., all services on the device are delivered over IP." —**Tze-Meng Tan**, Multimedia Development Corporation (MDeC), Malaysia, and director at OpenSOS; Internet pioneer as a member of the team that started Malaysia's largest Internet service provider (ISP), TMnet

"Demographic assumptions are OK; affordability and interoperability assumptions are much sketchier. The basket of 'typical goods' for those with terrestrial (broadband) access will be so vastly different than the services available over low-cost, handheld wireless that it will no longer be plausible to reckon the two as part of the same 'market' (think CB radio vs. telephony)." —**Tom Vest**, IP network architect and consultant, RIPE NCC Science Group, Office of the Chief Scientist; consultant for the Internet Society, and for OECD Economics and Statistics Division

"Internet access will not be carte blanche, but most likely specific data-oriented services like fish market data or microloan updates. Literacy and the predominance of Mandarin and English on the Internet will be the largest encumbrances followed by cost and regulation (censorship)." —**Todd Spraggins**, strategic architect, Nortel Carrier Networks; president and chairman of the board of directors of the Communications Platforms Trade Association

"I disagree with that statement because a mobile phone offers limited display and interactivity real estate to the user…I believe all common objects will be connected to the Internet—TV, watch, clothes, car, phone, music player, walls, lamps, tennis racquet ;)—and will all be gateways to the digital world in different fashions." —**Alexandre Winter**, cofounder and chief executive officer, LTU Technologies, a global leader in image search and recognition technologies

"Having visited Tanzania in the early 1990s, I was struck by the lack of physical infrastructure for landline phones. It was very clear that they were being leapfrogged by cellular communications. Therefore, it would not be strange at all that similar landline infrastructure for Internet access would be ignored for cellular access." —**Megan Holbrook**, partner, Kapow Inc., a site-design and development company; she has produced projects for Microsoft, Warner Bros., and Disney

"Power consumption will be a prime concern, and the mobile devices will be much of what people need to consume information, create and contribute, and generally communicate. Advances with solar power for small devices greatly lower the power need for the devices and the transmission towers." —**Thomas Vander Wal**, principal and senior consultant, InfoCloud Solutions Inc.; coined the term *folksonomy*; expert on tagging, the social Web, and social information use and reuse

"Mobile phones also replace our wallets, doubling as identification devices, credit cards, and car keys using RFID [radio frequency identification] tags or a similar technology." —**DJ Strouse**, international relations and computer science student, University of Southern California

"Mobile phones are, without doubt, the most useful technology to connect people around the world and particularly in developing countries. However, the success of the penetration of this technology remains to be seen in the years to come and will very much depend on the prices offered to users. Developing countries have some of the most expensive prices per minute. A competitive environment should be encouraged through public policies and laws to force mobile phone companies to not only reduce the prices but also to allow compatibility with other technologies such as IP voice in order to increase mobile phone penetration." —**Cristos Velasco**, director general of North American Consumer Project on E-Commerce; Mexico-based attorney and active in the Internet Governance Forum and Internet Society

"The mobile device (not phone only) will be the preferred ICT tool of choice for most people. Speech-recognition technology will have changed a lot of how we communicate." —**Robin Gunston**, consulting futurist for Mariri Consulting, a strategic and business planning company

"While I agree that cell phones will be the primary connection tool, that will be because of its relatively lower cost and broad abundance throughout the world. I believe people will still prefer to connect via computer rather than handheld device simply because of the more robust computing capabilities available, larger screen, and other convenience features they offer." —**Janie Graziani**, manager of new media and technology for the American Automobile Association

"I think this is true, however, I also think that a big breakthrough will have to do with expanding the phone interface, so that it can ergonomically match the keyboard and computer. One of the easiest/cheapest methods would be the development of cheap 'terminals' with keyboard and screens, which you connect to your phone that does the computing. Of course, the crucial thing for success will be the 'universal standards' part and the degree to which these standards will result from governmental/world organizations, or whether it will result from a monopolistic company that corners most of the market, so thus becomes the default standard. The latter scenario could be a disaster." —**Richard Hall**, professor of information science and technology and codirector of the Laboratory for Information Technology Evaluation, Missouri University of Science and Technology

"People are mobile; they want their Internet access to be mobile also. The key barrier to realizing this ideal world is the price of mobile data access. But it won't happen unless governments break up telecommunications monopolies and require compliance with universal protocols. These steps will ensure the real competition needed to drive down prices and increase access." —**Lea Shaver**, A2K (Access to Knowledge) program coordinator, the Information Society Project at Yale Law School; research includes developing a cross-national access to knowledge index

"While I think that OLPC is a great program, I don't see it succeeding to this extent. Given the different stages of cell infrastructure around the world, I don't see them converging on a single standard by 2020." —**Ben Spigel**, master's student in the department of geography, Ohio State University; researches micro-geographies of academic knowledge exchange

"It rather depends on what you mean by a mobile phone, of course. But if you consider the type of devices we have these days, then screens and functionality are very limited. Functionality issues may be solved by voice-based interfaces, but screen size will surely remain a problem, especially as the Web becomes more capable of delivering screen-based entertainment. So, no, I don't see people turning to mobiles to enjoy the Internet, although they will 'snack' online via mobiles." —**Jeremy Swinfen Green**, Telecom Express, an interactive marketing company

"I am a bit skeptical about the optimum economic output from many other diversified uses." —**Hakikur Rahman**, chairman, SchoolNet Foundation; coordinator of Sustainable Development Networking Programme in Bangladesh; active in Internet Society Board, South Asia Foundation

"The primacy of the mobile device ('phone') seems assured in the much shorter term. By 2020, we'll have global interoperability either from the top down (ITU-esque alignment of the systems) or from the bottom (phones capable of working on all major networks)." —**Alexander Halavais**, professor and social informatics researcher, Quinnipiac University; explores the ways in which social computing influences society

"I'm skeptical about the widespread adoption of technical standards, particularly with regard to telephony. Also—the term 'mobile phone' will probably be quite redundant imho [in my humble opinion], probably identifying you as an anachronism of the fixed, voice-only era. In 2020 my son (now 6 months old) will turn 13. He'll think 'phone' and 'mobile' are endearing terms (like we think of 'the wireless' or 'the terminal'...). With regard to mobility, I'm pretty sure that fixed displays (capable of greater size, etc.) will play a large part in communications too, but, as with the increased complexity of a/v communication, these fixed elements will be more integrated into a complex ecosystem of devices and protocols." —**Jeremy Yuille**, digital media coordinator at RMIT Communication Design and program manager at ACID in Melbourne, Australia; previously director at IXDA, the Interaction Design Association

"Mostly agree, but I don't think this in any way erodes the good work of the One Laptop Per Child project. Mobile phone space and Internet space still have very different qualities." —**Christine Satchell**, PhD, senior researcher, Institute for Creative Industries and Innovation, Queensland University of Technology

"This is likely to be the dominant means of Internet penetration in developing countries, but Internet access for most in the developed world will also have migrated to cell phones or cell phone–like devices. Laptop- and desktop-tethered Internet access will be mainly for offices and those who use it professionally." —**Jade Miller**, PhD student Annenberg School for Communication at the University of Southern California, with a research focus on global flows of information and culture

"This follows Nokia's ideal of the 'minicomputer' or 'computer in my pocket' notion that as handsets like the 'n' series, including the n95, were designed in the early 2000s to be a minicomputer that could be carried about on the person. Research that was carried out in 2005 in Japan by Ito et al. (2005) has shown how important the portability of this kind of technology—that relies on being portable and personable. Other research that has taken place in India by Jan Chipchase has shown how limited infrastructure in terms of broadband and wireless can encourage a 'leapfrogging' of technology, where the PC and laptop were not the primary method for Internet connection, as the mobile phone was a technology that was more easily available, affordable and reliable in this area." —**Maz Hardey,** social analyst, blogger, "defender of new media" completing a doctorate funded by the Economic Social Research Council in the U.K., based at the University of York

"While I'd agree that the 'majority' of people who have (personal/family, etc.) Internet access will have mobile access, I'm not sure it will be the majority of the world, simply, the majority of the connected world. For educational purposes, it's more likely to be computer (including laptops, etc.) based. Probably business will be mobile oriented (if people only have one way of accessing the Net)." —**Emma Duke-Williams**, lecturer in the School of Computing, and researcher, University of Portsmouth, U.K.; education blogger

"The statistics on the penetration of mobile handsets show that phones are a significant platform. If you include voice calls and SMS as a baseline data entry point, perfectly reasonable in answer to the stated query 'primary connection tool' that doesn't explicitly focus on 'data connection,' then this statement is already true, in that most people can receive very useful up-to-date information via SMS or voice-enabled services in most countries in the world. These services may well be hosted on platforms that provide Web, SMS, and voice access to the same back-end data. The same cannot be said of any form of computer (PDA, laptop…). The trend is clearly for more computing complexity in the phones, and richer interfaces, at least, in some segments of mobile handsets (business smartphones, and consumer media phones in particular). If and when voice telephony is affordable and of good quality over the data networks on these devices, then the voice will have become just another service on these personal communications devices; this has not quite happened yet. It will require the development of open SIP [session initiation protocol] stacks for handsets that utilise the full power of the inbuilt codecs. If all these phones were also IPv6 enabled, as was originally envisaged in the IMS strategy of the 3GPP and others, there is a huge potential for new innovative peer-to-peer services allowing these handsets to exchange information with each other." —**Micheál Ó Foghlú** research director, Telecommunications Software & Systems Group, Waterford Institute of Technology, Ireland Advisory Committee; member of W3C; blogger

"Although it is highly likely that a large proportion of people in the 'bottom' three-quarters of the world's population would have access to a mobile phone, it is highly unlikely in my opinion that they would be accessing the Internet. There are two primary reasons for this: (1) The capacity of the cheap mobile phones being planned is not that which is required for Internet access. See, for example, the Rs. 1000 phone to be launched by BSNL in India—these are pretty basic phones to be deployed over the next 5 years; (2) Even if the equipment were available, it is questionable whether the level of awareness and education required to use the Internet on a mobile phone would be able to be provided. In the first instance, people would have to consider using the Internet, and thus learning about it is a *necessity*. Will it be necessity, compared to food and basic literacy? The Internet is far more complex than transacting a conversation, which is a small extension of current cultural and individual practices." —**Amit Kelkar**, consultant and sociology researcher, Postmodern

"I only mostly agree if there are some breakthroughs in UI so that the limitations of screen size and entry are addressed. I do believe this is possible. We have to assume that the cost of the devices will approach zero as well." —**Robert J. Berger**, CTO for Cinch; expert on backbone networks, access networks, wireless networks, and innovative Web applications, a frequent speaker on next-generation networks

"Of course, this is not the mobile phone as we know it today, which has ergonomic problems for continued use of small keyboard and screen." —**Giulio Prisco**, chief executive officer of Metafuturing Second Life; formerly department head at the European Satellite Centre, analyst at European Space Agency, and an IT specialist for CERN

"In many poor, remote areas where individuals can't own phones, they still have access because of phone-sharing schemes. This nixes casual use, but still brings vast connectivity to services they find essential." —**Jerry Michalski**, founder and president, Sociate, a technology consulting firm; formerly managing editor of *Release 1.0*, Esther Dyson's newsletter, and cohost of the annual PC Forum

"Networked digital communications are extremely useful and are developing indigenously. People love their mobile phones, but the screens and keyboards are too small to be *the only connection*. Shared machines are really quite workable, even if not the preferred format of rich Westerners. As to *worldwide* 'universal standards and protocols,' that hasn't been true of mobile phones, TV, electrical standards, and by the way, when will the U.S. adopt the metric system? (Although conversion is always possible.)" —**Seth Finkelstein**, anticensorship activist and programmer, author of the Infothought blog and an EFF Pioneer Award winner

"Actually, I am intrigued by the possibility of what Mark R. Anderson has called the 'Carry-Along-PC,' combined with the various efforts to get cheap laptops to the world. I suspect the result will be a device that is a cross between today's cell phone and today's laptop." —**Rollie Cole**, director of technology policy, Sagamore Institute for Policy Research, a nonprofit, nonpartisan think tank headquartered in Indianapolis, IN

"For most of the world, I assume this will be true, due to current market penetration; however, I do exclude the U.S. from this prediction. While mobile technology has penetrated throughout the nation, I don't believe the U.S. will see anything like universal standards by 2020 because of the corporate monopolies with great power, who are using that power to make certain it doesn't happen. In the rest of the world, that power does not hold such absolute sway. But, at the same time, as the wonderful XO machine appears, India also is facing declining literacy rates, problems that, around the world due to the increase in poverty and slavery in the developing world, make it harder for populist computing to take hold. If your entire 18-hour day is spent in a sweatshop and you are locked in a 'dormitory' at night, when are you going to interact and build on a culture that might use the XO and the machines that come after it?" —**Christine Boese**, PhD, researcher and analyst for Avenue A-Razorfish and Microsoft

"Mobile telephones are likely to be more affordable and accessible than laptops or PCs, so it's feasible that more people will use this technology as a connection medium." —**Sam Ozay**, e-learning and e-communication specialist and solutions architect at Post-modern (Asia/Pacific); formerly general manager at European Language Centre

"I think the Google phone certainly has the potential to enter the expanding market of browserability of cell phones and interoperability with other platforms. Per-minute charge to browse has to be drastically reduced for most remote access phones to be online and that is dependent on critical mass use, which is connected with the need for using the Internet, something the cell phones itself won't be able to bring down." —**Shakib Ahsan**, MBA and MA in educational technology, now at Concordia University, Canada

"This will be true, in as much as the device will deal with phone calls. I doubt whether it is anything close to the traditional mobile phone we know today." —**Sam Smith**, Web interface developer, University of Manchester, U.K.

"Mostly agree, however, I expect even poor people to relish ownership of multiple devices, not just a phone." —**Brough Turner**, chief technology officer and cofounder of NMS Communications; oversees evolution of technology and product architectures

"The mobile will be the universal device, and mobile infrastructure is the only one deployed in all parts of the world. The multipurposes facet, and the ease of use of mobile, is making it the appreciated way to be connected. But the connections fees and the mobile's cost must decrease." —**Rafik Dammak**, software engineer, STMicroelectronics, Tunisia; DiploFoundation participant in the study of Internet

"Because the scenario is worded, 'for most people,' I would agree that the mobile phone is the primary connection. But, practically, that is true only because it will be the only connection for too many. The primary connection will be some device that provides a greater range of visual capability and be less portable. Further, the 'mobile phone' intended in the scenario will be replaced by something else that will have telephone capability as an adjunct." —**Don Heath**, Internet pioneer; former president and CEO of the Internet Society; member of U.S. State Department Advisory Committee on International Communication and Information Policy

"While this appears reasonable, I also note predictions that the bulk of Internet traffic will be between processes or devices (or user agents) rather than between people themselves. Therefore, it is *also* quite likely that the class 'people as users' will become an increasingly smaller fraction of Internet accessors. This does not contradict your prediction; it just qualifies it further." —**Steve Goldstein**, ICANN board member, retired from the National Science Foundation, where his job in the 1990s was to help diffuse the Internet globally

"I don't think that mobile phones will be the primary connection tool. It may be right that mobile networks will provide a substantial part of Internet connectivity, but I think the hardware industry will come up with some new appliances to compete with mobile phones. Treos and BlackBerrys will still be focused on upper scale segments, and some new accessible thin-client appliance will appear in the near future." —**Sebastian Ricciardi**, associate with Jauregui & Associates, a law firm in Buenos Aires; leader in the Argentina chapter of the Internet Society, formerly of ICANN's At-Large Advisory Committee

"Well, sure, wireless services will always be available in more places than wired, because the build-out is so much cheaper." —**John Levine**, founder of Taughannock Networks; a leader of the Internet Research Task Force's Anti-Spam Research Group and the Coalition Against Unsolicited Commercial Email

"Mobile phones are affordable, portable, and a virtual appendage for several generations. It is only logical that it will become the primary Internet connection." —**Hinda Feige Greenberg**, PhD, director of the information center for the Robert Wood Johnson Foundation, dedicated to improving health care for Americans

"The cell phone has become a primary tool for converging information technologies that is widely accessible and relatively easy to use. As the continuing rapid diffusion of cell phone technologies increases within modern society and the convergent depth of cell phone information services broadens, we will see the cell phone used as a primary access point for information services that are now accessed via the computer." —**Gary Kreps**, chair, department of communication, George Mason University; formerly founding chief, health communication and informatics branch, the National Cancer Institute

"By 2020, the phone will be melting away into the environment, its functions provided by distributed, intelligent components. People will talk to—and through—their cars, desks, etc." —**Greg Laudeman**, utilization catalyst and facilitator, community technology specialist, Georgia Tech Enterprise Innovation Institute

"As screens get bigger, everyone will use a mobile device to access the Internet. Everyone!" —**Dan Lynch**, founder of CyberCash, Inc. and Interop Company and an Internet pioneer; board member of Santa Fe Institute; director of computing for SRI International in the late 1970s

"Mobile devices have the potential to bring the power of the Internet off the desktop and out into the world. Will lead to further blurring of the digital and the analog realms. Phones are likely to become our way to access a variety of public and semi-public I/O [input/output] devices, as well as a readily available personal portal to online data and applications." —**Jim Witte**, professor of sociology, Clemson University; research is focused on differences between online and offline society, including a special interest in activity in Second Life

"I agree that the mobile phone will be the connection tool of choice for most by that time, but if the providers of those phones wish to achieve that lofty position, they will need to work together to create an open architecture that allows the phone to access different platforms on a per-use basis rather than a subscription. This à la carte methodology will enhance the experience for users and encourage users to try new platforms." —**Hal Widsten**, general manager, KWED/Seguin Daily News, Guadalupe Media Ltd.

"By 2020, I believe the 'bottom' three-quarters of the world's population will account for at least 60% of all people with Internet access. The convergence of traditional desktop or laptop computing will have transitioned to the next level: mobile computing in the form of, for want of a better term, mobile phone platform. Add-ons and accessories such as VR goggles integrated with the mobile phone will allow world-wide mobile computing to completely replace the desktop/laptop platform in many cases." —**Thomas Quilty**, president, BD Consulting and Investigations Inc., a firm that investigates and provides services tied to software piracy, trade-secret theft, and other high-tech concerns

"A hybrid computer and phone combination more akin to the Asus Eee (which has phone capability) will come into use. They will have more ways to input data but will include a full keyboard of some time and be capable of all laptop, as well as phone, functions." —**Bruce Turner**, director of planning services for a U.S. regional transportation commission; retired from U.S. military

"I agree that there will be a convergent computing/telephony device, but I'm not so sure they'll be called—or perhaps even recognized—as mobile phones." —**Aaron Schmidt,** Walking Paper Consulting, a blogger who writes about libraries, technology, and usability

"The 'mobile phone' people will be using as the primary access point in 2020 will be a completely different animal than today's mobile phone, or today's PC for that matter. It will be a very personalized device for entertainment, communications, productivity, location awareness, and financial transactions." —**Mark Youman**, principal, ICF International, a Washington, DC, consulting-services company that works with government and commercial clients

"I agree, but the phones will be regulated by the governments [in] many of the developing countries with chips that will control content and spy on the user as a way of keeping control…there will be a worldwide black market for 'stripped' phones, since the government will know if you try to tamper with one of the official phones." —**Garland T. McCoy**, founder, Technology Policy Institute, a think tank focused on the economics of innovation; formerly senior vice president at the Progress and Freedom Foundation

"Originally, mobile phones were much like text browsing in the early days of the Internet. Broadband has come a long way since then, and laptops are becoming smaller and more technology has converged with each other's platforms in mind. Don't be surprised if the communication devices as we know them today will be as simple as a watch with a hologram feature for viewing frames." —**Mark Terranova**, senior account manager, I Group Electronics, an electronic component distributor

"I think use of the Internet will be qualitatively different for people who access by phone or computer. Phone access will be used for quick transactions. The Internet has great information, but phones will never be able to display this information as well as computers from a human-factors viewpoint." —**Todd Wagner**, health economist, Health Economics Resource Center, Palo Alto, CA, part of the U.S. Veterans Administration; also involved with the Center for Health Care Evaluation

"Believe mobile 'devices' is a better term, since the evolution of devices is rapid. It may not be called a mobile phone, but it will be mobile, have multiple application capability, and work anywhere." —**John Murphy**, director of USA.gov technologies, General Services Administration of the U.S. government

"The ongoing hindrance with mobile phones is the size of the screen. That being the case, I can't accept the basic premise of your scenario—that a phone would be the primary Internet connection." —**Jill O'Neill**, director of planning and communication, National Federation of Abstracting and Information Services; author of the Infotoday blog; based in the Philadelphia area

"The cell phone is smaller and (usually) cheaper than a laptop. If, indeed, there are universal standards and interchangeability, it could be this powerful." —**Judith Siess**, president of Information Bridges International, Inc. and publisher and editor of the *One-Person Library* newsletter, author and blogger

"The keyboard is too small to type on comfortably on a mobile phone, and most people will continue to spend hours at work and home seated at their computers. You fail to mention whether the mobile phone will come down in cost, currently at $400 for a Nokia and more for other lines, only the wealthy can afford the phones—you don't explain how this cost gets lower in the future persuasively." —**Catherine A. Fitzpatrick**, Open Society Institute, Physicians for Human Rights; lecturer on humanitarian issues; formerly UN representative for International League of Human Rights

"Most of the U.S. will be wireless, if by mobile phone or other means. If you look at the numbers, the mobile phone is a technology that has quickly infiltrated every single aspect of society, whereas iPods, laptops and other items seem to still stink of digital divide. While the U.S. has not advanced as quickly using mobile phones for SMS and Web surfing, I'm sure we will be there soon." —**Tiffany Shackelford**, consultant who works with clients such as Phase 2 Technology, Stateline.org, Foneshow, WebbMedia, and Daily Me

"Short of some black swan (and who would rule that out!), anyone traveling overseas can observe that the cell phone is central to people's lives at the moment. For many people, cellular technology does an end run around other connectivity issues—building wired or Wi-Fi networks, the duke-it-outs between would-be providers—creating a de facto network of its own. It may not be a 'phone' by then, but yes, the computer will travel with us and probably replace the wallet and keys we now grab as we head out the door." —**Karen G. Schneider**, research and development, College Center for Library Automation, Tallahassee, FL; expert and thought leader in the library and technology community

"The physical interface is too small for this to be the primary connection tool, unless there is a way to 'unfold' the screen and make it larger and closer to the size of today's notebooks." —**Tim Terpstra**, cofounder of Teletekst Is Dood and cultural (media) entrepreneur, editor, researcher, and developer for The Generator/TodaysArt Festival, the Hague, The Netherlands

"I think we will be more likely device independent, with the gizmo we use to access the Net being less personalized and less of a 'thing.' Maybe a small earring, or tattoo with which one interfaces via galvanic skin responses or via eye movements...but interfacing with something connected as part of a larger central system. Would allow governmental control. Yes, standards and protocols should align worldwide." —**Kathryn Greenhill**, emerging technologies specialist, Murdoch University

"I am not sure that we can say there are just two tools to access Internet. In fact, we will have a lot of different tools to do that. Television, your car, laptop, mobile phone (more or less intelligent)..." —**Sebastien Bachollet**, president of the Internet Society of France, operates the European Global Event on Domain Names and Address Systems, known as EGENI, active participant in ICANN

"For many, information and entertainment channels will flow freely into powerful handheld devices. Users will create their own customized, ultra-personal news feeds, music channels, and video portals via phone interfaces. The iPhone hints at this now." —**Michael Stephens**, assistant professor, graduate school of library and information science, Dominican University, River Forest, Illinois; an expert on Library 2.0

"It's more likely that cheap, simple telephony will come to inexpensive mobile Internet devices than that more people will use dedicated mobile phones as we know them today for their main Internet usage. In addition, I believe device form factors will continue to evolve so that by 2020, the term 'mobile phone' may no longer be meaningful in the way it is today." —**Dave Coustan**, author of the Extraface blog and an independent consultant who formerly wrote EarthLink's official blog, Earthling, and Stuffo, a subbrand of HowStuffWorks.com

"In the medical/scientific field, 90% of the technology we'll be using in 15–20 years hasn't even been thought of yet. Therefore, while it may be the mobile phone that connects so many to the Internet in 2020…it may well be something much more technologically advanced, smaller, more powerful, and so user friendly—that uneducated people in undeveloped and underdeveloped countries will have access to it and be able to use it, easily." —**Dan Larson**, president and CEO of PKD Foundation, a nonprofit organization working for patient advocacy and education

"While I agree mobile phones will dominate Internet connections in the future, I believe there will be a significant space reserved for highly portable laptop or tablet devices that enable larger views of the incredible graphic presentations now found throughout the Web world." —**Bill Warren**, vice president of government relations, Walt Disney World; founding editor of the *Orlando Business Journal*

"While the idea of one laptop per child is a great ideal, the use of phones will prove to be much more possible." —**Ken Mitnick**, Down to Earth Software USA, Inc., providing software for real estate professionals

"Worldwide access among all incomes will not happen until there is a free channel for connection. The current connection model that revolves around telephone/mobile providers cannot work for people with no money. The network suggested by [Nicholas] Negroponte will connect people with no electricity." —**Buddy Scalera**, vice president for interactive content and market research for CommonHealth Qi, in charge of interactive online strategies, including social and viral marketing

"Communications will be the currency of the next decade, with telephony [being] the gold." —**Stan Felder**, president and chief executive officer, Felder Communications, a marketing and advertising firm in Grand Rapids, MI

"I disagree with the last statement. I believe the technology disparities will grow wider. Access to knowledge is a greater barrier than access to other resources, except nutrition. As technology becomes more advanced, the digital divide will become greater." —**Debbie Murray**, associate director, health education through extension, University of Kentucky

"I think this is mostly accurate; however, I still believe that laptops will still have a very large role to play in 2020, especially for the third-world countries. In America, England, and Japan and the likes, there will be a heavy shift to mobile technologies, with the advent of technologies like the iPhone playing a bigger role. The move to open standards will have finally taken hold back in 2015, and people will embrace the open standards by 2020 with little thought about it." —**David Newberger**, founder of Blackdot Ventures and systems administrator for Minneapolis Convention and Visitors Bureau, MN

"Look in today's public libraries for a clue as [to] the need for widespread public access to the Web. Many relatively low-paying jobs require an online application. Even the simplest tax situations are more easily addressed with the aid of online tax preparation software free on the Net to lower income people. Public librarians are becoming informal experts on how to access these services for their lower income customers. It's hard to believe that these functions could someday be done from a mobile phone, but I'll allow that it may happen." —**Jerry McCann**, vice president and director of the social marketing group at Carton Donofrio Partners Inc., marketing and advertising firm in the Washington, DC, area

"While it will take a while for mobile phones with significant computing power to become cheap enough for widespread distribution in developing countries, mobile phone will, indeed, be the primary means of communication, both via voice and via Web. This development is going to support—and be supported by—both the opportunity to be always conducting transactions online and by the opportunities inherent in localized services through the mobile Web. It should be interesting to see which role product piracy and the open source movement will play in this development." —**Peter Bihr**, freelance consultant on Web strategies, communities, blogging, and social media; based in Berlin, Germany

"[The] mobile phone seems to be materializing convergence better than any other gadgets. Laptops are a good rival, but mobile beats them in being cheaper and more practical by size and in terms of carriablity." —**Erkan Saka**, lecturer in media and communications systems at Istanbul Bilgi University

"Bricks and wires will be somewhat outmoded, and clicks and spires will take their place for much of the world. If telephony administrators can agree on standards and protocols, phones will increase in options and decrease in price as computers have, allowing more people across SES and national boundaries." —**Dorothy E. Finnegan**, associate professor of education, the College of William and Mary

"You are correct, it is the impact of mobile technology; however, the penetration of low-cost PCs and notebooks will be more difficult due to lack of many other priority resources in the poorer countries, like housing, sanitation, etc. Also the cooperation among telecom and Internet service providers will be the same as it is today: This means strong competition and only basic common protocols." —**João Miguel Rocha Filho**, director, DataOne, a provider of software for connecting to Linux; based in Brazil

"The mobile is profoundly liberating and equalizing for the individual, providing access to what is known by humankind. We can only begin to understand how this will transform human experience, but I think all the news is good." —**Judy Breck**, blogger at GoldenSwamp.com—aimed at "watching the global golden age of learning emerge from the open Internet"

"Open source will provide a popular mobile OS [operating system] that consumers can choose to install over corporate devices sold online or at retail locations." —**Don Kasprzak**, chief executive officer of Panaround.com, a Web-solutions design company; former system engineer at Apple Computer

"Consider the vast changes in mobile phone technology that have occurred in the past 15 years and the increased use of mobile phones across all age groups in developed countries. Expanded networks, connectivity, and affordability have also provided sweeping transformation in developing nations. These factors coupled with the pervasive cross-generational desire—in developed nations—to be connected and accessible any time/any place makes this scenario seem inevitable." —**Carol Schwartz**, distance learning coordinator, Northwest State Community College, Toledo, OH

"Although I strongly agree that the mobile phone could be the primary connection tool by 2020, who can predict the future? It is way too dynamic. However, looking at things like the Nokia N95 or the iPhone, you can see the power of a handheld to do just about everything, including being a computer, phone, camera, radio, video projector, GPS. Really, take all your gear now and wrap it into one device that you can stick in your pocket and that will be the future. We will look back at 2008 and say, 'Can you imagine, we needed a backpack to carry everything you have in the palm of your hand now.'" —**Leonard Witt**, associate professor in communication, Kennesaw State University, GA; research interest is citizen journalism and user-generated content; author of weblog PJNet.org

"The low-cost, high-tech mobile phone is more accessible and more practical than a computer and will prove to be an even more powerful force in the future." —**Joan Connell**, online editor, *The Nation* magazine, formerly an executive producer for MSNBC.com, senior editor for MSN and a Pulitzer finalist for her reporting

"I think 'phone' is a bit of a misnomer here. If one envisions something akin to the current BlackBerry, with a reasonably sized screen and keyboard/data entry functions that resemble a laptop, then yes, I think phones will be the primary means of communications. Already, we are seeing people use tools like PDA/smartphones in ways that resemble laptop computers, phones like PDAs, and laptops like phones. Portability and consolidation towards a single tool is happening now and will continue." —**Cameron Norman**, assistant professor in the department of public health sciences, the University of Toronto; actively engaged in use of the Internet to help tackle tough health issues, including work with the Centre for Global eHealth evaluation

"The direction is toward small, cheap, and mobile in technology. I don't see that changing." —**Alexis Chontos**, webmaster, the Art Institute of Pittsburgh

"Cell phones are currently in the majority of the students' hands that I work with at this medical center, and I see that increasing as the years go by, making the mobile phones sources for communication and information. While the younger eyes may be able to use the Internet on their mobile phones, the mobile industry will need to adapt to the millions of aging eyes—it is hard to use the current screens with bifocals, making accessing Internet content more difficult than using a desktop screen." —**Teresa Hartman**, associate professor and head of education, University of Nebraska Medical Center

"The mobile phone is too cumbersome—mostly because if its size—to do much of the professional work now done on the Internet. My hope is that by making wireless technology available to a larger part of the world, we can eliminate some of the communication barriers that are responsible for much of the misery and destruction in the world." —**Mike Samson**, interactive media writer and producer

"I mostly agree with certain reservations. I agree based on the fact that the whole world, i.e., second- and third-world countries, is included. However, using present technology, which by then could have changed significantly, we will still be restricted in ease of use and using the Web as an extension of one's work life. Instead, whilst the mobile element will be significant, I believe that the home network, especially in conjunction with in-house entertainment/workstation, as is already visible, will be the main stronghold of Internet usage. The mobile will act as the on-the-go, quick-check/mail/travel-entertainment module, updating seamlessly with the home Net." —**Robert Eller**, Concept Omega, a media marketing and communication company

"By 2020, mobile phone companies will have had to adjust to meet consumer demands for increased mobility, interoperability with competing networks, increased bandwidth, and lower prices. Advances in technology will lower the price for smartphones while increasing what they can do. This will result in most of the Internet being available on mobile phones, as well as a much richer multimedia experience. Because the essential Internet tools will be available via mobile phones, many people will have no need for home Internet connections." —**Jamie Richard Wilson**, journalist and freelance Web developer

"Our thinking about the Internet has always been based on a desktop concept. The recent spike in notebook sales and the expansion of Web-enabled mobile phones is just the beginning of our shift from home-based, digi-world connectivity to a truly mobile universe." —**Stephen Schur**, director of online communications, Ramapo College of New Jersey

"Additionally, I think it's likely that many of the functions of 'the Internet' will fade into the background; e.g., banking and credit card payment will no longer be accessed via Web sites but will be an integrated feature of mobile devices (as they increasingly are already in many LDCs)." —**Jacob Kramer-Duffield**, blogger and student, University of North Carolina School of Information and Library Science

"Relevant caveat: The road towards 'a relatively low price' is a bumpy one. European operators, for instance, had to be forced by the EU [European Union] to significantly lower their roaming tariffs, and this shows us that the industry always opts for 'concerted action' if given the opportunity. I believe that corporate attitude will not change in the next decade or so; something else has to in order to achieve such a scenario (perhaps by means of tough and 'citizenship-oriented' regulation…)." —**Luis Santos**, Universidade do Minho–Braga, Portugal

"I believe that as early as 2010, the majority of global Internet access will be via a mobile device. This trend will be accelerated by things like the ubiquity of wireless broadband in markets like the U.S. after February 2009." —**Chris Marriott**, vice president and general manager for Acxiom Digital, a marketing company

"With added connectivity through a range of computers (those infused into everyday things, those carried by choice, those accessed from stationary workstations, and so on), it is possible that the mobile phone will become the 'babble fish' of the 21st century—the tool that instantly translates any language (human, machine, or otherwise) to any other language." —**Ed Dieterle**, Harvard Graduate School of Education; research tied to handheld devices for ubiquitous learning

"Most people want a smartphone to access their own stored data (phone numbers and addresses), as well as to get point information from others (directions, text messages, Wikipedia). This means that these phones will grow in popularity and use because they are made for people to use connectivity in a simple, direct form. Laptops allow for more complex connectivity, but this will become increasingly marginalized." —**John Jordan**, associate professor of communications, University of Wisconsin–Milwaukee

"One laptop per child will fail, and most people in the world will still not have a reliable electricity source to power a phone, computer, or even a lightbulb. While telephones will certainly become more like computers (e.g., the iPhone), they will not be able to compete with more traditional computer configurations in terms of content creation. And it is in the area of content creation that the computer has had its greatest impact: Readers more and more are becoming writers—not content copiers, which is the traditional role for most writers, but originators and publishers of content." —**Dennis Baron**, professor of English and linguistics, University of Illinois, runs the Web of Language site and researches the technologies of communication

"While it's true that cell phones are gaining significant computing power, the more important issue is user interaction. Many online services could offer a great deal of functionality through a voice or cell phone touch pad interface; but the majority of online services are best interacted with via a mouse and full-sized keyboard. In terms of worldwide access, it's more likely that the digital divide will first be bridged by shared devices and connections, like Internet cafés, than it will by each individual owning a device, whether laptop or phone." —**Jay Neely,** social strategist in the process of founding News Armada, a Boston-based company working to advance Internet-based news and commentary and community online

"The mobile phone will be used daily/hourly/minutely as a Web client by most of the online population. But the idea of a 'primary Internet connection' will be increasingly obsolete. Many devices will connect to the Internet, for various reasons, and the very idea of consciously 'connecting to the Internet' will seem quaint in many parts of the world." —**Walt Dickie**, executive vice president and chief technology officer, C&R Research, one of the principal developers of CheckMetrix in the early '90s

"The mobile telephone will likely be a significant vehicle for the majority of the population to engage in personal information access and sharing. This will likely occur via text, orally, and one-to-one video. I am reluctant to accept that the small screen of a mobile phone or similar personal mobile access device will have the capabilities for multiperson interaction and collaboration as is now occurring with personal computers (Web-based conferencing, etc.). This maintains a distinction between personal interaction and professional use." —**Julie Anne Lytle**, PhD, educator and online learning coordinator, Episcopal Divinity School

"While it is probably true that these small, relatively cheap devices will find their way throughout the low-income areas of the world, they will probably not be much like phones as we know them. They will be capable of making calls, but will be much more integrated data-carrying devices. As such, in many cases, large notebook computers may become less necessary anyway." —**Sam McAfee**, chief technology officer and owner, RadicalFusion, a company that offers consulting on Web design and development, server-side programming, and database development

"New access to the Internet by large segments of societies without any previous experience or conceptions of what the Internet 'should' do will open up a whole new range of applications and services we can't even imagine today." —**Joe McCarthy**, principal instigator, MyStrands; formerly principal scientist at Nokia Research Center, Palo Alto, CA; he also has worked at Intel Research, Accenture Technology Labs, and Nokia

"Mobile phones are seeing an increasing trend towards innovative applications. The mobile has become a tool of convenience and empowerment." —**Syamant Sandhir**, director, Futurescape Netcom Pvt. Ltd., an experience-design and implementation company

"There will be tremendous pressure from the users to globalize communications and almost equal pressure by communication companies to maintain proprietary platforms that prevent the phone from being a global connection tool. I think we will also need to overcome the issue of reading a screen that is no bigger than a note card." —**Mark Fennell**, senior Web engineer, Athens Regional Medical Center, Athens, GA

"Log-file data today and year after year so far shows that mobile phone Internet access has not taken off in the U.S., even with the advent of the BlackBerry and iPhone. It is highly speculative that it will be the 'dominant' technology or that it will be desirable in the future. Costs will need to come down first. Currently, it is $80/month for iPhone service with decent Internet connectivity. This is in comparison to $35/month service otherwise. Equally valid arguments could be made that the TV will be the main connection point in 2020. But my gut says that, actually, it will be the laptop computer with its portability, the size of the keyboard and screen, and the preference for e-mail over talk." —**Theresa Maddix**, satisfaction research analyst, ForeSee Results

"The desktop computer will be largely gone by 2020, and people will all use some kind of mobile connection device. Whether that is a phone or a mini-laptop will depend on the user and applications they need." —**Lisa Carr**, director of strategy, Targetbase Interactive, health care strategist and writer

"The mobile phone/device blurs the lines between personal and professional time and users like that. They can use it during 'down time' to stay connected or manage their lives and those of their families. I don't see myself willing to give that up, and I think technology to support the ongoing adaptation of more Internet resources to mobile devices will be well received." —**Stephanie Geyer**, associate vice president, e-communications and Web strategy, Noel-Levitz, an operating division of Sallie Mae that consults on recruitment and marketing of higher education

"I agree with this insofar as I do not believe the personal computer will be the primary method of connection for most people; however, I also suspect that believing our only two options 13 years from now will be 'phone' or 'computer' is uncreative and fails to anticipate that the real answer will probably be 'something else entirely.'" —**Alexis Turner**, webmaster, Greenwood Publishing Group, New York

"Computers get smaller and fit almost everything you could want into a phone. Yes! This is already happening, but will happen more cheaply and, therefore, widely." —**Mariana Almeida**, product manager of Web products for health care, Kaiser Permanente

"Phone network providers cannot possibly make a profit by building out and maintaining the infrastructure required to bring this scenario to fruition. And that infrastructure is subject to vandalism, rendering it useless in situations ranging from political dissidence to terrorism." —**John Jobst**, IT specialist, U.S. Army Corps of Engineers

"It's the direction we're heading. We're more than likely to be using computing power outside the desktop in many or most cases." —**David Allen**, PhD, Temple University

"By 2020, one would hope the equivalent of today's iPhone will be affordable even in low-income economies." —**Brian T. Nakamoto**, cofounder of MrJoy Inc. and product-line manager for Everyone.net (a leading provider of outsourced e-mail solutions for individuals and companies around the world)

"Of course, by 2020 (or even 2010), the term 'mobile phone' will be an anachronism, as mobile devices increasingly become all-purpose connection tools, using innovative interface techniques to broaden their uses." —**Ivor Tossell**, technology columnist/journalist for the *Toronto Globe and Mail*, known as "the blogging journalist" and a social observer

"In total, I agree to the scenario, but I expect it will take more time to become reality, maybe till 2030." —**Jutta Croll**, managing director, Stiftung Digitale Chancen—Digital Opportunities Foundation, Berlin, Germany, promoting and supporting access and equal opportunities for all online

"Over the next decade, the mobile phone will be transformed by many entirely new applications and the voice interface will be exploited through voice recognition." —**Ed Steinmueller**, professor, science and technology policy research, University of Sussex; researches industrial structure of high-technology industries, coevolution of technology

"A further important influence to support this outcome is the increasingly critical need to reduce carbon footprints with mobile communications, enabling a reduction in travel." —**Tim Grafton**, market research director for UMR Research Ltd., a market research company based in New Zealand

"The mobile phone and wireless technology generally will certainly dominate for the next 5+ years. After that, new offshoots will evolve, almost certainly in ways we cannot imagine." —**Jasmine Sante**, Sante Strategies, independent Web strategy consultant in the Washington, DC, area

"Some intelligent cell phones are able right now to do as much as a computer is doing. I'm thinking about the N95 of Nokia. So it's safe for me to say that in a really near future, the cell phone will be pretty much THE device used to access information in our society, being on the Net or other proprietary networks." —**Bruno Guglielminetti**, producer for Radio-Canada, multimedia producer and blogger

"This is very true. The U.S. will remain behind, especially for older generations (40+ years old). Augmented reality functions will be commonplace, building out a set of mirror worlds." —**Bryan Alexander**, director of research at the National Institute for Technology and Liberal Education, blogger, expert on computer-mediated pedagogy; based in Ripton, VT

"The mobile 'phone' will indeed play such a role thanks to a great improvement in the ability to use them. Specific functions for use of the Internet will be shaping that use. For other types of functions, other devices will be used." —**Wim van de Donk**, professor of public administration and chairman of the Scientific Council for Government Policy in The Netherlands (WRR)

"The form factor, capabilities, and price of the 'mobile phone' of 2020 are hard to predict, of course. But I am sure that we won't think of it as a 'phone' any more than we think of today's GPS navigation devices as 'satellite radio receivers.' Mobile broadband is becoming widespread, and hopefully the U.S. will catch up with the rest of the world in terms of the price/performance ratio." —**Scott Brenner**, technologist, Web developer, consultant for clients ranging from Fortune 100 companies to small nonprofits

"Computers get smaller and become phones. Phones gain power and become Internet access devices. Phone computers like the iPhone are cheaper and more portable than other computers. They are already used for group access to information and communication in India and Africa. I expect to see much, much more of that." —**Paul Jones**, director of ibiblio.org at University of North Carolina–Chapel Hill; the original manager of SunSITE, one of the first Web sites in North America

"The move to a single primary form of wireless access to voice- and data-related services seems to me to be unstoppable. Whether we'll recognize it as what we think of today as a 'mobile phone' is questionable." —**Christopher Jacobs**, chief operating officer, Solutions for Progress, Inc.; formerly with Knowledge-Flow Inc. and Unisys Corp.

"Already today it is becoming hard to distinguish between a mobile phone and a computer. It is likely that this distinction will be obsolete in 2020, hence the question will not seem relevant at that time." —**Olav Anders Övrebö**, media blogger and former *Netzeitung* journalist, University of Bergen

"There is no doubt that the desktop PC and even traditional laptop will be surpassed by other, more lightweight devices—it is already happening. The only question is whether the access device will still be thought of as a mobile *phone*. It could be an 'ultra-portable' device which gets sold in channels more typical of the current lower end laptop market. It will also likely serve as a cell phone, but that may no longer be its primary purpose." —**John Eckman**, practice director, Next Generation Internet, Optaros, Inc., a professional services firm offering strategy, design, development, and consulting services tied to open-source software

"Mobile phones are spreading faster in developing countries than anything but TV now, notably, places that lack infrastructure for other kinds of broadband. So, upgrading cell phones to higher capacity will make more sense outside the OECD." —**Joseph Straubhaar**, professor, University of Texas–Austin; research interests include international communication and cultural theory, information sciences, and the digital divide

"I do not think cell phones will provide 'information in a portable, well-connected form at a relatively low price.' At least not by 2020. As it stands right now, connecting to the Internet from my BlackBerry sucks! Cumbersome, hard to read, hard to type. It offers me very little value; although I pay an extra $20 a month for Internet access, I'm not sure why. I guess for e-mail. And even that is a pain in the butt. That reminds me, I need to cancel that service. In any event, until they make cell phones from Mylar or something similar that can roll up or fold and put in my pocket or purse (yes, with touch features) and big keyboards and large screen, forget it! And that ain't happenin' by 2020." —**Virginia Bisek**, Web content developer and writer

"It's also possible that small, portable, handheld computers connected to a global, ubiquitous Wi-Fi connection will replace smartphones—i.e., the Nokia N810." —**Beth Gallaway**, Information Goddess Consulting, a Web 2.0 consultancy

"The mobile phone as we know it will change and 'smartphones' will become the *de rigueur*. There will still be roaming charges for global or country roaming, and the couple of mobile providers that remain will be very territorial about their networks. We'll see people eschewing buying a PC but investing in a mobile device that also connects to other devices within their life. E.g., instead of buying a movie on demand via a cable box/DVR or Apple TV. You'll do this via smartphone, and it's sent to the device you specify to watch." —**Chris Miller**, senior vice president, digital operations and new business for Element 79, an advertising agency

"Technically speaking, people will use 'phones,' albeit larger than current models, but the device will evolve into something like a 'comm.' device." —**Lynn Blumenstein**, senior editor, *Library Hotline, Reed Business Information*

"The term *mobile phone* is something of a misnomer. Advanced mobile phones should be viewed more as multipurpose, highly portable computing units." —**Daniel Fisher**, lecturer

"While I agree with regard to the mobile phone being a primary 'connection tool,' I think that it may contribute to a divide between people who are primarily just 'connecting' as brief communicators and as consumers and those who are substantial content creators." —**Jeremy J. Shapiro**, a professor of critical social theory at Fielding Graduate University with a research emphasis on the social and cultural effects of information technology and systems

"While I generally tend to agree with this, I think that the nature of the mobile phone will be significantly different than it is today. For example, I suspect that the screen will be much larger and roll out from the body of the device." —**Steven Hausman**, president, HausmanTech Consulting

"While the mobile phone will dominate, increasingly it will expand beyond voice communications into text, images, video, and other types of data." —**David F. Salisbury**, associate director for science and research communications, Vanderbilt University; formerly science and technology reporter for *Christian Science Monitor*

"The key here is the idea of the mobile phone being the primary Internet device. For most of the world, it will not. The mobile phone will be a peripheral to the main home computer just like televisions and dedicated portable music devices." —**Rob Boostrom**, PhD student, Southern Illinois University–Carbondale, creator of the Society for Word of Mouth, "a group for educators to share ideas about sharing ideas"

"My only reservation is the possible health and environmental effects of wireless phone/Internet transmission. For example, if people starve because the bee population is decimated by radiation, that could have an impact." —**Bruce Henry**, Concordia University, Montreal, Canada

"I don't agree with the part that it has a set of universal standards and protocols accepted by most operators internationally, but I do agree that the telephone will become the primary Internet connection." —**Christopher Brown**, strategist and managing editor of new media for the U.S. television program *America's Most Wanted* on FOX

"One by-product of this development will be a narrowing of the number of Web sites routinely visited by users and a significant increase in the market power of a small number of wireless companies (Verizon, Vodaphone, etc.) to funnel traffic to a limited number of core Web sites." —**Timothy J. McManus**, vice president of health care product management for Nuance Communications Inc., a software technology company known for speech-recognition work

"It strikes me that, although I do indeed mostly agree, this guarded response is not guarded because I doubt that in the current economic and technological climate it can become thus, but instead due to a concern that the economic (and thus technological) climate may change for the worse between now and 2020." —**Francis J. L. Osborn**, futurist and activist, philosophy department, University of Wales Lampeter (formerly St. David's University College)

"The mobile phone may be the primary connection in terms of number of hits, but not in significance. The quality of information that can be communicated via a cell phone is too low in terms of the amount of text that can be effectively read/reviewed, the quality of images that can be perceived, and the quality of sound. Perhaps it is wishful thinking on my part, but I believe that the most significant connections will still require skilled writing, images, and sound." —**Fred Ledley**, founder and chairman of Mygenome, professor of natural and applied sciences at Bentley College, Waltham, MA

"I foresee the use of a digital device in the future as being an integrated platform that allows access to an individual's digital repository of data and computing capability. The device will be part phone, part media player, part 'dumb' terminal enabling access, and possibly requiring biometric verification of a person's ID. Likely, the individual ownership of a particular device will hold less importance as access to the network will be gained by using any device on which a user's ID can be verified. All applications allowing access to a person's data, workspace, and networks will be reliant on this verification, not the ownership of a particular device. Portability of access devices like this will be important. The idea is one encompassed by the notion of 'ubiquitous computing.' Anyone's guess as to the timeline, but in 13 years, it will definitely be further along." —**Sam Punnett,** president, FAD Research Inc.; has worked in the field of interactive digital media since the 1980s, for the last 9 years on strategy, marketing, and e-business development

"Most people will probably be 'connected' via a mobile device that includes communications capabilities and some basic functions we currently associate with computers. However, these will not be 'mobile phones' per se, and they will probably not resemble current personal computers except superficially. They will, however, be able to be connected to powerful future personal computers, and they will also be able to be connected to powerful computer systems where people may have personal accounts. While a very large percentage of the world's population will have basic access, the 'digital gap' will actually increase as basic access will no longer really be sufficient. The gap between those with basic access and those with standard services in the near future will be like the gap between those who have a phone in the home which can receive calls from anyone but which is limited to emergency out-calls and those who have integrated phone-fax-Internet connections today." —**Benjamin M. Ben-Baruch**, senior market intelligence consultant and applied sociologist for Aquent, working at General Motors Corporation

"This is a glass half-full assessment, since there are a variety of terrible threats to human life on the planet that could prevent the beneficial spread of mobile connectivity: major pandemic, nuclear winter, famine, severe climate change, the return of the Inquisition in some form or other. But, it seems just as likely that we could continue to muddle through, skidding along from disaster to calamity like pinballs in a vast and blinking networked mobile metaverse." —**Havi Hoffman**, senior editor, product development, Yahoo; coauthor of *The Tech Buzz Game* and blogger

"By 2020, mobile phones would expand their functionality to include common business and personal organizational applications, from daily agenda to simple text messaging and news tickers. They will store personal information on a memory card, and function like an access key for entrances, transportation passes, maybe even credit cards. For security and privacy purposes, mobile phones will probably include a fingerprint scanner or similar biometric devices." —**Clement Chau**, research manager for the Developmental Technologies Research Group at Tufts University

"Whilst I agree overall, I do think the nature of the device itself will have shifted. I suspect that it will no longer be seen just as a mobile phone, more as a sort of mobile companion. Perhaps even there will be separate devices that can plug into each other, thus forming new objects on their fusion, such as speakers to become a mobile entertainment device, flexible screens to become video devices, projectors to become teaching aids, etc." —**Richard Osborne**, Web manager for the School of Education & Lifelong Learning, University of Exeter; research focus is e-learning

"The phone and computer are likely to converge, such that the phone has projected screens and keyboards." —**Mary Ann Allison**, principal, The Allison Group, has worked with Microsoft, Glasgow's Urban Learning Space, and other businesses, governments, and NGOs

"Mobile phone penetration accounts for 60% of the world's population in 2008, and with low-cost handset initiatives underway, penetration can extend beyond 80% in 2020." —**Michael Skarzynski**, managing director, Red Lion Technologies Inc.

"Easy and cheap Internet access via mobiles will be the standard in areas with young infrastructure. In the West, especially in the U.S., easy and cheap alternatives such as Wi-Fi combined with expensive closed balkanized mobile networks will limit actual Internet access via cell phone." —**Ted M. Coopman**, lecturer, San Jose State University

"Mobile phones have far greater reach than laptops (we see this currently), are portable, and have that 'anytime, anywhere' feature that computers don't have. I suspect that mobile calls will get cheaper and the obstacles we currently experience with international calling will disappear. I predict by 2020 most mobile phones will have Internet access and GPS capability.

People will use them for communicating and for gathering information." —**Rachel Kachur**, behavioral researcher, U.S. Centers for Disease Control and Prevention

"I do not see blanket interconnectivity as necessarily a positive scenario." —**Alex Don**, linguist and educator

"If you look at the speed at which we have already connected close to a billion people online, combined with power of mobile phone technology, combined with hope, then the dream of bringing everyone into the conversation can be a reality." —**Tiffany Shlain**, founder of the Webby Awards, cofounder of the International Academy of Digital Arts and Sciences; named one of *Newsweek*'s "Women Shaping the 21st Century"

"In purely statistical terms, the mobile will have to be the primary contact for the majority. However, in spite of improved interfaces, etc., it will (arguably) remain an unsatisfactory platform for accessing much Web information. Some form of PC—be it laptop or PDA—will retain an important role, simply because of the sheer size and accessibility of the available display area." —**Roderick White**, editor, *Admap* magazine, World Advertising Research Center

"I agree in general. However, for many people in the developed world, there will be several devices connected to the Internet, including the GPS in the car; home TV; screens throughout the home (kitchen, bedroom, etc.); home theatre and/or music center (formerly known as the stereo); etc. Without a giant leap in voice recognition, laptops and desktops will abound and all will be connected to the Internet. As devices shrink, so too do they grow. Look at how the viewing area of TV screens has grown as the size of MP3 players and phones has shrunk. Video will be on its way to becoming the dominant format on the Internet." —**Peter W. Van Ness**, president, Van Ness Group, a Web-development company; founded Personal Computer Solutions in 1983; cofounded StockPlan, Inc. and MyStockOptions.com

"We'll be using smart devices that are phone sized—yes—but the devices themselves will be more a portable laptop than what we now know as a phone, and satellite broadband access will be the way we connect—both to our telephone accounts, Internet, etc. So…not at all a telephone in the sense we think of it today." —**Sarah Houghton-Jan**, consultant for the Infopeople Project; digital futures manager, San Jose Public Library; author of the Librarian in Black technology blog

"The term 'Internet access' can be misleading and a poor determinant of digital inclusion. While I would agree that mobile telephony will be the tool of choice and that it is quite possible that access levels may reach 50%, I believe that actual 'accessing' levels will most likely be much lower than that." —**Nathan Botts**, community informatics research associate at the Kay Center for E-Health Research; working on the national health information network (NHIN) and outreach

"Today, the mobile phone mostly still feels and acts like telephony. But soon, like cars measured in horsepower, the phone part of the mobile phone will give way to more and more wand-like applications. And in this measure, the phone will become a bodily extension. We'll touch our device and get access and feedback. So the mobile phone will become our world window, through which we reach out and are reached globally." —**Barry K. Chudakov**, principal, the Chudakov Company, a marketing and advertising strategies creative consultant who has worked with many major corporations, including Microsoft and Disney

"Currently, there are many efforts under way to use mobile phones as a payment device. This will serve to increase people's reliance on mobile phones for Internet, payments, and communications." —**Denise Senecal**, research manager, Callahan and Associates, an Internet strategy consortium in the Washington, DC, area; expert on Internet and mobile banking strategies

"Your mobile phone is your PC, there are no longer two devices or concepts." —**Robert Grant**, chief executive officer, Voya-Care Inc., a medical connector company

"If universal standards are adopted, this scenario will be true." —**Dan Weingrod**, vice president for digital operations, Cronin and Co.; oversees creative online initiatives for integrated marketing communications company

"While mobile phones, however capable, will indeed qualify a person as connected and be of significant use because of their Internet connectivity, the real power of the Internet will still not be available to those with only a mobile phone to access it mostly due to the issue of input/output; still hard to enter data and a small screen hard to receive many types of data and then usefully do something with it. Those tasks will remain on a workstation that is at least a laptop or better capability." —**Shawn Kelly Apochromantic**, configurations manager, designer, technologist, futurist, General Atomics, and volunteer builder in Second Life

"Ninety percent of Internet access will be mobile; differences between mobiles phones, smartphones, PDA, small PC like the Eee from Asus will have disappeared, and we will carry with us 'MID' (mobile Internet devices)." —**Louis Naugès**, president, Revevol, an enterprise 2.0 company with offices in France, Spain, the U.K., and the U.S.; a founder of Microcost, an IT services and hardware company based in France

"From what I gather, most of the people in the world who access the Internet do so via mobile devices *already* here in 2008. I don't [know] what would happen to reverse this fact, i.e., I don't think that $150 laptops will entice people away from their highly functional portable devices." —**Hank Dearden**, director of business development, Digital Industry, Inc., a provider of technology services in the Washington, DC, area

"The mobile phone is changing the way we communicate and will continue to do so. It is the best device for most people around the world, as it is portable, less cost prohibitive than a laptop, and allows for both push and pull communication." —**L. Suzanne Suggs**, assistant professor of communication sciences, University of Lugano; research focuses on use of new media and messaging strategies to improve health status

"I see the phone as it evolves into something more than a phone, but a personal communications device more than a phone—the proverbial thin client to access all of what is currently on the Web." —**James Gorman**, principal, Working Technology Partners, a company offering technology solutions to businesses

"If the concept of 'mobile phone' means a battery-powered device that is always connected, then that will be a large enabler for the future. However, the current devices have focused their richness on direct communication (voice); entertainment (audio-video); and short-burst communication (e.g., text messages)—not on information access, interpretation, synthesis, and use. They may prove unsuitable for many purposes. Also, we need to better understand the future role of situated computing (connected devices built into other things like vehicles, appliances, even buildings)." —**Duane Degler**, user-centered designer and strategist for Design for Context, writer and editor for IPGems, focused on knowledge management, semantic integration and performance improvement

"Open global standards must be the technology norm. However, individuals will continue to connect only with their own cohort—open standards do not make open minds." —**Nancy W. Bauer**, chief executive officer and editor-in-chief, WomenMatter, Inc.

"Though the mobile phone will be the primary connection tool for most personal interactions, some sort of computer will remain the primary connection tool for most business interactions." —**Alan Webber**, senior analyst, Forrester Research, focused on business-to-business (B2B), e-government, and public sector linkages

"Certain specialized activities—professional correspondence, interactions with health care providers, scholarly research—will follow different paths, partly because of issues of confidentiality and privacy and partly because of the nature of the communication. In particular, interactions that imply length, analytical depth, and a generally more discursive mode will simply need more functionality (screen real estate, keyboard or mouse, functions, computational power) and are unlikely to migrate to the handheld devices." —**Amy Friedlander**, director of programs for the Council on Library and Information Resources, a nonprofit that services research and higher education

"Telephony under universal standards and protocols is the weakest part of this scenario." —**J. W. Huston**, president of Huston Consultancy and futurist

"In the developing and poorly developed world, mobile phones will continue to represent the most cost effective communications infrastructure." —**Jim Lucas**, Web manager, CACI, a provider of national security, defense, and intelligence-related solutions in the interests of the United States

"Beyond the fact that phones would seem to be more common devices than laptops or PCs in the developing world, I would assume that cell coverage would continue to be a better connectivity option than dial-up or any other possible form of broadband." —**Douglas Schulz**, managing editor for online publishing, America's Health Insurance Plans; formerly a Web team director for a biotech industry organization and manager of Internet and Web services at the Council of Better Business Bureaus

"The mobile phone will be the primary connection tool before 2020, especially outside the U.S. where the technologies are leapfrogging our own because they don't have the legacy infrastructure/mentality that we do." —**Kathryn K. Goldfarb**, president, KG Communications, an independent consultancy

"I accept that the processing power of mobile phones may be powerful enough to make this scenario possible. I have some reservations about the ability to create an adequate user interface on a mobile phone." —**Mike Langum**, Web developer, U.S. Office of Personnel Management

"I think online access through mobile phones will be ubiquitous by 2020, but I don't think it will be the primary connection tool." —**Jay Buys**, vice president for digital development, Fleishman Hillard, an international marketing and communications company

"The mobile telephone is one of the more disruptive technologies to arrive in the past 25 years. While the One Laptop Per Child is a pleasant pipe dream, the relative immediacy of mobile telephony and the relative low initial investment per user makes mobile Internet very much an obtainable goal in the developing world. Indeed, mobile telephony's relative independence from the traditional wired Internet infrastructure makes it even more appealing to developing countries. Two of the larger challenges going forward are the limits of current wireless bandwidth and competing wireless telephony (e.g., GSM vs. CDMA). These issues, however, are eminently surmountable." —**William Winton**, product manager, digital media, 1105 Government Information Group

"Just as marketers/corporate interests have driven much sustainable development across the world up until this point, they will have a similar effect in the transformation of mobile technologies. The ability to dig deeply into individual users and understand their consumer habits, detailed demographic information, and geographic movements is fast becoming the holy grail of marketing. Marketers will continue to drive for standardized protocols across the mobile spectrum and simultaneously absorb many of the costs involved, thereby driving barriers to access down for lower income populations." —**Kent Kirschner**, media specialist, Neighborhood America, inviting companies to understand the power of community-building online

"Unless there are substantial technology improvements to mobile sets, it will continue to be a mix of wired/wireless Internet connections. More home-based Internet appliances will be purchased." —**Mack B. Rhoades, Jr.**, Web services product manager, Michael Baker Corp.

"We're making children faster than laptops." —**Edward Lundquist**, senior science advisor, Alion Science and Technology; formerly of the Center for Security Strategies and Operations of Anteon Corp

"This changes the line between author and reader on the Internet. Telecom access is thumbing, or perhaps voice to text, both of which encourage first-draft product. New uses, new opportunities for aggregating information (news), less reasoned thought development leads to what?" —**Dick Davies**, partner, Project Management and Control, Inc.; past president of the Association of Information Technology Professionals

"More than adequate computational power and memory will fit into the cell phone form factor by 2020. The primary limitation will be display area. We may well see cybercafés providing large screens, rather than Wi-Fi connections. The power of OLPC will also have grown enormously, to the point where they will replace most current desktop and personal computers." —**James Jay Horning**, chief scientist, information systems security, SPARTA, Inc.; a former fellow at Xerox's Palo Alto Research Center; fellow of the ACM

"While the mobile phone might represent the primary connection tool for most people in the world, a drawback of the user interface will still mark a divide. Those with the kind of rudimentary connection now available via mobile phones will be able to access information but will not be able to construct informational 'hubs' as effortlessly as those with more robust tools." —**Kirk Munsell**, webmaster and science writer for the Jet Propulsion Laboratory

"The underlying question is whether the U.S. will join in the mobile phone revolution. Why hasn't the market force economy of the U.S. followed the other world economies in a more multifaceted integration of the mobile phone into daily use? And if it doesn't, will the U.S. lose the increased productivity stimulated by connectivity?" —**Michael Castengera**, senior lecturer at the University of Georgia's Grady College and president of Media Strategies and Tactics Inc., a media consulting firm

"The emerging cheap computers offered to disenfranchised children will encourage a whole generation to look for digital economic solutions to their poverty. Inefficient institutions will be replaced with (excessive?) technological enthusiasm and determinism, leading to a significant connection rate." —**Bertil Hatt,** researcher of Internet and social services, innovation valuation; employed by France Telecom and Orange (information technology and services industry) while completing PhD

"The cell phone will become the single entry point to the Net. The TV will likely remain the home-based access point, but the soon-to-be delivered OLCD and electronic paper could alter that aspect as well." —**Richard Hammond**, knowledge management team leader, United States Environmental Protection Agency; knowledge management expert currently examining the Semantic Web and RFID

"Considering the difficulty involved in installing and upgrading physical delivery systems, it only makes sense that telephony and wireless connectivity be the desired norm in 2020." —**Patti Nelson**, a webmaster who works on U.S. government sites

"The human interface to an Internet-capable device must have a very significant technology breakthrough. Laptop and desktop computers of today are barely adequate, and new phones like the iPhone are just toys. Voice recognition and eye tracking on a screen will need to become reliable interface methods. Direct interface through brainwave activity will hopefully become at least experimental." —**Dixon Hutchinson**, software engineer

"It's convenient and cheap and easier to carry around than a laptop, and if it gets lost, damaged, or stolen, is easier and cheaper to replace. In developing nations, the mobile phone is leapfrogging laptops and fixed-line Net access—skipping a generation." —**Helen Keegan**, founder of Beep Marketing, a self-employed consultant and a judge for the Webby Awards

"There are two factors that will probably see Internet access via the mobile phone achieve higher penetration. In developed countries, Internet access via mobile phones will probably achieve higher total penetration on account of being an individual, rather than household, technology item. Many people are likely to have a mobile Internet access device they will use as a complement to their home-/work-based access solutions. The ability to be 'online—anywhere, anytime' via such devices is likely to see them achieve widespread adoption. In less affluent nations, the price of Internet-enabled handsets and actual data charges will obviously have a big impact on their take-up. But something that might also tip the scale in favor of mobile via handsets is power requirements. Browsing the Internet on a small screen (handset) is probably going to require less power than on the big screen, and this may make a difference in some parts of the world." —**Heath Gibson**, manager of research and market analysis, Big Pond, a competitive intelligence company and provider of broadband customer Web sites in Australia

ANONYMOUS COMMENTS

The following section includes a selection of contributions to the discussion of the evolution of mobile Internet communications from survey participants who chose to remain anonymous. Among them:

"This question confuses connection with inclusion and, thus, appears somewhat political. Most people will use their mobile phones ('personal communicators') for access, but that says nothing about their economic status. In general, access by mobile phone will remain the vehicle of choice for upper income users, while most others will rely on PCs and thin clients."

"With increased interconnectivity via mobile telephony, national and regional bodies come under increasing scrutiny and pressure to regulate both access and fraud. Popular and industrial calls for a global coordination of design opens wider the fissures between their respective allies, the UN NGO communities, and the states. Education turns its focus toward producing language and cultural interpreters attached to every field of study."

"Given the lack of infrastructure in less developed countries, devices that operate in free-standing mode will bypass more traditional equipment."

"It will be as different from today's PDAs as the Apple iPhone is from a standard GSM phone. Calling such a 'telephone' very much trivializes the difference."

"While I could expect the mobile phone to be 'primary' in the amount of use for Internet, I expect the desktop to largely still 'feel' primary. The mobile phone access would be used for mundane tasks like getting directions, facts, news, etc. The desktop would be used for entertainment and work."

"With Asia's huge populations catching up with the West, this seems like it will happen."

"Most phones are likely to have cameras, so visual communications will also be possible."

"Laptops per child will not be an interesting issue. Everyone will have some mobility device, including children."

"Personal mobile computing will drive the technologies, which will drive the use. People want support in communicating, and to do it when they want, how they want, and when they want."

"China and India will adopt the Web en masse. The nature of the Web will change; it will cease to be dominated by the U.S."

"By 2020, mobile phones will be one (ubiquitous) pathway to the Net, but far from the only one."

"With wireless communications embedded in more objects, the connectivity is likely to be through multiple routes."

"Small portable 'tablets' will be the primary interface such as the Apple iTouch...The primary communication will be using tools like MySpace, Facebook, etc."

"Just consider the iPhone and how it has started to change things in just 6 months…13 years from now, one marvels to imagine what such devices will be capable of."

"The phone may no longer be recognizable as descended from the invention by A. Bell."

"There is an issue of the killer application; what would that be?"

"Within another 15 years, it will be another mechanism (with embedded telephony), but not a mobile phone as we know it today."

"The mobile 'Internet' as currently offered by most major wireless carriers is a highly crippled, closed system that is not the Internet that most users are familiar with."

"To suggest that mobile devices will be used for more involved computing (word processing, spreadsheet work, presentations, etc.) is not practical. There are two kinds of 'connectivity' work—work while stationary and work 'on the go'—mobile devices will be for work 'on the go.'"

"Cell phones are diverse, it is so, but their use in developing nations is quite unlike that in the developed world. The phones are also not well suited to the task. Low-cost computing clients is a functionally more appealing way to go."

"Google will be a contributor to the universal standards and protocols mentioned. Open-source software is continuously evolving and becoming much more accessible."

"Phones will be at the center of complex and heated debates about using them as ID and locational devices. These phones will also be more autonomous, helping (with rules) manage the flow of messages to and from the device."

"For the wealthier nations, I would still see multiple platforms—phone, computer, TV, STB [set-top box on a television], car, etc. Also, age plays a factor, as the population ages in advanced countries, smaller screens are hard to read and see, even for the technically literate."

"The cell phone is too small. With Wi-Fi, WiMAX, and other wireless (not cellular) protocols maturing, other devices (small laptops, etc.) will overtake cell phones as Internet communications devices."

"All electronic devices are converging. We now have mobile Internet devices with 5" screens that are A/V capable."

"Telephony will never hit universal standards in 12 years. We don't even have cell service in Vermont."

"Corporate interests will likely slow this process considerably."

"Monthly fees will remain beyond the budgets of low-income families. Using a telephone in different regions will always be an expensive/exclusive good."

"By 2020, there will be some new technology that combines mobility, communication, and content. The mobile phone will be transformed in some way."

"The primary Internet connection will be Wi-Fi devices like iPod Touch with microphones. Skype will be on all the phones, and VoIP will have almost completely replaced cell phones."

"Geolocation, etc., will be done on cell phones, and perhaps there will be more of those than computers."

"Camera, instead of a mobile phone."

"Thirteen years out is unknowable—inconceivable. See Ray Kurzweil."

"Yes. Cost, size, location."

"Without significant advances in small-screen interaction technologies, interaction will remain on smaller and less expensive laptops with mobile wireless. With VoIP, these may even replace mobile phones for business people."

"Agree. It's portable and at the point of need. It will be cheap."

"A computer requires literacy, which is not the case for a mobile phone. Illiteracy is one of the big issues in poorer countries."

"The implementation and expansion of WiMAX and later technologies will change the way we communicate, allowing us to switch seamlessly between cellular and VoIP communication. An extension of this is the centrality of the cell phone, particularly in developing countries."

"Internet access and applicable charges will require standardization and lower costs to reach the 'bottom' of the world's population."

"Individuals will store their medical records on them and be able to use them for financial transactions. Phones will also have antitheft and antiloss devices because they will be so vital for communication."

"'Wired' computing will be much like the portable computer—something we'll look back at with disbelief. Mobile devices in much of the world are already the primary connection."

"I don't think we'll have universal standards for telephony."

"All will need Internet access in order to conduct business and purchase necessities. Internet technology will keep us connected at all times."

"Not requiring elaborate building-to-building infrastructure will be the only way to dramatically increase access."

"This depends upon more miniaturization of laptop-like features, reasonably sized keyboards, or touch pads, and the nonproprietary cooperation of those developers who currently now operate without regard for open-source standards."

"Laptops will be improved (portability and ease of use/access), and this will be the primary connection."

"This means that cell phone companies will need to be regulated to stop charging by data size."

"Even as Web sites design expressly for handhelds, there is still so much data to squeeze into a small format. Condensing it with an effective information architecture creates a deeper Web site with more clicks to get to the intended target. In short, the small format slows the speed of the Internet."

"Phones may need to be a bit larger to accommodate the eyesight of the aging population."

"Business is too focused on squeezing computer technology into phone-sized products. The focus should be on expanding phone technology to computer-like power with mobile ease of use."

"Our human need to connect will increase as technology erodes our personal contact."

"The 'mobile phone' of 2020 will have evolved so much it will have only a passing resemblance to the device we call a phone."

"One question is whether the mobile phone will represent an individual, in which case verification/authentication for secure transactions will occur at the device level, or whether mobile phones may still represent groups (family/villages/etc.), in which case, secure transactions will still need to be verified repeatedly."

"I've been amazed, when I travel internationally, at the extent to which people use mobile phones almost seamlessly in their lives."

"The percent of the bottom three-quarters who have Internet access might be higher than 50% by 2020."

"By 2020, the concept of a device that is identified primarily with one application (i.e., telephony) will be laughable… Nevertheless, the basic premise of this scenario—that a small, integrated device will serve as the primary link to our networks —is likely true."

"While lots of little interactions with the Internet will go to cells, more comprehensive info and entertainment will not. I doubt universal standards, even by 2020."

"Our mobile phones will be used for telephony, as well as shopping—with credit card swipe ability."

"The current popularity of the iPhone demonstrates the direction we may be heading. Cell phones have proliferated massively, and they have become very powerful. Unfortunately, humans haven't figured out how to balance between being in constant contact AND having a personal life."

"The mobile Internet will grow and grow, but only for certain contents, while the online communication will still be in charge of big displays that will substitute our actual TV sets."

"In Japan, this is already happening."

"The advance of battery technology will enable this fact."

"The Telco's will corner the market and eliminate the current Web neutrality, and the 'bottom' three-quarters will find illegal or alternative ways of accessing these services. So yes, I mostly agree that the future is mobile, but I don't think it'll fit within a utopian bell curve where access and quality are consistent."

"It will be interesting to see the impact of mobile communications on the capacity of authoritarian regimes to restrict expression on the Internet."

"The current GSM and 3G phones already cause health problems. Data connections are even riskier than voice connections."

"Mobile phone will be mobile Internet…it goes into a social interface, a global community of storytellers."

"I wonder how the impact of the market structure in telephony will affect these developments vs. the Internet where the infrastructure was not 'owned' and the system not segmented by multinationals' ownership."

"The expanding markets of India and China will contribute greatly to an increase from 30% in 2007."

"Some type of small or mobile computing platform—like the iTouch or iPhone—will become the foundation, and telephony will just be one application added to it. The telephone application, even in the mobile environment, will be liberated from the device and become a download for whatever small computer people want to use it on."

"Small, relatively cheap, agreed-upon standards for interoperability—it works. Will we design services that accommodate the limitations inherent in the lower end of the technology scale, though?"

"Telephony still requires access to a cash economy, and there are no incentives to keep prices low."

"In developing countries, mobile Internet access by phone is, and most likely, will continue to be very expansive. Most people will have access at cybercafés and other similar locations."

"In Africa and Japan, it is the dominant networked device."

"3G or equivalent access will leapfrog wired technologies."

"To really tackle the digital divide, it is vital that people have personalised and privatised access and that they have the skills to work with these technologies in ways that will benefit them directly. The way the market is currently set up and the way in which it is linked to education is still below standard in most countries."

"Portable devices will be the leading form of Internet access... Talking on them will be a minor, secondary function."

"The summary does not address the quality of information and the fact [that] the users are bombarded with ads. The mobile phone may be the primary connection tool, but for certain usages only."

"People want information quickly and briefly wherever they are."

"Small multifunctional devices will be used to connect people with greater frequency than larger PCs/laptops—especially as hardware capabilities expand inversely proportional to size."

"The various telephony companies have built huge bases of support and profit without compatibility. If there is an impetus for cooperation and a single standard in the future, it will most likely be in Europe and Asia as opposed to North America."

"First, OLPC and similar initiatives are not going to be so successful. Second, the mobile devices that will be used in 2020 will be quite different from the mobile phone as we know them. Third, seamless roaming may continue to be a serious headache. Fourth, cost will continue to remain above purchase power of the majority of people."

“It won’t happen, especially in poorer regions, unless we ensure adequate competition among multiple carriers.”

“Wireless telephony could fall prey to the same concerns that afflict wireline telephone and Internet—big companies with power to control the rate and possibly the direction of technological innovation…governments may impose constraints on a cautionary principle basis.”

“The rest of the world except the U.S. will be on this model. The U.S. telephone industry will have a hard time with universal standards and protocols.”

“Somewhere between the MacAir and the iPhone and new foldable screen technology is the opportunity for a new kind of phone.”

“Boundaries between mobile phones and PCs will collapse as computing power migrates to the Web and innovators develop more ultra-portable devices…allowing access to a fuller roster of Web-based applications than is currently or conceivably possible on a phone-sized device.”

“With the increasing capability and decreasing cost of flash memory, it would be unrealistic to not expect the further development of the smartphone—but most likely not in the form we’re currently used to seeing it in.”

“Unfortunately, human nature will dominate…we will keep the best stuff for the ‘haves’ and try to keep the ‘have-nots’ from getting cheap access.”

“Interoperability will be the norm.”

“Much of this will depend on the openness of mobile networks, which, in fact, appears better in developing countries than established areas. It may be due to the fact that they are not willing to deal with the complexity of walled gardens compared to their cost. Another critical factor will be display technologies such as roll-up display, HUD, micro projection, etc.; small displays cannot depict a lot of information nor easy to understand. A way to provide a large display (without going as far as eye implants or goggles) while still maintaining a small form factor would be a key.”

"By 2020, the mobile phone will have become the primary lifeline for most of the world's population—and the recent growth in mobile adoption in the third world indicates that cheap handsets and reliable mobile service can bring first-class access to information and communications tools to populations that were previously inaccessible."

"The mobile phone will be the ultimate one-to-one personal communication, entertainment, and information tool."

"Interconnection rights between carriers in the U.S. have been fought over for the past decade. I see little to make me believe domestic carriers, let alone international carriers, will solve that problem in another 13 years."

"It's hard to imagine that major corporations will assist the bottom to gain access to the Internet. Someone along the way will need to foot the bill."

"This is already occurring. More teenagers in China view the Chinese-language version of 'American Idol'—'Super Girls'—on a mobile device than on either TV or CPUs. The real question will be what governments will do to limit the access to search criteria, etc."

"Getting the telecoms to agree on a common standard will be the biggest hurdle. Nice to think that market forces would drive this."

"Innovative pricing models will contribute to access."

"There will be a revolution in hardware, and we may not be able to now understand the device and the protocols that will arise to support persistent and fast information exchange between individuals."

"With the technology, education, and economic explosion in China and India, this prediction is likely to be low."

"Phones will be more universal because they are a leapfrog technology—and because they do not require the power and infrastructure that other Internet connections require."

"I am not convinced there will be universal standards and protocols. This has not happened in existing landline telephony, video formats, software packages, nor even in basic electricity supplies. There are too many vested interests in maintaining competition."

"Telephony standards will never be that seamless."

"The marketplace for telephone service will be impeded by regulatory, corporate, and technical barriers."

"Universal standards are almost sure to emerge, since there is such a strong economic incentive for telecommunications providers to offer seamless service across geographic boundaries."

"The world will be better connected through cell phone technology, but does that make a better world?"

ENDNOTES

1. See *World Information Society Report 2007: Beyond WSIS*, published by International Telecommunication Union and the United Nations Conference on Trade and Development in May 2007(b), available at http://www.itu.int/osg/spu/publications/worldinformationsociety/2007/report.html.
2. See *The Role of Mobile Phones in Sustainable Rural Poverty Reduction*, published by the World Bank ICT Policy Division Global Information and Communications Department on June 15, 2008, available at http://siteresources.worldbank.org/EXTINFORMATIONANDCOMMUNICATIONANDTECHNOLOGIES/Resources/The_Role_of_Mobile_Phones_in_Sustainable_Rural_Poverty_Reduction_June_2008.pdf.
3. See the report *Mobile Access to Data and Information*, by John Horrigan, published by the Pew Internet & American Life Project in March 2008, available at http://www.pewinternet.org/pdfs/PIP_Mobile.Data.Access.pdf.
4. See *World Information Society Report 2007: Beyond WSIS*, published by International Telecommunication Union and the United Nations Conference on Trade and Development in May 2007(b), available at http://www.itu.int/osg/spu/publications/worldinformationsociety/2007/report.html.
5. From "The Next 4 Billion" by Allen Hammond, William J. Kramer, Rob Katz, Julia Tran, and Courtland Walker, available at http://www.wri.org/publication/the-next-4-billion. Drawing on income data from

110 countries and standardized expenditure data from 36 countries across the globe, it assesses the market opportunities represented by 4 billion individuals who make up the economic "base" of the world population pyramid (BOP). The analysis provides a quantitative assessment and characterization of BOP markets at several levels.

6. See "The Impact of Telecoms on Economic Growth in Developing Countries" by Leonard Waverman, Meloria Meschi, and Melvyn Fuss, featured in the Economics Focus section of *The Economist*, March 12, 2005, available at http://www.london.edu/assets/documents/PDF/L_Waverman_Telecoms_Growth_in_Dev_Countries.pdf.
7. From "Shaping Policies for the Future of the Internet Economy," by the Organisation for Economic Cooperation and Development Committee for Information, Computer, and Communications Policy, March 28, 2008, available at http://www.oecd.org/dataoecd/1/29/40821707.pdf.

Scenario 2

The Internet and the Evolution of Social Tolerance

Prediction: *Social tolerance has advanced significantly due in great part to the Internet. In 2020 people are more tolerant than they are today, thanks to wider exposure to others and their views that has been brought about by the Internet and other information and communication technologies. The greater tolerance shows up in several metrics, including declining levels of violence, lower levels of sectarian strife, and reduced incidence of overt acts of bigotry and hate crimes.*

Expert Respondents' Reactions	
Mostly Agree	32%
Mostly Disagree	56%
Did Not Respond	13%

All Respondents' Reactions	
Mostly Agree	33%
Mostly Disagree	55%
Did Not Respond	11%

Note. Since results are based on a nonrandom sample, a margin of error cannot be computed. The "prediction" was composed to elicit responses and is not a formal forecast.

OVERVIEW OF RESPONDENTS' REACTIONS: A majority mostly disagreed with the proposed future. Many said while there is no doubt the Internet is expanding the potential for people to come to a better understanding of one another, it also expands the potential for bigotry, hate, and terrorism; thus, tolerance will not see net gains. They believe that the natural human tendencies to congregate with like-minded allies and to act in tribes are too potent to be overcome by technology tools that expand communication and the flow of information. Still, about a third mostly agreed with the premise, optimistic that gains will be made, while adding the qualifier that negative agendas will always also be well served by advances in communications technologies.

More than half of the respondents mostly disagreed with the idea that the Internet will help inspire a significant increase in social tolerance. A representative response came from **Adam Peake**, a policy analyst for the Center for Global Communications and a leader in the United Nations–facilitated World Summit on the Information Society and Internet Governance Forum. "Not in mankind's nature," he wrote. "The first global satellite linkup was [in] 1967, BBC's 'Our World: The Beatles "All You Need Is Love,"' and we still have war, genocide, and assassination (Lennon's poignantly)."

An anonymous respondent wrote wistfully about the ideal of greater tolerance: "If only." Others wrote, "No way!" "We hope," "Dream on," and, "The Internet won't change our nature; we'll still find ways to hate each other, the Internet will just help us."

Another respondent wrote, "Tensions in the real world are likely to exacerbate by 2020, and there is no reason why real people would become less aggressive just because they have a Facebook account." Another expressed the opinion of many when she wrote, "Internet has nothing to do with social tolerance; it comes from education; if Internet advances education, yes, it will indirectly contribute to social tolerance." And another commented, "Tolerance is in decline, at both a 'local' and a global level."

Jamais Cascio, the founder of Open the Future, active in the Institute for Ethics and Emerging Technologies, commented, "Sadly, there's little evidence that greater observational exposure to one's 'enemies' automatically reduces hostility and increases tolerance. In many cases, it does the opposite, especially if that observational exposure is controlled or manipulated in some way."

The same line of reasoning was followed by **Alex Halavais**, a professor and social informatics researcher at Quinnipiac University. "Wider exposure to different views does not guarantee more tolerance," he wrote, "and there are plenty of opportunities for people to use the Internet to encourage factionalism and ignorance."

Fred Baker, Cisco Systems Fellow, Internet Society and IETF leader, and an architect of the Internet, wrote, "Human nature will not have changed. There will be wider understanding of viewpoints, but tolerance of fundamental disagreement will not have improved."

An anonymous respondent commented, "People are people. The Internet is only technology—and just the most recent technology in a long line of advances that have 'made the world smaller.' The groups that are marginalized and persecuted may be different in 2020 than they are now, but when people get scared or greedy, they will still find 'others' to scapegoat or hate."

And **Tom Vest**, an IP network architect for RIPE NCC Science Group, expert on Internet protocol policy, and consultant for the Organization for Economic Cooperation and Development, commented, "Absent some major external shock, affective education on the kind of global scale necessary to make this one come true will take much longer than 15 years. On average, people will not be much more tolerant/intolerant (or educated/ignorant) than they are today."

An anonymous participant responded, "Rewiring our neurons and DNA is not going to be easily facilitated by user-friendly landing pages."

Philip Lu, vice president and manager of research analysis for Wells Fargo Bank Internet Services, commented, "Just as social networking has allowed people to become more interconnected, this will also allow those with extreme views (who would otherwise be isolated) to connect to their 'kindred' spirits elsewhere. Therefore, I am not optimistic that violence will go down."

Several survey participants pointed to an overload of input as a reason this scenario rings false.

"The barrage of overstimulation makes it harder for people to grasp what they're seeing, and then they will regress back to base human nature, where a competitive us-vs.-them attitude always crops up," wrote an anonymous participant. "This *won't* ever go away. Perhaps it will subside to a degree, but it won't make a large-scale difference. However, I believe there's a case to be made that this is true one generation past 2020—as kids grow up with more of a world view, perhaps they will be able to advance to the environment that is described."

Clay Shirky, author of *Here Comes Everybody*, a book about the ramifications of the new forms of social interaction enabled by emerging technology, responded, "The Net's ability to enhance the sense of in-group membership will enhance fragmentation of previously large, multiethnic polities. (Consider that there are secession movements in Scotland and Belgium.) There may be lower levels of sectarian strife, but only in the same way and for the same reason that there are lower levels of sectarian strife in the former Yugoslavia today, relative to 1997."

And **Frederic Litto**, president of the Distance Learning Association of Brazil, wrote, "Much to the contrary, all our advancement in knowledge about evolution, human cognition, and medical diagnostics and treatment have done little to reduce human stupidity, hate, and violence. We may advance indefinitely into new worlds of technological competence and globalized knowledge about one another, but there's no guarantee that universal education, sophisticated flows of communication, and international organizations attempting to reduce intolerance and acts against peace will be entirely successful. This reminds me of Henry Thoreau's famous retort (1870s?) when told that the first long-distance telephone lines had been put into place, linking the inhabitants of the states of Georgia and Vermont: 'All well and good; but what if the peoples of Georgia and Vermont have nothing to say to each other?'"

Some Say the Internet Will Accelerate or Expand Fragmentation and Reinforce Prejudices

A number of respondents said the Internet's capabilities enhance the opportunities for people with ill will and violent agendas.

"Are you kidding?" responded **Dan Larson**, CEO of the PKD Foundation. "The more open and free people are to pass on their inner feelings about things/people, especially under the anonymity of the Internet—will only foster more and more vitriol and bigotry."

Many expressed concerns over the use of networked communications to further the goals of groups that sometimes leverage the differences between themselves and others to gain unity. "Globally, a more significant 'us-versus-them' attitude will prevail," responded **Scott Brenner**, a U.S.-based technology consultant.

"I see more anger in society, more carelessness, less regard for rules of civility and behavior," wrote **Alexis Chontos**, webmaster for the Art Institute of Pittsburgh. "There will be greater crime, an increase in the 'you owe us' mentality, less tolerance, more sectarianism, more hate crimes (religion against religion)."

Steve Goldstein, an ICANN board member and Internet pioneer who worked with the National Science Foundation to help diffuse the Internet globally, wrote, "The Internet will continue to be used as a tool for supporting intolerance, revolution, terrorism, and sectarian hatred, as is painfully evident today."

Fred Ledley, founder and chairman of Mygenome, was even more certain of the negatives. "The Internet is a danger to social tolerance," he wrote. "The easy distribution of hate and propaganda through the Internet allows dissemination of hateful material that would not previously have received attention. Worse, it makes it harder to appreciate what is fringe behavior by a small number of individuals, and what represents a true movement or organization. The prevalence of anti-Semitic propaganda on the Web is a frightening example of what the Web can sustain."

The propagation of propaganda and lies is a concern for **Bruce Turner**, director of planning services for a U.S. regional transportation commission. "Bad info drives out good, and the degree of intolerance will rise as superficial examinations of nonissues become more and more the order of the day," he commented. "Bigots and governments spoofing as knowledgeable experts will make the information suspect and largely ignored. Bigotry and hate crimes will be facilitated for the remaining fringe who pay attention."

An anonymous respondent wrote, "The anonymity of the Internet does not encourage tolerance in most cases, but rather allows people to react in ways that would be deemed inappropriate offline, because they do not have to face the same sort of immediate or devastating consequences." And another wrote, "Take a look at forums and comments on a newspaper Web site, and you will notice that the Web can also be an open gate to hatred messages and intolerance."

Jim Horning, chief scientist for information security at SPARTA, Inc. and a founder of InterTrust's Strategic Technologies and Architectural Research Laboratory, responded, "Tribes will be defined by social enclaves on the Internet, rather than by geography or kinship, but the world will be more fragmented and less tolerant since one's real-world surroundings will not have the homogeneity of one's online clan."

Fernando Barrio, senior lecturer and leader in the MA program for e-business regulation at London Metropolitan University, said inadequacies in education play a significant role. "Due to the problems in general education, the exposure to what is different without proper guidance will exacerbate sectarian violence, incidence of acts of bigotry, and hate crimes."

Bernardo Huberman, senior fellow and director of the Social Computing Lab at HP Laboratories, commented, "Have you been on the Internet? It allows people to find their own insular communities that are outside the criticisms of others. See: furies." An anonymous participant added, "There will be more tolerance on a whole, which will only aggravate extremists even more." And another added, "By bringing people of every background together, the immediate effect is more and bloodier wars, perhaps not on the battlefield, but certainly in social movements and politics."

Sean Steele, CEO of infoLock Technologies, a security company, wrote, "Online bigotry and hate crimes are rampant from those nation-states that do not enforce limitations on or [do] condone such activities."

T. J. McManus, a vice president with Nuance Communications, Inc., emphasized the fact that some governments control content. "Unless the dominant nondemocratic governments that exist today (Iran, Saudi Arabia, China, countries in Africa, etc.) change, restricted Internet access might have the opposite effect, i.e., provide a different channel for creating and maintaining sectarian differences."

An anonymous respondent added, "We will see the growth of extremism and extremist communities that will help to make the world bleaker...Governments will try to limit access to stuff they disagree with. And the pressure is for more authoritarian responses, sometimes for excellent motives."

Many shared the view that people will spend less time in face-to-face communications, and that this will damage their ability to have empathy and relate well to others. "Insofar as the virtual world permits less actual interaction, then individuals with dangerous biases will have no cause to question their beliefs," wrote one anonymous contributor.

Another anonymous participant commented, "Evidence already shows that the Internet, in fact, normalizes socially unacceptable attitudes for those who hold them because they can seek out communities who share the same ideologies."

And an anonymous respondent wrote, "'Tolerance' is a Western concept, and the Web may cease to be primarily Western."

Many Respond That the Internet Will Contribute to the Expansion of Tolerance *and* Intolerance

Many mostly disagreed with the scenario because the Internet, like all technologies, serves both good and evil human motives equally well. "This bifurcates," wrote an anonymous participant. "Most people are tolerant, but the pockets of extremists and ideologues are more toxic and more extreme."

"Although I believe the Internet is a net positive for tolerance and sociability, its impact will be gradual, even generational, and although positive on balance, it will also contribute to the cohesion and separateness of intolerant (and worse) subgroups," responded **Tom Hughes**, COO at The Connors Group, a financial markets information company.

"Polarization will continue and the people on the extremes will be less tolerant of those opposite them," wrote **Don Heath**, Internet pioneer and former president and CEO of the Internet Society. "At the same time, within homogenous groups (religious, political, social, financial, etc.) greater tolerance will likely occur…I hope I am wrong."

William Winton, project manager for digital media at the 1105 Government Information Group, wrote, "The Internet is a two-edged sword. Its openness and ease of communication have also fostered the rise in online Jihadists, resurrected a flagging neo-Nazi movement, and enable all sorts of intolerant movements, ideas, and people to flourish online. The jury will probably still be sequestered in 2020 as to whether the Internet has fostered 'tolerance' or merely 'siloed' hate."

Some respondents commented that the expanded communications capabilities offered by the Internet will serve to accentuate the world's economic and religious divisions. "Unfortunately, more egalitarian distribution of information will show others lifestyles they have no reasonable chance of attaining or that they might object to on religious grounds and that will cause intense resentment," wrote **Jim Lucas**, Web manager at CACI, a provider of national intelligence solutions for the U.S. government. "On balance, expanded communications better illuminate human nature but don't change it."

An anonymous respondent wrote, "With the gap between rich and poor in all countries increasing, we have seen levels of violence increase. The gap is only going to rise, leading to increased resentment and, thus, more violence. This rise in violence will be especially high amongst young men who realise that they will never earn more than the minimum wage."

Shawn Kelly Apochromantic, a manager, designer, and futurist for General Atomics and a volunteer builder in the online synthetic world Second Life, commented, "The developed and nonextremist peoples of the world will continue to improve, but the isolated and extremely intolerant will continue to be a source of hatred...As the levels of education and exposure improve, the pools extreme groups can draw upon for narrow-minded followings will diminish."

Richard Osborne, Web manager for the School of Education & Lifelong Learning at the University of Exeter, responded, "Knowing more about someone online could just as easily lead to less tolerance as opposed to more; because you can read their views more fully, you might find this enhances your dislike."

Berlin-based Web strategist **Peter Bihr** said the positives will come "within certain strata of society...Within the more highly educated crowd, the Internet might increase social tolerance for the reasons put forth in the 2020 scenario."

"This is one of those dialectical things," responded **Jeremy Shapiro**, a professor of critical social theory at Fielding Graduate University. "Both social tolerance and intolerance will advance. That appears to be part of the dialectic of modernization." And an anonymous respondent wrote, "Perhaps in the wired elite there will be less strife, but those who generally lean toward picking up a brick to solve a problem will continue to do so."

Some Say the Internet Is Making a Positive Difference, Allowing People to Come to a Better Understanding

Many respondents agreed with the scenario. "I do see a long, slow road of improvement," wrote **Paul Jones**, director of ibiblio.org, based at the University of North Carolina–Chapel Hill. An anonymous participant commented, "Levels of sectarian strife and overt bigotry and hate crimes will peak after 2020 (not before) in response to this wider exposure and increased public presence of cultural minorities." Another anonymous respondent wrote, "The more you know about the 'other,' the less extreme your views of them will be, the more open to differences you will be; history confirms this, I believe." And another wrote, "This is the only hope for world peace."

"Knowing how others really are helps develop tolerance; this will take time, but connectedness will accelerate tolerance," responded **Dan Lynch**, Internet pioneer and board member for the Santa Fe Institute. An anonymous respondent agreed, writing, "Ultimately, the Internet's ability to increase individual awareness, education, and access to information will overcome organised social and cultural intolerance."

"One can only hope," wrote Buzzmachine.com blogger **Jeff Jarvis**. "I wouldn't go so far as predicting world peace through the Internet. Sadly, there will always be fanatics and criminals...But I do, at least, believe that the Internet's ability to bridge nations and divides and bring together individuals can only be positive."

An anonymous respondent wrote that the benefits will transfer to those who are not online: "Tolerance is just one feature of an increased global spirituality that has already started emerging. Lack of access to the Internet doesn't keep one from this, since the ubiquity of the Internet will spread this spirituality through offline social contacts, as well as via offline contacts."

Another anonymous respondent wrote about a recategorization of human culture. "It's harder to have bigoted notions about other kinds of people when you play Facebook poker with them every day. The Internet, by its very nature, brings diverse people together as they use it to organize themselves according to categories that have nothing to do with race, religion, class, or country."

"Access to information will increase cultural, social, and intellectual tolerance among people who have access," responded **Clement Chau**, manager for the Developmental Technologies Research Group at Tufts University. "Because of this, we shall see that the control and access of information will become the primary concern for governments worldwide."

Jamie Richard Wilson, a journalist and freelance Web developer, commented, "When people can put names, faces, and experiences to people on the other side of social barriers, they will begin to understand and empathize with others. Greater access to information will make it difficult for governments and organizations to tell people what to think. Diversity will be seen as a positive asset, and more people will seek opposing views to their own ideals for perspective."

"Increased access to information about different people will enhance our understanding of different cultures and promote greater intercultural sensitivity," wrote **Gary Kreps**, chair of the department of communication at George Mason University. "People will recognize similarities in values and goals and use these shared values as a basis for coordination and cooperation."

Joe McCarthy, self-described "principal instigator" at MyStrands, formerly principal scientist at Nokia Research Center in Palo Alto, CA, wrote, "Yochai Benkler's book, *The Wealth of Networks*, shows how the Internet can help transform economics and society, and enable more people to be both self-sufficient and entrepreneurial. As more people are able to truly engage in this increasingly inclusive economy, there will be less violence. We'll all come to see that 'everyone's a customer'...and that everyone's a potential trading partner (on an individual, not just a national, stage)."

"I believe that as Derrick de Kerckhove so aptly named it, the Internet has created a global, connected intelligence," wrote **Barry Chudakov**, principal of the Chudakov Company, a marketing strategies firm. "And while this connecting can be used to foment hate and divisiveness, the larger use of the Internet is to create intelligent communities. Further, one can encounter voices within these communities that build awareness of wider views than one may have known before. So it is the community-building, the focusing of shared interest, that has the potential, at least, to allow more and varied voices to be heard. Whether this will indeed result in greater tolerance and declining levels of violence and strife…let's just say there is great potential for that to happen."

An anonymous participant equated information with power, writing, "Much of the intergroup hatred of the past has been fomented by leaders who wish to derive power by uniting one group of people against a perceived common enemy. This includes divisions along political, ideological, racial, and religious lines, among others. Much of this promotion of hatred is achieved via disinformation campaigns. As communication technologies facilitate more direct communication amongst ordinary people and make it easier to debunk planted myths, people will realize they have more in common than in contention and be more accepting of their differences. The same technologies make it easier to spread disinformation, but the balance will tilt in favor of the truth."

Another anonymous respondent commented, "This [scenario] will be true, as people use the power of the Internet to solve worldwide social problems such as poverty, hunger, sickness, and lack of education. As there is less need for violence and dishonesty to survive, people will become more trusting of one another." Another asked, "When you were 10, could you play chess online with someone across the world?"

A reference to the effectiveness of online exposure for issues came from **Gerard LaFond**, founder of red TANGENT, a strategies firm. "The Internet, by 2020, with its army of bloggers, cell phones with video cameras, and citizen journalists, shines a bright light on differences between people and cultures. This has an effect of creating tolerance and also brings swift condemnation when hate and bigotry is illuminated by the Web's reach."

An anonymous respondent who mostly agreed with the scenario wrote, "This cannot happen, however, if the content management is strictly enforced. To me, making information free and available to everyone is how the social tolerance will increase. If we create a 'haves vs. have-nots' society with respect to content and info, violence will continue to grow."

Do Our Tools Shape Us, or Do We Shape Our Tools? The Question of Technological Determinism

This question drew the attention of several respondents who are attuned to the concept known as *technological determinism*. A dominant view holds that advances in technology are the driving force behind social change and that they carry inherent effects—that our tools are vital to how we act and who we are. This view is referred to as technological determinism by those who argue against it—they say technological innovation is mostly shaped by society through the influence of economic, political, and cultural motivations.

"It would be marvelous if this were to happen, but be wary of attributing deterministic effects to the Internet and other ICTs, never mind assuming they will change human nature in this short a time scale," wrote **Victoria Nash**, of the Oxford Internet Institute and formerly a fellow at the Institute of Public Policy Research.

An anonymous respondent commented, "The Internet is a means of communicating with no inherent positive or negative value. Do we believe that the telephone advanced social tolerance in the 20th century?" Another wrote, "Technology won't change fundamental human nature."

Benjamin M. Ben-Baruch, senior market intelligence consultant and applied sociologist for Aquent, wrote, "First, I disagree with the notion that social tolerance has advanced or increased. Second, I disagree with the notion that either technology or education tends to increase tolerance. There is, as far as I can discern, no body of evidence that supports such notions. To the extent that evidence exists, it supports the notion that both education and technology can be used to increase tolerance but only under conditions that are unlikely to be replicated broadly across large populations (at least in the foreseeable future)."

"To credit the Internet would be overly technologically deterministic," responded **Christine Boese**, information architect for Avenue A-Razorfish. "There are aspects of both greater and lesser social tolerance online. If the technology tends to lead cultures in any particular direction, it is leading to greater polarization of extremes and less of the middle. Does greater tolerance constitute the middle? Not in this case. The extremes find support for their views online, more so than in the less connected, face-to-face world, so bigots find their views reinforced and even the far extremes of social relativists find their views reinforced…Is everyone really entitled to his or her own opinion, or are there very real and socially constructed methods to evaluate whether some opinions and views are, indeed, superior to others? I believe the latter. Perhaps we should all go back and read that dated study by William Perry on the intellectual development of Harvard undergraduates in the homogenous 1950s."

An anonymous respondent commented, "It is not inevitable that KMD (knowledge-enabled mass destruction), which appears viable with emerging technologies, will not become more probable in 2020 than it is now, since emerging technologies will rely more upon deep technical knowledge, not rare physical resources, as has been the case with WMDs [weapons of mass destruction]."

Many Note Other Concerns: Economy, Environment, Religion, Global Population

Many respondents expressed concerns about world population growth, the limits of resources and problems with sustainability, sectarian strife, and other major issues. "It could be argued that in the next decade, environmental, economic, and population pressures will put so much strain on society that we'll be doing quite well just to be as socially tolerant in 2020 as we are now," responded **Michael Edson**, director of Web and new-media strategy for the Smithsonian Institution.

Olav A. Øvrebø, editor of the Web magazine *Vox Publica*, wrote, "This is too heavy a burden to lay on online communication. There are so many other variables in society that influence tolerance. Hopefully, online communication can contribute positively, though."

Kathee Brewer, a freelance journalist and consultant from Texas, wrote, "Few people are willing to give up core beliefs (partially out of fear and partially because they can see no way to reconcile diametrically opposed tenets—particularly religious). Religion, in particular, will continue to divide people, because it has to drive wedges between them in order to retain control."

Roderick White, editor of *AdMap* magazine and the World Advertising Center, wrote, "Sorry, I'm a pessimist here. A reasonably informed guess says that the influence of the Internet will be unable to overcome the pressures created by a combination of a growing global population and climate change, which together are going to create enormous pressures on populations to migrate and amplify the existing tensions that sour relations between races and nations."

Jane Sarasohn-Kahn, founder of THINK-Health, pointed to economics as a lynchpin. "This scenario is highly dependent on the health of the economy and whether incomes and access to social institutions are more equitably distributed than they are today," she responded. "Economic development and distribution of wealth is a key wild card for 2020."

An anonymous respondent commented, "Serious economic and environmental changes will have to be affected globally before this is possible. We can only tolerate each other if we're not fighting over resources."

Sam McAfee, CTO and owner of RadicalFusion, a design, programming, and development company, also expressed deep concern about world economics. "The questions of the next 12 years for the economic well-being of most of the world's poor are how deep and how long the recession facing the U.S. will be, how much a U.S. recession will drag the rest of the world economy down with it, and how governments throughout the world will deal with massive increases of poverty in their own countries created by a global economic slump," he wrote. "No amount of cheap access to the Internet is going to alleviate the suffering that will come with a downturn of this magnitude."

An anonymous respondent added, "The information area has increased the stratification between rich and poor. We now have super-rich and strongly poor—and that gap seems to continue to widen. No amount of interaction on the Net is going to compensate for that amount of alienation. And if the information era becomes distorted where it is used by those in power to maintain their power and quell dissent—as we have seen in countries such as Burma—alienation will increase."

Cambria Ravenhill, manager of national channel planning at TELUS Communications, commented, "The road to such social tolerance will be violent. The driving force behind this tolerance is the ability to form new communities and connections outside of traditional boundaries and tribal norms. The stakeholders of the status quo will fight this new order long and hard—it's already happening. The growth of extremism and polarization of religion is one example; a second is the recent increase in hate crimes. By 2020, these should have declined significantly, with one side effect being the possible marginalization of religion in the newly tolerant society."

One anonymous contributor said authoritarian regimes will use the Internet to extend control through conflict, writing, "The Internet will be used for the continuation of social strife...The vast amounts of information collected about people and groups will make people vulnerable to policing—unless we start defining acceptable practices and limits now."

Havi Hoffman, senior editor in product development and a blogger for Yahoo, wrote, "Hatred and sectarian strife will continue to flare across the planet before we humans exorcise tribalism, intolerance, and xenophobia. Recently, I heard Kevin Sites speak about the common denominator he found in the dozens of hot zones he visited over a year of travel, covering conflict. Under the tribalism, bigotry, and ideology, he found greed, hunger, fierce competition for limited resources. I don't think we'll have the resource and distribution inequities solved in the next 12 years. I can see the possibility of tolerance growing—especially as young people (the digital natives) grow up, talking and meeting and trading effortlessly and effectively in virtual spaces—the next global youth culture may embrace networks of terror and strife, as well as communities of work, play, science, craft, art, sport, gaming, and every flavor of collaboration. Even what we cannot imagine."

An anonymous participant commented that most people are not online and cannot be influenced by Internet content in a positive or negative way. "This is only plausible if the idea of digital identity is an important aspect of an individual life," the respondent wrote. "I foresee a large proportion of people having no vested stake in digital life. For these people, any laudable ideals associated with digital cultures have little effect on attitudes and behaviours...This is Social Identity 101, folks."

And **Ilima Kane**, program manager for Klein Buendel, a Colorado-based health firm, commented, "I think this (scenario) will hold true for the types of violence and bigotry we are familiar with today but, unfortunately, will be replaced with new forms of intolerance on issues we can't even imagine today."

The Potential for New Communication Tools to Bring People Together to Serve the Common Good

The vast majority of respondents to the survey equated the word *tolerance* in the scenario with an understanding of differences—if not to grant complete respect for them, at least to exhibit a willingness to withhold any overt negative reaction to them. Many respondents equated it with the ideal of a world at peace or mostly at peace.

Only a few survey participants took issue with the language used in the scenario, but "tolerance" as a concept is open to question and worthy of contentious discussion in and of itself. "Tolerance is a myth," wrote one anonymous respondent. "What most people define as tolerance is acceptance of their behavior and adherence to their concept of others' 'bad behavior.' At no point does anyone bother to explore why they feel the need to judge others. They simply score them on how close to their own norms the other comes."

There was some discussion of the fact that in the first years after they are invented, new technologies are usually pronounced to be world changing.

Frank Thomas, a *Future of the Internet III* survey respondent who did not divulge personal information about his workplace or other identification, wrote, "Has this scenario text been copied from the telegram that Queen Victoria sent in 1866 [*sic*] when the first transatlantic cable was opened? Technology enlightens society, the eternal American dream. First the telegraph, then the same ideology with the telephone, now with the Internet."

The social tolerance scenario was presented in this *Future of the Internet* survey to see if people have reason to believe that the latest communications technology is likely to be more successful at enhancing global harmony than past innovations. Such technologies have been hailed throughout history as tools that might help humans overcome some of their differences or, at least, recognize and respectfully withhold overt negative responses to them.

In the earliest years after the telegraph, radio, telephone, and television were innovated and brought to public use, all were pronounced likely to change the way humans communicate, help ease rivalries, and contribute to world peace.

When the first transatlantic cable was built from England to the United States and President Buchanan and Queen Victoria exchanged messages in 1858, Buchanan wrote, "May the Atlantic telegraph, under the blessing of heaven, prove to be a bond of perpetual peace and friendship between the kindred nations, and an instrument destined by Divine Providence to diffuse religion, civilization, liberty, and law throughout the world."[1] Authors Charles F. Briggs and Augustus Maverick wrote in their 1858 book *The Story of the Telegraph*, "Of all the marvelous achievements of modern science the electric telegraph is transcendentally the greatest and most serviceable to mankind...The whole earth will be belted with the electric current, palpitating with human thoughts and emotions...This binds together by a vital cord all the nations of the earth. It is impossible that old prejudices and hostilities should longer exist."[2]

Electrical Review writer John J. Carty projected in his Prophets column (really an early blog—a personal column) in 1891, "A system of telephony without wires seems one of the interesting possibilities...

Someday we will build up a world telephone system, making necessary to all peoples the use of a common language or common understanding of languages, which will join all the people of the earth into one brotherhood. There will be heard throughout the earth a great voice coming out of the ether which will proclaim, 'Peace on earth, good will towards men.'"[3]

Ithiel de Sola Pool wrote in his 1983 book *Forecasting the Telephone*—a study of early projections of the potential of the phone—that people predicted the telephone would further democracy and make the world smaller, increasing contact between peoples of all nations, thus, fostering world peace. [4]

The same has been said of the Internet since its earliest days of mass adoption in the 1990s. Jean Jipguep, chairman of the board of the International Telecommunication Union, said in the keynote address of the Internet Society's 1995 International Internetworking Conference, "It is clear that this can be vitally important for economic growth, fostering democracy, and promoting the exchange of ideas and knowledge of other cultures, which favors peace."[5]

There is no doubt that the Internet has allowed an unbelievable level of worldwide collaboration; the network itself is proof of that. It is a collaboration of many people, led by its chief architects—the global members of the Internet Engineering Task Force.

Clay Shirky, a respondent in the *Future of the Internet III* survey, just published the book *Here Comes Everybody*, a look at how human systems are impacted by technology and how technology is impacted by human systems. Among the examples he noted of the "wisdom of crowds" and their influence on outcomes is the work of Internet activists to make public the intolerant remarks made by U.S. Senator Trent Lott and radio personality Don Imus. Online evidence of their offensive remarks and the ensuing outcry from the general public led them to lose their positions of trust. In his book, Shirky also illustrated the ways in which terror organizations coordinate their activities online. He pointed out how both spontaneous and coordinated online networks present opportunities and threats.

Martin Carnoy and Manuel Castells have written about the revolution in the evolution of the actions and forms of governments and governance in an age of emerging collaborative media in their article "Globalization, the Knowledge Society and the Network State: Poulantzas at the Millennium." They wrote that thanks to new ways of communicating, "the dramatic expansion of non-governmental organizations around the world, most of them subsidized and supported by the state, can be interpreted as the extension of the state into civil society, in an effort to diffuse conflict and increase legitimacy by shifting resources and responsibility to the grassroots...What emerges is a new form of the state...the state of the Information Age is a network state...Legitimacy through decentralization and citizen participation in non-governmental organizations seems to be the new frontier of the state in the 21st century."[6]

The involvement of more participants in active roles in human organizations is expected to raise their levels of engagement in all actions, in turn raising the likelihood that they will be more deeply involved in acts of tolerance and acts of intolerance, depending upon their motivations and goals.

Internet negatives, in addition to crime, terror, and child victimization, include the rising number of cases of young people posting evidence of violent or cruel acts online in a bid for fame. In 2008 six girls and two boys in Polk County, FL, filmed the beating of a cheerleader specifically so they could post it on the Web. Three New York teens were arrested after they were accused of beating a girl and posting videos of it on MySpace and YouTube. A video of the beating of a 12-year-old from Brevard, FL, was posted on PhotoBucket. A series of videos posted on YouTube by different cruel pranksters from locations across the United States involved innocent victims who had been hurt and had the damage done against them posted for an international audience. Sometimes such evidence can be used to capture and prosecute those who deserve to be held accountable, as it was in most of the cases listed here.

Whether one agrees that there will be an overall net gain in tolerance or not, it is definite that innumerable sites on the Internet have been tailored to educate people about differences. Among the online messages that are raising issues in ways that can make a difference is a series of nine videos released by the United Nations Educational, Scientific and Cultural Organization (UNESCO) in 2008 to address topics tied to human rights, tolerance, and ways to work against discrimination. People from around the globe were invited to submit proposals to produce the videos. More than 100 proposals were submitted from 50 countries. The selected works were produced and made available on DVD and online.[7] Even malicious online content produced by those who many might call hate mongers can be used to educate people about the stark realities of humanity.

In 2008 the United Nations and Google agreed to work together on an online mapping project established to turn the world's attention to refugees throughout the world and to humanitarian efforts to provide aid. The outreach program utilizes Google Earth and Google Maps to spotlight projects being undertaken by various aid agencies. The groups can overlay audio, video, and text information on sites that explain their work. The project will continue to expand but initially included a look at the refugee camps in Darfur, Iraq, and Colombia. Pop-up windows explain the specific needs in each location to highlight the humanitarian crisis and allow logistical planning.

In 2007 Google teamed with the U.S. Holocaust Museum to launch the Genocide Prevention Mapping Initiative, including layers of information about people at risk for or affected by genocide, including those in Darfur today and those in Rwanda in 1994, when between 500,000 and 1 million people were killed.

Activists are using the Internet as a tool to try to pressure governments and groups to work for human rights and greater tolerance of freedom and allowances for differences. An example is the organized protests conducted along the route of the Olympic torch run prior to the 2008 Olympics in Beijing. The Save Darfur Coalition, a group that leverages the social network Facebook among other new-media tools, launched an e-mail campaign asking people to go to the route of the torch run in San Francisco and demonstrate to "extinguish the flames of genocide in Darfur."

The largest known protest to date organized on a social network came when people used Facebook to rally between 100,000 and 1 million people in 165 cities around the world in support of the Colombians' protest of the criminal acts of the Revolutionary Armed Forces of Colombia, known as FARC. No official count of participants could be reliably recorded. People were urged to take to the streets to protest the kidnappings constantly being committed by the rebel group. They wore white t-shirts to symbolize peace. The largest protests were staged in Latin American capitals, but rallies also took place in France, Hungary, Spain, India, Italy, the United States, and other countries.

While social networks are bringing diverse people together to work for causes—yet another example is Facebook's 2007 group A Million Strong for Tibet: End the Violence—they also allow people to categorize and sort themselves more easily, often into splinter groups that do not always work for the common good.

The launch of the Encyclopedia of Life (EOL) in 2008 increases human knowledge about differences. The EOL (http://www.eol.org) is a work in progress, bringing together contributors from across the globe in a celebration of the diversity on earth that will thoroughly document all of the 1.8 million known species on the planet.

It has been said that we fear that which we do not understand. The Internet is affording people the ability to share more information. It is also teaching people to understand what they must fear.

ADDITIONAL RESPONSES

Many other participants in the survey shared valuable input on the social tolerance question. Among them:

> "This 'tolerance' will be mostly fake tolerance, necessary to be integrated to the community...Underground networks will (and already) exist where people join anonymously, and the level of social tolerance yielded will be close to zero." —**Alexandre Winter**, cofounder and chief executive officer, LTU Technologies, a global leader in image search and recognition technologies

"Tolerance seems to rise and fall, in part in response to critical events; so it seems unlikely that 'tolerance' will reduce violence and strife, although by chance, 2020 might be a moment in which some public figure captures global attention and distracts us for a bit." —**Oscar Gandy**, author, activist, retired emeritus professor of communication, University of Pennsylvania

"The real change agent for social tolerance is the massive social engineering programs that are pushing how people are taught to perceive. That may change as our social needs change. I suspect social tolerance may be a temporary fashion like an avocado refrigerator. At that time, we may need something different like, 'the ability to do complete work.'" —**Dick Davies**, partner, Project Management and Control, Inc.; past president of the Association of Information Technology Professionals

"I think this is a 'risk area.' Right now, I see more danger of further siloing and confirmation of our isolated thinking than an opening up to other ideas." –**Lawrence Swiader**, chief information officer, U.S. Holocaust Memorial Museum

"It may be exactly the opposite. Young people accept each other when spending time together—whether because of court order or economic advantage. This may explain somewhat why the traditional elite schools also are the most integrated and diverse ones." —**Michael Botein**, professor of law and founding director of the Media Center at New York University Law School; consultant to the FCC

"Five years ago, I would have agreed that technology would encourage greater tolerance. I now think that unintermediated, personal, face-to-face contact, focused on a common goal of some kind, still is needed to break down these barriers. Otherwise 'the other' remains strange and feared." —**Susan Crawford**, founder of OneWebDay, celebrated each September 22; ICANN board member; associate professor, Cardozo Law School; visiting professor, Yale Law School

"Peer-to-peer network communications (whether Internet or anything else) can readily serve as mechanisms for intolerance, as has been ably demonstrated. Arguably, mass television media may help enhance social tolerance." —**Anthony M. Rutkowski**, cofounder of the Internet Society and a founding trustee;

longtime leader in International Telecommunication Union; vice president for regulatory affairs, VeriSign

"Can anybody reading this actually bet anything meaningful on declining violence, sectarian strife, bigotry, and hate? Whatever the growth of the Internet has done, it certainly hasn't solved these age-old problems." —**Howard Rheingold**, Internet sociologist and author; one of the first to illuminate virtual communities; author of *Virtual Reality*, *Smart Mobs*, and *Virtual Community*

"As much as I wish this scenario were true, I'm afraid that it will take a lot more than mere technology to tame human nature. The base instincts that cause violence, sectarian strife, and hate crimes are based on a fundamental 'fear of the other,' which is deeply rooted and will take generations to curb through widespread educational programs, reinforced by daily practice, peer pressure, and law enforcement. There may be some progress towards social tolerance by 2020, but the significant advances described by this scenario strike me as wishful thinking. At best, we may see very limited progress in highly developed and affluent societies." —**Fabrice Florin**, executive director, NewsTrust.net, a nonprofit social news network that allows people to rate the news on quality

"Let's not overstate the importance of media and communication. The Internet will not change human nature. We know from the 20 or so years of online culture that conflicts are rather accelerated. The Internet is not an ideal platform for tolerance and conflict resolution. There is a lot of distortion on the line. The critical point here is the real-time nature of online communication, which prevents people from reflecting on what they do." —**Geert Lovink**, professor and expert on culture, sociology and the Internet; based in Amsterdam; author of *Dark Fiber* and *Uncanny Networks*; responsible for the Institute of Network Cultures

"Tolerance is a personal decision. My personal experience has been that the more insular or small the community, the more gossip or judgment. A worldwide community should lead to more acceptance of diversity and recognition that on key human issues we are more the same than different, and that the differences for 90+% of people just are not that big a deal." —**Ed Lyell**, professor of business and economics, Adams State College,

Regis University, San Luis Valley Board of Educational Services; pioneer in issues regarding Internet and education

"Our natural inclinations are to become more tolerant when we learn more about others. Especially, if the world grows more prosperous. Note also that most of our myths—movies and such—preach messages of tolerance and admiration of eccentricity." —**David Brin**, futurist and author whose 1998 nonfiction book *The Transparent Society: Will Technology Force Us to Choose Between Privacy and Freedom?* identified key issues of concern

"Oh please, you're killing me. Exposure to other cultures, people, and ideas has increased in the 20th century, but the Internet is hardly the root of it, just an extension of it. Next you'll say that Silicon Valley invented the Internet." —**Tom Jennings**, University of California–Irvine, creator of FidoNet, the first message and file-networking system online; builder of *Wired* magazine's first online presence

"Bandwidth probably increases intolerance in people over 25 but increases tolerance in people under 25. By 2020, the proportion of people who grew up in a high-bandwidth world will be sufficient to push world culture in the direction described." —**Fred Hapgood,** technology author and consultant; freelance writer in technology and science; in the 1990s, he took on the role of moderator of the Nanosystems Interest Group at MIT

"I wish I saw an increasing level of tolerance coming about because of the Internet, but I believe it's just the opposite. Individuals are able to seek out communities that bolster their own limited viewpoints—and ignore the larger world or anything that runs counter." —**Dian Schaffhauser**, writer and editor for CampusTechnology.com, THEJournal.com, *Redmond Magazine*, *Computerworld*, and Web Worker Daily; founder of Sourcingmag.com

"This is highly related to the [first scenario]. Unless substantial accountability mechanisms are created, cowards will continue to be encouraged and empowered by the anonymity that the Internet provides. This is not necessarily the best way to develop the Internet. Creating accountability at the cost of all other attributes would destroy the Internet. Starting small—fixing e-mail, for instance: Getting rid of spammers by authenticating e-mail senders might be a great place to start. Without these basic measures in place, I don't believe that the Internet will help tolerance overcome bigotry." —**Ross Rader**, director of retail services, Tucows Inc.; works with ICANN in the Registrars Constituency, part of the Generic Names Supporting Organization (GNSO)

"In spite of advances in communications over the previous 3 centuries, the levels of crime and violence have not decreased, merely changed form." —**Adrian Schofield**, manager of the applied research unit, Johannesburg Centre for Software Engineering, South Africa; leader in the World Information Technology and Services Alliance

"I don't think the ambient level of tolerance in any given society has much direct correlation with the Internet." —**Milton Mueller**, professor, Syracuse University and Technology University of Delft; expert on Internet governance, technology policy; active in ICANN's Non-Commercial User's Constituency

"My sense is that online (as well as offline) people generally gather with those like themselves rather than seek out new people and new opinions, so I don't expect there to be wider exposure to others and their views." —**Steve Jones**, professor of communication and associate dean of liberal arts and sciences, University of Illinois–Chicago, cofounder of the Association of Internet Researchers

"The Internet has enabled people 25 and under to gain self-knowledge about challenging personal issues such as sexuality, ethics, morals, and values at a much earlier age by exposing them to—and connecting them with—a broader community of both peers and experienced adults. This self-knowledge, in turn, leads, for most, to the greater social tolerance described here...but it also polarizes to the edges, supporting hate speech and bullying among more narrow segments." —**Susan Mernit**, independent consultant and former senior director for product development, Yahoo; blogger; previously a vice president at America Online (AOL)

"Amnesty International U.K. From an international perspective, it's difficult to agree or disagree—different trends are being witnessed at the same time. On the one hand, the Internet is creating a huge space for freedom of speech, assembly, and association. It has given people a means with which to change society and culture. That is positive. But the fear this has sparked has led to crackdowns and attempts to exert more control on the part of governments and companies. In some parts of the world, there is great tension between the Internet as a progressive force and as a means of better social control. I imagine that the Internet will continue to be a battleground until 2020, and who ultimately 'wins' will depend on how the Internet is developed and who exerts ultimate control over it." —**Nick Dearden**, campaigns manager, Amnesty International, the human rights organization

"My sense is that the wide diversity of views online allows people to find their niche and protect their own views rather than learn and grow as a person from exposure to others. My thinking follows that of Cass Sunstein, who wrote in Republic.com that the Internet might dramatically increase the possibilities for people to hear 'echoes of their own voices and to wall themselves off from others.' While the rise of personal blogs and the opening up of news articles for reader comments has helped bring multiple voices into play, I fear most still feel most comfortable listening to the voices they agree with." —**Michael Zimmer**, PhD and resident fellow at the Information Society Project at Yale Law School; research includes social and cultural dimensions of new information technologies

"Alternatively, increased personalization, increased capabilities to engage with geographically dispersed communities of interest, etc., actually lead to a turn inward. We find those with views similar to our own, wherever they may be, and only communicate with them. Our contact with those who are different declines, and we no longer even have access to the 'different' views of mainstream media news outlets." —**Paul Miller**, technology evangelist on the senior management team at Talis, a company delivering human-centric Web applications, based in Birmingham, U.K.; a blogger for ZDNet

"*Very strongly* disagree with this one. If anything, what Cass Sunstein called "The Daily Me" captures the very strong tendency to exploit CMC [computer-mediated communication] technologies to largely find and reinforce agreement with one's own views, prejudices, etc., rather than to seek out strongly different views. The same goes, unfortunately, with regard to culture. At the risk of oversimplifying, especially the consumer/commodification orientation of the Net fosters a kind of consumer self-interest—one that requires us to make everything 'Other' easily comprehensible and consumable (sometimes literally): This means, however, stripping away from 'the Other' everything that defines it precisely as radically different. This means, in turn, that, in the absence of careful teaching and structured experiences, the default setting of Internet use is *not* to encounter the Other as Other, and thereby come to learn genuine tolerance and understanding for radical difference. On the contrary, as the Net allows me to more and more surround myself with 'the daily me'—I am required less and less to be tolerant of difference. So, my dismal prediction (hope I'm wrong) is that, again, with some exceptions at the margins, mainstream uses of the Internet will foster less, not more, tolerance and understanding across radical cultural differences." —**Charles Ess**, a professor of philosophy and religion and researcher on online culture and ethics, Drury University, Springfield, MO, and active leader of the Association of Internet Researchers

"Yes, there is more tolerance at some levels, but I am not optimistic about a decline in sectarian strife." —**David W. Maher**, senior vice president for law and policy, the Public Interest Registry, the Internet top-level domain registry; formerly vice president of public policy for the Internet Society

"Yet the ability to be part of segregated communities by being less isolated geographically will be even greater, so cultural or other identification could enable greater divisiveness." —**John C. Abell**, new-media project director, Committee of Concerned Journalists

"The Internet will be an enlightening tool to introduce people to different cultures and traditions, but mainly teach how to be more tolerant and accept people from other origins the way they are!" —**Hanane Boujemi**, ICT researcher for DiploFoundation, working on educating people about Internet policy and Internet governance, Malta

"I hope social tolerance will improve. I see more positive changes with the millennial generation, now age 6 to 26." —**Janet D. Cohen**, blogger, futurist and trend analyst

"Greater levels of personal and societal information and transparency may just make it easier to discriminate." —**Scott Smith**, principal, Changeist LLC; consultant, writer, and futurist; formerly with Yankee Group, Current Analysis, and Jupiter Communications; expert on global influence of technology

"Although some still engage in scapegoating, more people now see humanity as all being 'in the same boat' under the challenges of climate change, peak oil, and economic collapse." —**Cliff Figallo**, social innovator and original member of the first online community—The WELL, now of AdaptLocal.org; expert in fitting and implementing social Web applications to groups

"Most conflicts come from myopic perceptions that do not see the other side of the coin. With the Internet (thanks to social networks, tourism Web sites, blogs, and YouTube), more people have access to information about other (including opposing) views and cultures. Tolerance will be on the rise, and more people will break beyond myopia by seeing the parts of the world they are unable to visit physically through online opportunities. At least many people now know that Africa is not a country. :)" —**Gbenga Sesan**, Internet for development consultant, Paradigm Initiative, Nigeria; his work is tied to the use of ICTs in socioeconomic transformation, focusing on underserved groups

"I mostly agree, as a matter of my optimism and idealism, but would be careful to treat the Internet as a unifying cultural influence towards tolerance—current studies (Putnam, danah boyd, et al.) show that the Internet can result in online communities with less diversity and tolerance than offline." —**Michele Perras**, artist, consultant, researcher, and futurist with Interactive Ontario

"I totally agree with this. I also predict that 'shock' sites such as rotten.com and the more extreme porn sites will see traffic slack way off as they become just one more part of the enormous library of humanity. Violence, bigotry, and hate crimes often stem from individuals who are disconnected and marginalized. In a world of near-perfect interconnectedness and transparency, these things decrease." —**Josh Quittner**, executive editor, *Fortune* magazine; formerly editor of Business 2.0, Time.com, and technology editor of *Time* magazine; also formerly at *Newsday*; freelancer for *Wired* magazine

"'If we could read the secret history of our enemies, we should find in each man's life sorrow and suffering enough to disarm all hostility.' —Henry Wadsworth Longfellow. When our linkages are beyond the family, beyond the tribe, beyond the nation-state, our peer-to-peer relations will trump the hierarchies who have brought us war after war that we do not want." —**Hamish MacEwan**, consultant, Open ICT, New Zealand

"This is very true. Though the Internet allows intolerance to be broadcast more easily, it also allows it to be argued against in a more public forum." —**Megan Holbrook**, partner, Kapow, Inc., a site-design and development company; she has produced projects for Microsoft, Warner Bros., and Disney

"I mostly agree that social tolerance has advanced significantly, but I really don't think that the Internet will be the greatest cause. It more likely due to—again—inevitable cultural shifts brought about by demographic changes bubbling through neighborhoods, educational institutions, commerce, and community affairs. Also, it looks like we are headed for hard economic times in the country that may extend well into the period leading up to 2020, and facing that as a nation may well bring us together, provided we get past the scapegoating of 'the others.'" —**Peter Eckart**, director of health information technology, Illinois Public Health Institute

"Social tolerance will be reduced in some societies but will not be fully achieved in many others. The diversity of cultures and religions, as well as social and political problems, will remain no matter how developed the Internet is in 2020." —**Cristos Velasco**, director general of North American Consumer Project on E-Commerce; Mexico-based attorney and active in the Internet Governance Forum and Internet Society

"Wider exposure of views to others is not, in and of itself, going to bring about social tolerance. If social tolerance is advanced in 2020, it might be aided by information and communication technologies, but the reduction of violence, sectarian strife, bigotry, etc., is more closely linked to belief systems, perception of well-being, and reduction in perceived inequality or victimization." —**Rich Miller**, chief executive officer, Replicate Technologies, Inc.; an Internet pioneer with ARPAnet who implemented some of the first computer messaging and conferencing systems

"This scenario has no basis in current trends, thus it is unlikely to eventuate—there are major value shifts in society that affect these types of metrics that are independent of the proliferation of ICT." —**Robin Gunston**, consulting futurist for Mariri Consulting, a strategic and business planning company

"I wish this was true, but there are just way too many places in the world where people are suffering due to various bases of social discrimination such as sex, race, political alliances, etc., all around power, who has it and who abuses it. There will be more information and communication technologies online in the future, and people will become enlightened by others through time, but changing the way people fight in countries around the world will not necessarily have a correlation in terms of impact to wider exposure to others and their views. Hate crimes are deeply embedded in the pathologies of people's minds and ways of behaving, often separate from the influence of the technologies online. These technologies could play a role in changing people before they produce hate crimes, but I don't believe greater tolerance will show up via a relationship to communication technology usage online." —**Joanna Sharpe**, senior marketing manager, Microsoft

"The television and telegraph were each once hailed as a technology that would bring global peace. Marx suggested the railway would bring about the communist utopia. Some ancient thinker, no doubt, suggested a similar idea about the smoke signal." —**Charles Kenny**, senior economist for the World Bank; expert on technology and economics and author of many reports, including *ICT: Promises, Opportunities and Dangers for the Rural Future*

"The Internet is actually collapsing the size of the group one interacts with, so if this trend continues, we'll see more polarization rather than tolerance. This will be due to the diminishing exposure to alternative ideas and concepts." —**Bernardo Huberman**, senior fellow and director of the Social Computing Lab, HP Laboratories; consulting professor in the department of applied physics at Stanford University

"Learning more about people of other cultures usually does create feelings of acceptance. I hope you're right." —**Janie Graziani**, manager of new media and technology for the American Automobile Association

"I actually believe this is true, based on my experience with the Internet, but that may be partly a wishful feeling." —**Richard Hall**, professor of information science and technology and codirector of the Laboratory for Information Technology Evaluation, Missouri University of Science and Technology

"People are people and just won't change that much, especially in so short a period of time. The Internet can breed intolerance and spread lies in the same way that it can breed understanding and spread knowledge." —**Jeremy Swinfen Green**, Telecom Express, an interactive marketing company

"I'd really like to believe this, but the strife described here is largely brought about by resource, rather than knowledge, inequalities. If you add domestic, open-source nanotech fabrication into the mix, then I'd sway more in the 'agree' direction." —**Jeremy Yuille**, digital media coordinator at RMIT Communication Design and program manager at ACID in Melbourne, Australia; previously director at IXDA, the Interaction Design Association

"I think that tolerance will be more prevalent, but not that it will have much to do with the Internet." —**Julian Hopkins**, social scientist and PhD candidate at Monash University, Malaysia

"The Internet is the left wing's form of talk-back radio." —**Christine Satchell**, PhD, senior researcher, Institute for Creative Industries and Innovation, Queensland University of Technology

"I disagree. Online, people will connect with others like themselves and can choose not to be exposed to dissimilar others if they don't want to—there will be no great change in either direction in terms of social tolerance." —**Jade Miller**, PhD student, Annenberg School for Communication at the University of Southern California, with a research focus on global flows of information and culture

"Actually, I wanted to put 'indifferent' instead of mostly agree/disagree. The scenario is too technologically deterministic. Whilst declines in the levels of violence, criminal activity, and sectarian strife will be positive steps forward by 2020, these will not be 'because of the Internet.' Moreover, these rates are likely to fluctuate as technologies account for 'new' incidences of criminal behavior that are not yet possible in 2007.

However, in terms of building communities and a sense of togetherness, new social media—that includes the Internet—will represent the most popular point of access for people to communicate across." —**Maz Hardey,** social analyst, blogger, "defender of new media" completing a doctorate funded by the Economic Social Research Council in the U.K., based at the University of York

"The Internet will help, but fundamentalist groups use the Internet for their needs today as well, and they will continue doing so in the future." —**Iddo Genuth**, founder and chief editor of *The Future of Things*, a science and technology e-magazine

"No. While I don't think WWIII is imminent, I think the Net will allow like-minded individuals to gather and to emphasize the difference. While it's not in the least related, climate change could well lead to water/food shortages, especially in the majority world, and that could well have a major impact on things like tolerance of others, etc. I'd like to think that the current levels of Islamophobia will have declined, but..." —**Emma Duke-Williams**, lecturer in the School of Computing and researcher, University of Portsmouth, U.K.; education blogger

"The Internet and the Web together democratize the flow of information. This has many consequences, including much fairly useless information being circulated, but also it makes it much harder to create walled-off knowledge. Extremist groups can use the same technologies to promote their ideologies very effectively, but, on balance, free flowing information does lead to more tolerance." —**Micheál Ó Foghlú** research director, Telecommunications Software & Systems Group, Waterford Institute of Technology, Ireland advisory committee; member of W3C; blogger

"Although I agree in the most part that a greater tolerance for behavior may be present, I feel that the more identities merge, the more they separate. Sameness often leads to a romanticism of individuality and more specific identity. As the philosophers Deleuze and Guattari said, 'Smooth space is being constantly translated, transversed into striated space; striated space is always being reversed, returned into smooth space' (Deleuze & Guattari, 1987, p. 474, *A Thousand Plateaus*)." —**Amit Kelkar**, consultant and sociology researcher, Postmodern

"This really could go either way. A lot depends on what happens with other aspects of society in the next few years." —**Robert J. Berger**, CTO for Cinch; expert on backbone networks, access networks, wireless networks, and innovative Web applications, a frequent speaker on next-generation networks

"There is no evidence whatsoever that more communication makes people more tolerant. On the contrary, many people pick and choose from the wide menu of the Internet the particular content that reinforces their prejudices. The facility to form groups is used to associate with people 'like me' more often than for discovering new horizons." —**Christian Huitema**, distinguished engineer, Microsoft Corporation, pioneering Internet engineer (on the Internet Architecture Board 1991–1996, Internet Society leader 1995–2001)

"Well I would *like* this scenario, but it looks like wishful thinking. Unfortunately, current trends are in the opposite direction. What I see happening is an increase of entropy, with both tolerance and intolerance rising simultaneously." —**Giulio Prisco**, chief executive officer of Metafuturing Second Life; formerly department head at the European Satellite Centre, analyst at European Space Agency, and an IT specialist for CERN

"This one's hard to call because so many people seem to be closing themselves away in gated communities, real and virtual. Yet any small chink in that armor allows all sorts of things to pour through. The many places that people interact online allow very different people to get to know one another as friends before discovering they are supposed to be enemies. This leads to more tolerance." —**Jerry Michalski**, founder and president, Sociate, a technology consulting firm; formerly managing editor of *Release 1.0*, Esther Dyson's newsletter, and cohost of the annual PC Forum

"My positive answer is a wishful thinking. I hope the Internet can promote and enhance acceptance to cultural diversity." —**Gilda Olinto**, researcher, Brazilian Institute for Information in Science and Technology

"Human nature will *not* change significantly in 13 years, regardless of the cause." —**Rollie Cole**, director of technology policy, Sagamore Institute for Policy Research, a nonprofit, nonpartisan think tank headquartered in Indianapolis, IN

"A greater exposure to others and their views may incite a greater social tolerance, but I do not agree that this can be taken to mean people will become more accepting of the views of others. This will only open the way for a greater degeneration of society in which nothing is surprising and indifference is dominant." —**Sam Ozay**, e-learning and e-communication specialist and solutions architect at Postmodern (Asia/Pacific); formerly general manager at European Language Centre

"Mostly agree, but I'd add an indirect path—Internet-based communication facilitates trade, and trade reduces war and sectarian violence." —**Brough Turner**, chief technology officer and cofounder of NMS Communications; oversees evolution of technology and product architectures

"People will be accustomed to share opinions and discuss in an open and free way even if some minority of fanatics won't accept this fact, and they will still try to defend their ideas in an intolerant way." —**Rafik Dammak**, software engineer, STMicroelectronics, Tunisia; DiploFoundation participant in the study of Internet

"The Net makes it much easier to retreat into a world where the only people you talk to are people who think just like you do." —**John Levine**, founder of Taughannock Networks; a leader of the Internet Research Task Force's Anti-Spam Research Group and the Coalition Against Unsolicited Commercial Email

"I'm not sure how much the Internet matters in this area and whether it will be a positive or a negative enabler. The Internet can be used as a tool for bigotry and divisiveness as much as it may foster greater awareness and understanding of human diversity." —**Greg Laudeman**, utilization catalyst and facilitator, community technology specialist, Georgia Tech Enterprise Innovation Institute

"'Sectarian strife' will continue at levels comparable to those in the first decade of the 21st century. The increased ability to communicate through various media—broadcast, text, IM, image—will help affinity groups recruit followers to a central, and perhaps exclusive, message. This message is just as likely to promote health and happiness as it is bigotry and subversion of external values." —**C. R. Roberts**, Internet journalist based in Vancouver, Canada

"But the line between social tolerance and indifference is a thin one." —**Jim Witte**, professor of sociology, Clemson University; research is focused on differences between online and offline society, including a special interest in activity in Second Life

"People seem to be much more likely to use the Internet to reinforce their preconceptions and to isolate themselves in an environment that does not challenge beliefs and prejudices. Tolerance arises from human contact, not Web surfing." —**Brian Dunbar**, Internet services manager, NASA office of public affairs

"Social tolerance will be achieved when it is no longer profitable for people to believe otherwise, and the Internet has nothing to do with that." —**Hal Widsten**, general manager, KWED/Seguin Daily News, Guadalupe Media Ltd.

"The Internet is great, but I don't have this much faith in its powers. This scenario seems possible on an intellectual level, but looking at humanity's history…" —**Aaron Schmidt,** Walking Paper Consulting, a blogger who writes about libraries, technology, and usability

"Like any other technology, the Internet is and will be used for hate mongering, as well as for bringing a wider vision for tolerance." —**Mark Youman**, principal, ICF International, a Washington, DC, consulting-services company that works with government and commercial clients

"The world is small after all!" —**Mark Terranova**, senior account manager, I Group Electronics, an electronic component distributor

"People will continue to hear what they want to hear. Today on the Internet there are relatively impartial and honest journalistic sites, but I bet FOX News, which is not balanced, gets more traffic. I don't expect this to change." —**Todd Wagner**, health economist, Health Economics Resource Center, Palo Alto, CA, part of the U.S. Veterans Administration; also involved with the Center for Health Care Evaluation

"While the Internet may enable the spread of tolerance, it will not be the catalyst. The Internet has, just as equally, the potential to increase intolerance. Some other sequence of events will need to happen in our relationships with one another in order to fundamentally change the seemingly instinctive intolerance we have today." —**Chris Bell**, vice president and director of worldwide marketing for iTunes, Apple

"We keep hoping this will happen, and it might, but not necessarily due to the influence of the Internet." —**Jill O'Neill**, director of planning and communication, National Federation of Abstracting and Information Services; author of the Infotoday blog; based in the Philadelphia area

"This will never happen. Ethnic groups elsewhere in the world (Sunni vs. Shiite, Bosnian vs. Serb) have been fighting for centuries, and this is not likely to change, just as black vs. white, etc., in the U.S. The bigger problem is affluence vs. poverty. The gap is growing and probably will continue to do so. What I personally would like to see is a cap on top executive earnings, based on the salary of the lowest paid employee. For example, if the lowest paid employee earns minimum wage—$7 or $14,000 per year—then a cap at a hefty multiplier of 200 would give the CEO $2.8 million. I suppose there could be a way around it, but if the government assesses a confiscatory tax (95%) on the excess, this might work. And stop giving the zillionaires deductions for their excessive homes—anything over 3,000 square feet or $1 million should not qualify. Give the poor a gasoline credit equal to 90% of the cost of gas so they can afford to work. Make *everyone* who receives government assistance (handouts) work—the mothers could be trained as childcare workers (which would give them better parenting skills) and have them work to provide daycare for others who have to or want to work. And no, I am *not* a Democrat. I'm a red-blooded, 60-year-old, female, Jewish Republican. I just think things ought to be fair." —**Judith Siess**, president of Information Bridges International, Inc. and publisher and editor of the *One-Person Library* newsletter, author and blogger

"Every indication we've seen of the growth of the Internet —online bulletin boards, fora, games, virtual worlds—have shown us that people grow far more intolerant, as they gain more access to tools of expression and anonymity, and far more aggressive and hateful. Instead, we will see more and more hate crimes and bigotry being planned and incited online and expressing themselves more in real life." —**Catherine A. Fitzpatrick**, Open Society Institute, Physicians for Human Rights; lecturer on humanitarian issues; formerly UN representative for International League of Human Rights

"I have to believe this. I wouldn't be involved in technology if I didn't want this to be the case." —**Tiffany Shackelford**, consultant who works with clients such as Phase 2 Technology, Stateline.org, Foneshow, WebbMedia, and Daily Me

"If there is one really bright spot, it is that though the Net can clearly coalesce communities of hate and violence, the Web can also expose people to new ideas and build tolerance. It's not the *only* reason this is happening, but it's a good outcome." —**Karen G. Schneider**, research and development, College Center for Library Automation, Tallahassee, FL; expert and thought leader in the library and technology community

"Contradictory trends will continue to coexist. More detailed knowledge of others will bring tolerance among those with affinities and a grouping of some of them to evermore viciously attack others and band against them. While the Internet is able to foster greater knowledge and respect, these don't happen on their own. They need strong leadership and will, but also a consistency across cultures and classes that doesn't seem, to me, able to prevail over the ever-growing levels of hate and intolerance in our societies. The buildup of tolerance is happening in groups comparatively smaller than the buildup of hate and takes a larger moral and intellectual effort. Hate scales, goodwill does not seem to—certainly not at an equal rate. The level of invective, mistrust, and rapid radicalization of views that we see online (whether flame wars among fans of sports teams or among partisans of political views and organizations or among genders, ethnicities, nationalities, etc.) and the abuse of others' information and images give pause to optimism." —**Alejandro Pisanty**, director of computer services at the Universidad Nacional Autónoma de México; active in leadership with Internet Society, ICANN, and the Internet Governance Forum

"Will stay the same. People remain people regardless of the technologies in place." —**Kathryn Greenhill**, emerging technologies specialist, Murdoch University

"In fact, I don't know, but I hope." —**Sebastien Bachollet**, president of the Internet Society of France, operates the European Global Event on Domain Names and Address Systems, known as EGENI, active participant in ICANN

"I fear that this will remain a dream. The same thing was said several times in the past decades, but reality has been different. What we could say is that there will be parts of the world where social tolerance would have advanced significantly, but I can't see the disappearance of phenomena like bigotry and hate." —**Roberto Gaetano**, ICANN board member; also responsible for software development for International Atomic Energy Agency; an active participant in the ICANN policy-making process

"I agree 100%. The Internet is electronically joining all of the world's people together at an amazing rate. I can't think of anything more exciting for my children's future than to be a part of it." —**Bill Warren**, vice president of government relations, Walt Disney World; founding editor of the *Orlando Business Journal*

"I think this an extremely optimistic scenario due to the time involved, but nonetheless it's very probable. Internet means more access to free information." —**Tiago Casagrande**, works with social communications and new technologies for verbeat

"There will always be a new crop of social intolerance, especially as the world gets smaller." —**Stan Felder**, president and chief executive officer, Felder Communications, a marketing and advertising firm in Grand Rapids, MI

"For the most part, people will become more tolerant of others. There will be some people who will be the opposite of this though, I believe, because they will use the Internet to find things that support their view of the world. So when racial, religious, or other hate crimes do occur, they will be more pronounced than they are now." —**David Newberger**, founder of Blackdot Ventures and a systems administrator in Minnesota

"I agree in the name of the rule that we always expected a better future. Communication tools may be used to spread violence, prejudice, and so on. You need not to go to Africa to see ethnic mass killing, just look for Yugoslavia in Europe." —**João Miguel Rocha Filho**, director, DataOne, a provider of software for connecting to Linux; based in Brazil

"We are a culture that looks to violence as entertainment, and young children are exposed to violence through both the news media and primetime programming. Many, if not most, parents seem to be unwilling or unable to monitor their children's viewing habits or encourage a lack of television altogether. While I would heartily applaud a change in direction, I can't see a reduction in violence, strife, bigotry, or hate crimes when the frequent presentation of violent images is still the order of the day." —**Wendy Cook**, Web administrator working for the state of Texas

"As much as we will be able to learn about social tolerance with advances in communication and education, we will never witness declining levels of violence, lower levels of sectarian strife, and reduced incidence of overt acts of bigotry and hate crimes, sadly." —**Don Kasprzak**, chief executive officer of Panaround.com, a Web-solutions design company; former system engineer at Apple Computer

"So far, this has been true. But we have not fully explored as a society potential downsides of what happens online." —**Joan Connell**, online editor, *The Nation* magazine, formerly an executive producer for MSNBC.com, senior editor for MSN and a Pulitzer finalist for her reporting

"We will be more tolerant, but intolerance and its cousins, bigotry, hatred, unfair discrimination, will live on." —**Cameron Norman**, assistant professor in the department of public health sciences, the University of Toronto; actively engaged in use of the Internet to help tackle tough health issues, director of evaluation with the Centre for Global eHealth Innovation

"Although greater connectivity will allow certain subversive groups to virtually congeal, the amount of information available for individuals to research will promote greater tolerance through understanding." —**Peter Kim**, senior analyst, Forrester Research, Boston; specializes in e-strategy and management, social marketing, blogs; recent reports include *Microblogging for Marketers*

"I hope that this is the case and that we, as a world, will perceive intolerance as much as a negative as we do rape." —**Robert Eller**, Concept Omega, a media marketing and communication company

"The increased availability of (randomly scattered and ungraded) knowledge is no more than just that—an available variable value collection of 'stuff.' Knowledge results from an appropriation process that is grounded on a cultural predisposition to learn. Knowledge must be apprehended within context, and context can only be assured by others. Using any of those anecdotal 'average American' studies/quizzes as an example, we very easily conclude that knowledge has very little to do with the availability of contents (in newspapers, libraries, or the Net)." —**Luis Santos**, Universidade do Minho–Braga, Portugal

"By its nature, the Internet provides opportunities for users to find answers to their questions. Increasing open-mindedness will not happen spontaneously. More likely, specific searches will allow people to validate their current beliefs as they search for more information about their current interests." —**Ed Dieterle**, Harvard Graduate School of Education; research tied to handheld devices for ubiquitous learning

"Sadly, this likely will not become true. The presumptions needed in order for it to be true are that there be a finite number of sites that people visit and that these sites have valid and valuable information. The more likely scenario is that the number of such sites will be equaled by sites that post spurious information, hate speech, and other forms of bigotry. People will look up and gravitate toward sites that confirm their opinions more than they will seek out sites that challenge their beliefs and experiences. This is why the role of universities will only become more important, as people will not only desire but will, in fact, need the insights of those who can study such discourse and help evaluate its cultural and communal significance and worth." —**John Jordan**, associate professor of communications, University of Wisconsin–Milwaukee

"The technology is neutral in terms of 'tolerance' metrics. While exposure to the new and different can certainly bring about acceptance, it may also result in a rejection of the new and different, a reinforcement of current prejudices. It's not all *West Side Story* out there, as Shakespeare knew even before Lenny Bernstein picked up the theme." —**Dennis Baron**, professor of English and linguistics, University of Illinois, runs the Web of Language site and researches the technologies of communication

"While visual media such as television and online video do much to promote tolerance (the show *30 Days* is a specific attempt at this), I disagree that the effects will be as pronounced as in the prediction. While I have high hopes for advancements in realism made in human interaction via a virtual medium, I believe that even by 2020, the most impressive of those advancements will have just been made, if at all, and will still be going through an adoption period in which much misuse, due to perceived anonymity online, occurs. Some experiments in Second Life have already shown that racial bigotry not only occurs in virtual worlds but, under some circumstances, becomes even more pronounced. The most important technological advancements that can be made to increase tolerance are those that increase the number of opportunities for (a) recurring interactions with people from different cultures and (b) immersion within a different culture. Therefore, technological advancements that increase opportunities for foreign travel, that place people from different cultures in a position of repeated interaction (such as international offices or organizations made possible by advanced communications), or that increase exposure to different cultures during early childhood education will have a much greater effect than advances made in broadcast or recreational communications." —**Jay Neely**, social strategist in the process of founding News Armada, a Boston-based company working to advance Internet-based news and commentary and community online

"If anything, the Internet seems far more effective at helping political fragmentation, rather than cohesion. The Internet is a medium of small groups with narrow interests, not large ones with broadly shared concerns." —**Anthony Townsend**, research director for the Technology Horizons Program of the Institute for the Future, providing long-range forecasts on technology; he is also a cofounder of NYCwireless

"People are people. From the ancients until now, there has been a huge increase in the levels of exposure to different viewpoints—but no change in tolerance that I'm aware of. I see no reason this will change with greater access to the Internet." —**David Lee King**, digital branch and services manager for the Topeka and Shawnee County (KS) Public Library

"Rather than increasing social tolerance, the Internet will have become a perpetuating factor in sectarian strife, bigotry, and hate crimes. Individuals will gravitate toward sites and other individuals espousing their views. For extremists, this will allow their views to seem normative, increasing the likelihood that they will act on them." —**Naomi L. Lacy**, assistant professor, research division, University of Nebraska Medical Center

"As society adapts to increased contact with other parts of the world, conflict is likely to increase—conflict of beliefs, thoughts, and cultures. It will, however, slowly begin to subside towards 2020. Not sure it will be a more tolerant society." —**Syamant Sandhir**, director, Futurescape Netcom Pvt. Ltd., an experience-design and implementation company

"I have seen no data to support the closing of the digital divide. In fact, I only see data in support of its widening. Instead, two societies emerge, one with Internet access and one set of values, and another, the have-nots, who don't necessarily speak English or read and have limited access to the technology of the day." —**Theresa Maddix**, satisfaction research analyst, ForeSee Results

"While tolerance is growing, the ability for people to be anonymous—and virtually say what they want without fear of exposure—is still a threat. It's a form of bullying that can restrict many from truly expressing their views." —**Mike Driehorst**, messaging strategist; leads social media for Hanson Inc., an interactive communications and video production company in Ohio

"The Internet is an echo chamber. People do not tend to willingly expose themselves to different viewpoints when using the Internet; they tend to expose themselves to similar viewpoints, thus causing their views to become more extreme. If anything, the exact opposite of what you predict is a very real danger, although I am not *quite* pessimistic enough to believe things will get worse than they already are." —**Alexis Turner**, webmaster, Greenwood Publishing Group, New York

"Sadly, the Internet will feed and foster more divergent, out-of-the-mainstream hateful beliefs. People with seeds of hate will find each other more easily and nourish one another's beliefs." —**Mariana Almeida**, product manager of Web products for health care, Kaiser Permanente

"Let's hope so. But remember, knowledge is not virtue." —**Don Ranly**, PhD, professor emeritus, University of Missouri School of Journalism

"I agree overall. However, it looks like there will increasingly become pockets of ignorance online where groups live in ignorant walled gardens reinforced by self-generated and self-selected misinformation." —**Brian T. Nakamoto**, cofounder of MrJoy Inc. and product-line manager for Everyone.net (a leading provider of outsourced e-mail solutions for individuals and companies around the world)

"Despite its potential, the Internet is not a place where people go to be exposed to different views. More often than not, it's a place where people go to find people who agree with the things they already believe. The increasing balkanization of the Internet along geographic lines will only perpetuate this. (On the up side, improved translation software might prove a great help to those who actively seek out cross-cultural understanding.)" —**Ivor Tossell**, technology columnist/journalist for the *Toronto Globe and Mail*, known as "the blogging journalist" and a social observer

"The Internet supports both tolerance and intolerance—by increasing proximity, division rather than consensus is more likely." —**Ed Steinmueller**, professor, science and technology policy research, University of Sussex; researches industrial structure of high-technology industries, coevolution of technology

"I fail to see the causal link between the capacity for people to acquire more information about others and becoming more tolerant. Why, indeed, would people become less tolerant as they became aware of the range of extreme and ill-informed views that exist? Have the Al-Qaeda executions on the Internet made us more tolerant of that behavior?" —**Tim Grafton**, market research director for UMR Research Ltd., a market research company based in New Zealand

"I hope this will be true. However, there is still the constancy of crime, fraud, and general meanness, which usually keeps up with the latest technology/social innovation." —**Jennifer Jarratt**, principal, Leading Futurists LLC; works with formalized methodologies to assess and interpret potential futures

"Violence will continue. More content is a neutral thing, affording both a greater sense of shared humanity and reasons for hatred." —**Bryan Alexander**, director of research, National Institute for Technology and Liberal Education, blogger, expert on computer-mediated pedagogy; based in Ripton, VT

"Our culture seems to be sacrificing our moral standards on the altar of tolerance. I don't think greater tolerance for the sins of others will lead to less violence. We will simply be no longer labeling actions that are wrong as violent and unacceptable. Furthermore, those who do stand up for traditional Judeo-Christian morals and values will face harassment and persecution." —**Jana Vanderslice**, counselor and technoethics researcher

"I hope this will be true but fear that a further balkanization will be the more probable scenario." —**Wim van de Donk**, professor of public administration and chairman of the Scientific Council for Government Policy in The Netherlands (WRR)

"I think that this will be true for many, but as can already be seen, the Internet allows various disaffected groups, who would in the past never have come together, to 'meet' and grow stronger through this union." —**Ruth Martin**, National Chengchi University, Taiwan

"The Net overall would seem to foster a greater social tolerance, but I don't think it will make a measurable difference in 13 years, given the many sources of intolerance (including some who use the Net for that purpose)." —**Earl Babbie**, professor emeritus and writer, BEinc, prolific author of books on social research

"We'll see fewer incidents but larger and more violent ones. Those who use ethnicity/race issues as a means to power will continue to press forward, now with ways to mobilize larger groups." —**Christopher Jacobs**, chief operating officer, Solutions for Progress, Inc.; formerly with KnowledgeFlow Inc. and Unisys Corp.

"Great degrees of connection between people from different perspectives; smaller global village; more tolerance. At least one hopes so." —**John Eckman**, practice director, Next Generation Internet, Optaros, Inc., a professional services firm offering strategy, design, development, and consulting services tied to open source software

"The more you know, the more you realize you don't know. The Internet could be making people a lot more insecure about what they know and don't know. So maybe this insecurity could result in more people staying home and trying to get smarter (which takes a lot of time) and producing less violence." —**Virginia Bisek**, Web content developer and writer

"While I believe that 'social tolerance has advanced significantly' generally, in 2020 the use of the Internet and communication technologies will bring about more organized acts of violence, terrorism, and sectarian strife. The acts of violence will be bigger, and perhaps more frequent, as perpetrators vie for the best YouTube video and worldwide exposure." —**Woody Degan**, chief executive officer and operations director, Memphis Sound Entertainment; Consumer First Consulting, IT Consulting

"The general population will embrace this, but the Internet and other digital tools will enable bigotry, hate, and religious groups to continue their intolerant beliefs underground." —**Chris Miller**, senior vice president, digital operations and new business for Element 79, an advertising agency

"I agree partly because I hope this will be the case. Social change brings on a corresponding backlash that must be dealt with in other venues—schools, houses of worship?" —**Lynn Blumenstein**, senior editor, *Library Hotline, Reed Business Information*

"There are no real signs that toleration is increasing on a global level. The Internet seems to serve both the cause of greater tolerance and greater intolerance at the same time and will undoubtedly continue to do so." —**Daniel Fisher**, lecturer

"I don't think that it is at all clear that the Internet is influencing people to be more tolerant." —**David F. Salisbury**, associate director for science and research communications, Vanderbilt University; formerly science and technology reporter for *Christian Science Monitor*

"The social tolerance will occur in part due to the vast networks of weak ties that people will maintain through social networking sites." —**Rob Boostrom**, PhD student, Southern Illinois University–Carbondale, creator of the Society for Word of Mouth, "a group for educators to share ideas about sharing ideas"

"Wider exposure can make those already receptive to new ideas more tolerant; however, the way in which those not receptive to other ideas grow more violent and intolerant appears to be an 'equal and opposite reaction.' Witness Al-Qaeda recruiting via the Internet, and beheadings being carried out specifically for the mass media." —**Francis J. L. Osborn**, futurist and activist, philosophy department, University of Wales Lampeter (formerly St. David's University College)

"The perpetrators of the acts described in the scenario are the acts of those who are not going to be affected by something like Internet-promoted social tolerance. While the existence of new means of disseminating information and opinion are significant in moving things incrementally along, the Internet is just as capable of serving those looking to create disharmony. I believe the benefits of the Internet will be important and positive in the long haul, but they will be realized only after generations have passed." —**Sam Punnett,** president, FAD Research Inc.; has worked in the field of interactive digital media since the 1980s, for the last 9 years on strategy, marketing, and e-business development

"The Internet exposes us to many ideas and things we wouldn't normally be exposed to, and as a result, over time, it will increase tolerance because people will increasingly connect as global communities." —**Jonathan Dube**, president of Online News Association, director of digital media at CBC News, the Canadian Broadcasting Corporation, publisher of CyberJournalist.net

"It's highly doubtful that Internet, which reinforces social and educational differences (if not creates them), will increase social tolerance." —**Yves Froidevaux**, Swiss Federal Statistical Office

"The world is likely to split into several different forms of society/civilization: One large group will be characterized as described in this scenario…but not all groups. Increased specialization in a more complex environment is an evolutionary pattern." —**Mary Ann Allison**, principal, The Allison Group, has worked with Microsoft, Glasgow's Urban Learning Space, and other businesses, governments, and NGOs

"Tribalism and identity-related group divisions may become psychologically more inviting or advantageous; loyalty to one's tribe may be rewarded more than an easy-going tolerance of difference...perhaps." —**Alex Don**, linguist and educator

"I don't believe that these statistics have anything to do with the Internet. People are people, period. Perhaps after a few generations, we will see the greater ease of communication and global community resulting in more tolerance, but I have a hard time believing we're seeing that right now, so early in the Web's history." —**Sarah Houghton-Jan**, consultant for the Info-people Project; digital futures manager, San Jose Public Library; author of the Librarian in Black technology blog

"Tolerance or nontolerance are too deeply rooted in people's psyche to change significantly within 10 years." —**Ulrich Spalthoff**, director of advanced technology, Alcatel-Lucent, Paris; expert in solid state science, semiconductor materials, optical fiber communications

"Exposure to other points of view should increase tolerance, as long as those points of view are portrayed from a level-headed perspective. Luckily, we've seen self-leveling phenomena in well-moderated social sites and forums where participants will correct any skewed data put forth early in the thread." —**Jason Stoddard**, managing partner/strategy at Centric/Agency of Change, an interactive strategies company; he is also a popular speaker on social media and virtual worlds

"We will have a new set of intolerant norms replacing the old ones." —**Robert Grant**, chief executive officer, VoyaCare, Inc.

"Unfortunately, increasing exposure of individuals to each other allows extremists with a minority point of view to become even more radicalized and potentially violent. This, along with desensitization of previous trend mentioned, may not lead to greater tolerance." —**Dan Weingrod**, vice president for digital operations, Cronin and Co.; oversees creative online initiatives for integrated marketing communications company

"There is already a disturbing trend in Internet. Making it easy to create groups with very specific views is allowing people to stay within groups that share these views. If it's too easy to share only with 'same-view' people, we may lose our tolerance." —**Louis Naugès**, president, Revevol, an enterprise 2.0 company with offices in France, Spain, the U.K., and the U.S.; a founder of Microcost, an IT services and hardware company based in France

"I hope this scenario comes to pass, but I'm not sure of it at all." —**Hank Dearden**, director of business development, Digital Industry, Inc., a provider of technology services in the Washington, DC, area

"Unfortunately, access to the Internet and ICTs will not reduce crime and improve tolerance of others to the extent this scenario suggests. This scenario will be true only if we also reduce poverty and provide safe access to clean water, nutritious food, education, and health resources." —**L. Suzanne Suggs**, assistant professor of communication sciences, University of Lugano; research focuses on use of new media and messaging strategies to improve health status

"The Internet is a tool for communications. Human nature remains the same: While it can promote peace, it can also be used to promote hate (see Al-Qaeda's use of it). I see the violence only going down in proportion to demographic (lower birthrates means fewer males between 16 and 34) and social (higher standards of living) conditions than as a result of the Internet." —**James Gorman**, principal, Working Technology Partners, a company offering technology solutions to businesses

"Historically, integration of people at a social, community, travel, and trade level should have had a similar effect. I'm not sure we're seeing the evidence." —**Duane Degler**, user-centered designer and strategist for Design for Context, writer and editor for IPGems, focused on knowledge management, semantic integration, and performance improvement

"Hate crimes come out of childhood programming. Individuals who are brought up to weigh evidence and not prejudge others will use the Internet for openness—the others will use it to connect with other hate mongers." —**Nancy W. Bauer**, chief executive officer and editor-in-chief, WomenMatter, Inc.

"This scenario is devoid of myriad real-world issues, notably, global population growth and competition for scarce resources that are likely to breed discontent." —**J. W. Huston**, president of Huston Consultancy and futurist

"Technology can work both ways. Text messaging was recently used to incite violence in Kenya. On the other hand, exposure to 'the other' will increase acceptance and understanding." —**Kathryn K. Goldfarb**, president, KG Communications, an independent consultancy

"This 'tolerance' will be mostly fake tolerance, necessary to be integrated to the community…Underground networks will (and already) exist where people join anonymously, and the level of social tolerance yielded will be close to zero." —**Alexandre Winter**, cofounder and chief executive officer, LTU Technologies, a global leader in image search and recognition technologies

"Tolerance comes from a sense that the 'other' is not a danger. I think most people (and peoples) are relatively benign. Therefore, knowledge of 'others' always trumps prejudice. I sure hope this one comes true." —**Mike Langum**, Web developer, U.S. Office of Personnel Management

"The Internet brings wider exposure but sadly does not teach tolerance. It'll take much more than a few Web sites to reduce levels of crime and strife." —**Jay Buys**, vice president for digital development, Fleishman Hillard, an international marketing and communications company

"This will be true, and we will also see communities aligning themselves in the physical world more closely in ideological terms." —**Kent Kirschner**, media specialist, Neighborhood America, inviting companies to understand the power of community building online

"Sadly, being more informed does not breed tolerance. People will be more informed, but that does not mean they will be more tolerant. It could even have the opposite effect of intolerance. The information on the Internet is a mix of fact and fiction. This scenario assumes otherwise." —**Mack B. Rhoades, Jr.**, Web services product manager, Michael Baker Corp.

"It is mainly a hope that social tolerance will improve as people make connections with other people across economic, social, and geographic borders. My major misgiving is that the Internet also allows more narrow niche developments that may also foster separation." —**Michael Castengera**, senior lecturer at the University of Georgia's Grady College and president of Media Strategies and Tactics Inc., a media consulting firm

"The Internet has nothing to do with socialization and attitude reproduction by parents, the true cause of intolerance and hate. Positive exposures to cultural differences must happen in person to be effective. The Internet still allows social distance that cannot be avoided when face to face." —**Melanie McCalmont**, owner-operator of Synchronous Publications

"National pride appears a clumsy electoral argument, as many election results are influenced by foreign analysts, and many important decision are taken outside of national authority. The largest ethnic and national groups among the 2020 newborn is 'mixed race' and 'double national.' Not knowing the key elements of the 10 largest religions is considered a concerning closed mind." —**Bertil Hatt,** researcher of Internet and social services, innovation valuation; employed by France Telecom and Orange (information technology and services industry) while completing PhD

"I fear that while technology allows for broader understanding, it also allows focused extremist groups to form more readily. These groups will often have a negative impact, conduct violent acts, and perpetrate hate crimes." —**Robert H. Rich**, PhD, strategic planning and evaluation, American Chemical Society

"Constantly increasing stresses on society as we migrate from oil-based to alternate-based economies will continue to provide plenty of opportunity for misunderstanding and hatred. What may be true is that skin color will not be the determining factor. Other factors such as the haves and the have-nots will be the determining factor." —**Richard Hammond**, knowledge management team leader, United States Environmental Protection Agency; knowledge management expert currently examining the Semantic Web and RFID

"This is the great unknown. Religious fanaticism will continue, particularly in the Christian and Muslim portions of the world. There will always be those that are not smart enough or too lazy to reach out and learn about those different from themselves. Decreasing the percentage of people who fall into that category is the goal of a modern society." —**Dixon Hutchinson**, software engineer

"If the Internet changes a society for good, why are we not now seeing greater social tolerance in proportion to Internet usage?" —**Marco Rivera**, Internet specialist, Vistronix, Inc., an information management, technological solutions, and managed services firm serving federal, state, and local agencies

"Not everyone is part of this digital age. Some will deliberately exclude themselves from it. Negative behavior will still persist, and although there will be improvements in social tolerance, there will still be social intolerance, much of it present on the Internet as it is today." —**Helen Keegan**, founder of Beep Marketing, a self-employed consultant and a judge for the Webby Awards

"I optimistically agree that the Internet is connecting more people and more people to more information. An enlightened population is a more tolerant one." —**Elizabeth Talerman**, chief operating officer, Talerman+Partners LLC, a marketing integration company; previously senior vice president of marketing at Martha Stewart Living Omnimedia

ANONYMOUS COMMENTS

The following section includes a selection of contributions to the discussion of the future of the Internet and social tolerance from survey participants who chose to remain anonymous. Among them:

"The Internet allows minorities to interact with each other, giving them a greater voice."

"Civilization and technology are not correlated positively; in fact, they have been found to be inversely correlated. The Nazis regarded technological advance as one of their heights of achievement, but they were hardly 'love thy brethren' kinds."

"Social engineering is so 20th century. Besides, 'greater tolerance' usually means indifference. This is completely the wrong track."

"Absolutely disagree and would argue that it has helped advance terrorism, militants, hate groups, and further Balkanize society."

"There is a loss of a shared experience with fragmented media. People may only read things that fit their prior views."

"The more information made available to more people, the more those individuals will learn about other cultures/people different from themselves. Tolerance levels for different cultures and people will increase due to the increase in knowledge and information."

"The Internet is dividing us more. Despite Web 2.0 being about community, I have yet to see much of a trend of being inclusive and bringing people from different walks of life together. Instead, people are finding like-minded people online."

"As the world's poor get connected and find out how bad they have it compared to rich nations, they demand more equity."

"Because information can be shared so quickly and is available on demand, it is easier to incite racial hatred and to describe how to commit crimes, etc. (c.f. pedophiles grooming young children via chat rooms, how to disable burglar alarms, etc.). The opportunities are just there for a different way of committing crime."

"The Internet cuts both ways. Visit an extremist site sometime—there are thousands to choose from, and they are not going away."

"Tolerance is decreasing in many parts of the world, particularly in the Middle East. This is a long-term trend unlikely to be changed by the Internet."

"People will become more aware of differences but no more tolerant."

"The Internet has a proven ability to create communities. But communities can be exclusive—people can choose to tune out (e.g., the Internet makes homeschooling easier for parents who wish to shield their kids from public schools' broad curriculum)...There's a good chance that we'll see the Internet used as a way to 'tune out' in the future."

"Bigots can use the Internet just as easily as nonbigots."

"We'll see more tolerance from person to person, but unfortunately, I don't think this will lead to less violence and strife on a larger scale."

"The Internet exposes people to other people's points of view, culture, etc., and that promotes understanding and, therefore, tolerance. As more people get connected to the Internet, this tolerance will reach more and more people. However, extremists will always exist and will try to subvert the Internet for their purposes; but the tolerant majority will prevail."

"The Internet becomes a tool for the fringe and extremists to have a platform and recruit. Conversely, technology will make it easier to ignore such groups and enable authorities to ensure they do no more than just 'talk.'"

"If this happens, it will occur because of increased levels of education, not the Internet, and will take place in wealthier, developed countries where this trend is already evident."

"On the other hand, radicalism also increases as the disaffected fringes of society find it easier to collect and coordinate. Isolated violence may decrease, but terrorism and organized violence may not."

"Education and knowledge bring down walls."

"People tend to only talk to their own 'kind.' Right wings with right wings, left wings with left wings, etc. There's not that much crossover. See also John Kelly's research, Columbia University, NY [mapping the content of blogs]."

"Unless different social norms can be brought to bear, there will remain an absence of civil discourse that can promote understanding and tolerance."

"The Internet will continue to be a tool for those who encourage intolerance and hate, which will slow this progress. Ultimately, however, the Internet's ability to increase individual awareness, education, and access to information will overcome organised social and cultural intolerance."

"We are living in [a] speedy age. It's far from tolerance. It will only change the format of intolerance, like massive hacking."

"Greater tolerance can only occur through additional efforts to create understanding and valuing of difference. Exposure to difference will only generate a modicum of tolerance and, unless reinforced by people with the wisdom to foster it, can lead to greater conflict. Education is necessary."

"The Internet will enable 'Democracy 2.0'—next-generation democracy."

"The people who are tolerant today will be tolerant of more variation in 2020, but the intolerant people of today will be even less tolerant in 2020—they will find more variations to hate and be more convinced that they need to be separated and protected from all those variations."

"There is always going to be strife and disagreement, and devious minds will have faster ways of coming together."

"Monkey wrench politics are only likely to rise over the next 15 years."

"Access without education can breed more intolerance (e.g., all the rumors circulated via e-mail) and give platforms for the fringe radicals (left or right) to have a strong voice."

"There will be a divide between extremes."

"The Internet will not decrease the number of poor people in the world, and, as the populations increase, the divide between rich and poor will increase."

"One could argue today that the trend is, in some ways, toward decreasing social tolerance, at least as exercised by the government on the people. Societal tolerance between individuals apart from the government, yes, probably has advanced."

"There has been no noticeable decrease in bigotry since the Internet began, and it is unlikely to start to decrease now. The Internet is a safe place for bigots to fester and encourage each other. Sad, but true."

"Tolerance may increase, but it won't be because of the Internet. Radicals of all sorts have connected online, and it's possible to tailor your media diet to match and reinforce your viewpoint. This will only get easier in the future. Increasing educational and economic opportunities will have greater impact."

"Tolerance is increasing and will continue to increase, but the pace will remain slow, and the increase will primarily be in younger generations."

"People across the globe are feeling more disconnected from meaningful relationships and seem to be more prone to violence, not less."

"As the Internet will become readily available on portable computing systems such as mobile phones, people would very well connect and socialise with each other through the Net. Especially on portals for friend finding, dating, and sharing similar interests. But this will inversely affect their normal in-society, face-to-face socialising skills. This may make them short tempered and more intolerable to nuances of the world."

"Social tolerance progresses and recedes. There is a constant ebb and flow to this dynamic process. At times, it may seem as if two steps are being taken forward, and then intolerance will rear its ugly head again, with one step backward. I am not sure where things will stand at a single point in time (2020), but technology is no guarantee that social tolerance will improve."

"'Niching,' for lack of a better term, will draw like to like. For example, there are online groups for homeschooling parents, support fora for parents with children of rare diseases. We are more widespread, but people are still drawn to those like them…As people are drawn to 'like,' I could also foresee that there would be an insular approach in cultures where differences are not tolerated."

"Millions of years of evolution have led us to become a violent species with strong allegiance to our own groups. This will not change significantly in 13 years—maybe in 1,300 or 13,000 years or even longer."

"Our sense of claiming territory will adjust to different boundaries."

"The deterioration of religion and social values will lead to [the] uprising of more crime and violence and more government control."

"The world continues to be a violent place, and it is hard to imagine the world without sectarian strife."

"Sometimes I think that the only thing that could really get humans (and their governments, in particular) to cooperate and play nice is a threat from the outside."

"The Internet allows you to understand someone's race or creed in a whole new light (thank you, Wikipedia and Google). There will always be hate speech, but the current viewpoint of the Internet will likely prevail, which is race/creed/nationality/gender/age don't matter. It is what you can do and how well you can do it. There will still be hate speech, but it will be taken to represent the actions of someone, not a race/creed/blah. Though its origins will be negative, its use won't be."

"Awareness does not equal tolerance."

"Can this be envisioned in those nations and regions dominated by more restrictive religious and social creeds?"

"The Internet is neutral in its morality and, if I am correct, still has pornography as the leading single use. Minority views and opinions may find a voice, but this does not mean peace and harmony!"

"As fast as positive images can be transmitted, negative images can also be portrayed. The media will have a larger influence in our lives."

"From person to person, we have an excellent chance of becoming one human race of equals."

"Exposure to different cultures and views through a variety of media, including traditional ones, is the main contributor, with the Internet being one player among many."

"We tend not to be exposed to variety, but rather only seek out that which is most like ourselves. We join only similar enclaves and [become], thus, more narrow in our views."

"The Internet narrows the world for most people, as they tend to communicate online mostly with those like themselves."

"The clash between conservative social forces and nonconservatives will only become more pronounced, especially as natural resources become increasingly scarce."

"The ability to speak what is on your mind is truly powerful. That is one of the gifts of the Internet. Learning the diplomacy of expression and how to play well with others is another Internet gift."

"New tolerance will only apply to connected and user audiences. While a smaller percentage of people might be left unconnected by 2020, the potential for that group to harbor even more violent, paranoid, and reactionary segments may only increase. It seems likely that the threats and actions of bigoted violence will still be with us and may, perhaps, only increase in intensity."

"Tolerance has not increased so far due to technology. However, intolerance has become more disguised."

"It will be easier to find out information about different cultures, etc., but it will also be easier to anonymously bully people."

"The culture of celebrating diversity based on ethnic heritage and the ability to connect with like-minded people will create more prejudice and hate between people of different origins and beliefs."

"Tolerance will erode as more people spend time in online echo chambers with like-minded folks."

"Calephs and Imams will not allow people to question Islam, and if they do, they will be punished. Governments will not open their doors to outside influences; they will crush all attempts to do so. Those with free thought will become pariahs and refugees from the 'true' believers."

"Social tolerance will increase, but not because of wider exposure to others and their views, but because people must adapt."

"Human nature and tribalism will tend towards the opposite outcome, more intolerance and hatred as more people become uncomfortably close."

"When people actually know their neighbor (the need of) fear/protection will decrease."

"People who are tolerant may become more tolerant. People who are violent may become more violent."

"The potential for demagoguery also increases with uncensored mass information capabilities."

"The Internet is a destroyer of tolerance because the quality of the information allows easy manipulation, populism, and the result is the diminishing of tolerance."

"Open communication makes us all brothers, as such differences blur. We take care of the needy, discipline in the commons those who are aberrant. Many explore polymorphous perversity in their own evolutionary trajectories. Few get stuck in any one of them, but most pass through quite varied ones and become more accepting."

"The social isolation that increased electronic vs. personal one-to-one interaction will lead to fewer social skills and less ability to deal with and cope on a personal level with anything that feels uncomfortable. Violence and strife are an individual action; tolerance is a feeling. They do not necessarily equate."

"Internet, technology will only make things worse."

"The world will be forced into collaboration based on mutual survival needs."

"On the Internet, anything can be normalised, reinforced, furthered, and even the most heinous perspectives can find support and encouragement. Anyone who thinks that the Internet gives people exposure to other ideas hasn't looked at how people use it. And the sophistication of search and folksonomies and Semantic Web, etc., will mean that people can increasingly zero in on specific stuff related to their 'community of one' rather than that messy, diverse stuff one normally encounters in the street, classroom, workplace, or 'real' environment."

"While overall tolerance may increase, extremism will become more extreme. Technology will also allow a tiny group of radical thinkers to have greater impact than they could have in previous ages."

"Technology does not change basic human behavior, that is, the result of long-term evolution. And not everyone agrees that tolerance is good (or that evolution exists, for that matter)."

"Violence prevails because many in power have found that conflict between groups can be used to strengthen their personal positions. This will not change."

"The world can be a vicious place, and the Internet as well. That will never end completely."

"Values-based dialogue will slowly progress and morals-based dialogue will diminish."

"While exposure to other ideas and lifestyles can increase tolerance, the anonymous dog-pack mentality allows people to be even more effectively hateful and divisive."

"It will advance, but not by a factor of more than 10%."

"The Internet and Web facilities have often had a negative effect in our society. Such activities as gang recruiting, sexual predator interaction, identity theft, and terroristic intercommunications have expanded in our society because of the Internet's openness and low cost of entry for people with bad intentions."

"So far, none of these scenarios allow for the basis of human nature being inherently selfish. These scenarios are assuming that human beings are inherently good."

"I'm not seeing this at all. People are becoming more 'anonymous,' more able to lash out at others, forgetting how to interact with others in person, staying inside, playing Wii instead of going outside."

"The only thing that will happen is that many will become more blasé and cynical."

"Tolerance comes with better education, and better education has very little to do with Internet exposure."

"What a wonderful utopian idea! I don't think the telephone or printing press made intolerance disappear. Nor will the Internet."

"People are less and less forced to acknowledge people who differ in religion, skin color, nationality, and political viewpoint. This is one of the scariest outcomes of personalization. It used to be everyone listened to the few radio stations or watched the few TV stations. You were then exposed to different things. Today, you can read only blogs that agree with you, read only editorials you agree with, listen to only music you like, etc. Furthermore, it's very easy to set up extremist Web sites."

"Tolerance is an evolving concept, and people will always be intolerant...As long as there are two people with unique egos, social tolerance will never be achieved."

"People are likely to flock more and more towards opinions and people who are like their own. People will become more skilled at filtering out those things that they do not like to be confronted with. Information overload will lead to more selective use of sources. This scenario might be slightly different in those parts of the world where freedom of speech is suppressed and people have a clear need to hear outside opinions and will be exposed to sources they would normally not be exposed to."

"The Internet not only mixes viewpoints between groups but also starts providing logical nonliner and semantic frameworks for discussion. Large religious and irrational divides still remain, however."

"This is the great leveling effect of broader communication."

"In more than one million years, human behavior did not improve. The rise of threats to survival in the real world, e.g., access to water and energy, will cancel any trend toward smoothening relations thanks to whatever facility."

"The Internet, particularly the Web and its associated technologies, has allowed people to narrowcast information to their own values and attitudes far more easily. Selective attention will still be key in 2020 in understanding how new media works."

"Tolerance will improve due to a greater understanding through the Internet. But human nature is human nature and has been throughout time."

"The amount of public or impersonal vitriol has seemingly increased with the decrease in civility and psychic pressure of local social norms."

"Cyber violence has increased. The emotional damage is just as bad in online environments as in face-to-face environments."

"Social tolerance will always be constrained within the bounds of human nature, but as we connect with others online, we find common ground we did not have before."

"As long as the Internet remains open, it will provide a free and unfiltered exchange of information. This will allow people to assess multiple sources of information before forming perceptions. However, it's a stretch to expect the Internet to reduce violence. Terrorists are adept at using the Internet to encourage violence."

"In third-world countries, the Internet can be used to disseminate information and increase understanding and knowledge about other people and other ideas."

"Our nature is to surround ourselves with people who agree with us."

"This is the most likely scenario of all that I have read in this survey; however, it also assumes that governments and radical elements of religion do not undo all the good that can be accomplished online."

"Digital divide, language barriers, Internet dropouts, dictators in many countries will probably slow down the social tolerance advancement. In some countries, a parallel Internet network could be owned and controlled by governments who will not allow their citizens to navigate freely on the real Internet."

"The ease of disseminating new ideas and building off others is creating a greater tolerance in the Western world. It is yet to be seen what will happen in the Middle East and Asia."

"Other factors (such as TV) began to shift social tolerance before www."

"What is more likely, especially in the short term, is that access to information will be divisive. Cultures will retreat to maintain and protect traditions. The extremes at either end of the social spectrums will excel faster than the middle where there's more of an institutional lag. As the left and right become increasingly vocal, a greater divisiveness will result."

"The increase of population size, prevalence of inexpensive travel and communications technologies, and increasing interdependence of nation-states will drive tolerance (due to proximity) more than the Internet."

"Tolerance is still a fairly personal, localized thing. The Internet will not radically change tolerance levels, at least not by 2020."

"Hate crimes may decrease due to other trends in culture, but more traditional tribal areas that are cut out of the predominantly Asian and Western Internet culture will continue to perpetrate violence from past wrongs."

"No reason to believe that technology will transform human nature (e.g., suspicion of 'others'). Fault lines of intolerance might shift and opportunities to find common cause might grow, but ultimately, conflict will prevail (e.g., across communities vs. within communities)."

"I don't think the Internet will make much of a difference when it comes to sectarian strife in tribal territories. The conflicts go back too far and have more to do with power than they do with differences."

"The oneness of humanity is a fact that will become more and more obvious to more and more people."

"There will be much social strife as the education gap increases."

"Social, political, and religious trends will continue to move in the direction of intolerance as the percentage of the population that adheres to a conservative belief system increases."

"I agree, however, spectacular terrorist violence is unlikely to be reduced by this broad tolerance."

"The Internet helps people find other like-minded people and groups and does little to expose people to new opinions."

"I am not convinced the Internet breeds tolerance. It might ultimately prove, in some areas, to be a tool to spread hatred. However, as races and cultures interbreed, tolerance should increase as distinctions decrease. This issue is too much influenced by the state of the economy to predict at this point."

"Sadly, despite the greater availability of information and differing viewpoints, tolerance does not seem to be increasing. Instead, people are often choosing to gather with like-minded individuals online."

ENDNOTES

1. See "History of the Atlantic Cable & Undersea Communications: From the First Submarine Cable of 1850 to the Worldwide Fiber Optic Network; Cabot Strait Cable and 1857–58 Atlantic Cables," by Bill Glover, available at http://www.atlantic-cable.com/Cables/1857-58Atlantic/index.htm.
2. See *The Story of the Telegraph and a History of the Great Atlantic Cable*, by Charles F. Briggs and Augustus Maverick (New York: Rudd and Carleton, 1858, pp. 21–22).
3. From "The Prophet's Column," a series of articles by John Joseph Carty that appeared in the *Electrical Review* in 1891. For more on Carty, see the 1936 National Academy of Sciences biography by Frank B. Jewett, available at http://www.archive.org/stream/nationalacademyo009315mbp/nationalacademyo009315mbp_djvu.txt.
4. From *Forecasting the Telephone: A Retrospective Technology Assessment*, by Ithiel de Sola Pool (Norwood, NJ: Ablex Publishing, 1983).
5. See *The Global Telecommunication Infrastructure and the Information Society*, a statement by Jean Jipguep delivered at the June 1995 Internet Society INET '95 conference, available at http://www.isoc.org/HMP/PLENARY/L1-6/html/paper.html.
6. See "Globalization, the Knowledge Society and the Network State: Poulantzas at the Millennium," by Martin Carnoy and Manuel Castells, in *Global Networks*, volume 1, issue 1 (2001, pp. 1–18).

7. See *ICT-Enhanced Public Service Broadcasting*, published by UNESCO in 2008, available at http://www.unesco-ci.org/cgi-bin/media/page.cgi?g=Media_Development%2FICT-enhanced_Public_Service_Broadcasting%2Findex.html;d=1.

SCENARIO 3

THE EVOLUTION OF IP LAW AND COPYRIGHT PROTECTION

PREDICTION: *Content control through copyright protection technology dominates. In 2020 strict content controls are in place thanks to the efforts of legislatures, courts, the technology industry, and media companies. Those who use copyrighted materials are automatically billed by content owners, and Internet service providers automatically notify authorities when they identify clients who try to subvert this system. Protesters rarely prevail when they make claims that this interferes with free speech and stifles innovation.*

Expert Respondents' Reactions	
Mostly Agree	31%
Mostly Disagree	60%
Did Not Respond	9%

All Respondents' Reactions	
Mostly Agree	31%
Mostly Disagree	61%
Did Not Respond	8%

Note. Since results are based on a nonrandom sample, a margin of error cannot be computed. The "prediction" was composed to elicit responses and is not a formal forecast.

Respondents were presented with a brief set of information outlining the status quo of the issue 2007 that prefaced this scenario. It read:

> Major content producers such as the world's music and film businesses are lobbying governments and courts to protect their assets. Digital rights management (DRM) is one of the umbrella terms used to describe various technologies being developed to help copyright holders control access to digital products and prevent copying. Its intent is to assure that content creators maintain control of their work and are rewarded with appropriate compensation. Opponents of DRM say its language and approach are geared toward forcing public acceptance of intellectual monopolies. They argue that the movement toward assigning ownership of everything stifles innovation and competition, saying DRM is actually "digital restrictions management," and IP stands for "intellectual protectionism" and "intellectual poverty."

OVERVIEW OF RESPONDENTS' REACTIONS: A number of predictors used the phrase "the horse is out of the barn," implying the old paradigm of intellectual property (IP) protection is ineffective. Others talked of the "arms race," implying that those who wish to access information without regard to law continue to find ways to circumvent IP-control attempts. Still others referenced continued coexistence, suggesting that in the future, content owners will sometimes expect monetary payment but will sometimes offer their content free or in exchange for attention or other action. The varied themes among the "mostly disagree" responses to this scenario were dominated by two points: Regulators will not arrive at universally accepted policy, and people everywhere will

continue to circumvent IP structures if regulatory guidelines are not enforced globally. Several respondents said they think the future of IP is "up to China." Many respondents said "cracking" technology will stay ahead of IP-control technology. Some said new economic models will be developed to deal with new realities of digital, online content. A number of respondents predicted the following: To gain a sizeable audience, most content will have to be offered for "free"; regulation will be layered, and concepts such as Creative Commons will prosper. Those who mostly agreed with the scenario said content will be privatized and kept under the control of governments, media, and/or telecommunications firms. Some said content control is being reasserted by currently entrenched institutions through hardware and software restrictions including nongenerative, "tethered" devices such as smartphones, TV, Netflix, and so forth that allow people to only access content through IP-controlled gateways to the Internet.

A significant majority disagreed with the idea of a dominant and successful copyright protection system by the year 2020. Some people's remarks echo Electronic Frontier Foundation cofounder John Perry Barlow's 1994 essay on the changing nature of "ownership": "Humanity now seems bent on creating a world economy primarily based on goods that take no material form. In doing so, we may be eliminating any predictable connection between creators and a fair reward for the utility or pleasure others may find in their works. Without that connection, and without a fundamental change in consciousness to accommodate its loss, we are building our future on furor, litigation, and institutionalized evasion of payment except in response to raw force…We're going to have to look at information as though we'd never seen the stuff before…The economy of the future will be based on relationship rather than possession. It will be continuous rather than sequential. And finally, in the years to come, most human exchange will be virtual rather than physical, consisting not of stuff but the stuff of which dreams are made. Our future business will be conducted in a world made more of verbs than nouns."[1]

Some respondents noted that it is human nature to desire to acquire at no cost those things for which others pay a price. And some warned that extreme management of IP rights would not be worth the trade-off of the potential inherent in free and open communications networks.

"Digital rights management is fool's gold," wrote **Michael Botein**, founding director of the Media Law Center at New York University Law School. "Many people want IP protection, but everyone wants to steal. Regardless of the legal mechanisms so far—e.g., automatic damages, compulsory copyrights—many people would prefer the illegal route, perhaps because it runs up their adrenaline."

"The dominant business plan will be access to attention, rather than access to content, so this scenario seems rather unlikely," responded **Oscar Gandy**, author, activist, and emeritus professor of communication at the University of Pennsylvania.

Brad Templeton, chairman of the Electronic Frontier Foundation, commented, "While people will try to do this, it is so technologically intractable as to not succeed. Copying data is the natural state of computers; we would have to try to compromise them too much to support this regime."

An anonymous respondent wrote, "First, DRM as ownership silos is unmanageable; second, free production is likely to grow; third the fight against IPR's [intellectual property rights] 'violators' will continue to be legal gesticulation, as it is not enforceable in many places where it is the most active." Another commented, "Money has the predominant voice in government these days, and it's confused short-term profit with long-term gain." Another noted, "Examples like Radiohead's distribution of DRM-free music for any price will also help change 'hearts and minds' to the possibilities of distributing content without strict IP controls. Money can be made, markets can be developed, and everyone can come out ahead." And another predicted, "Plagiarism will be rampant, with global content available to all and various national laws allowing willy-nilly infringement. Everything on the Web is viewed as free—and most will not care, hoping each copycat offers 15 minutes of fame."

Some See Changes Coming at the Hardware Level

Several experts noted the ways in which digital rights management is already being exercised at an accelerating pace through the introduction of digital-information-access appliances or devices, like the iPhone, that are closed systems.

Steve Jones, cofounder of the Association of Internet Researchers and editor of *New Media & Society*, commented, "While I applaud the efforts of DRM opponents, I am discouraged by the progress DRM seems to continue to make in hardware as much as in software. Having purchased an iPhone, I was delighted when Apple updated its software to allow custom ringtones, only to discover that I needed to pay for a ringtone via the iTunes Music Store even though the ringtone I wanted to use was one in which I own the copyright!"

Social media researcher **danah boyd** of Harvard University's Berkman Center for Internet and Society referred to the work of her colleague, Jonathan Zittrain. In his 2008 book, *The Future of the Internet—And How to Stop It*, he described the pros and cons of "generative" (wide open to innovation and contribution, with everything shared by all) and "nongenerative" technologies (controlled, proprietary systems like cell phones and digital video recorders [DVRs]). "While the media and public are talking about DRM at a software level, the reality is playing out at a hardware level," boyd wrote. "Non-generative technologies are being produced, restricting users from nearly everything, often to protect DRM. New mobile handsets and Intel-based hardware are having DRM backed into the circuitry. This is a problem and, if this continues, strict controls are somewhat possible. While we are marching toward this scenario at a fast pace, I think that we'll see a disruption before 2020. I'm not sure what the disruption will be. Ad hoc mesh networks? Foreign-produced technologies brought in on the black market? As long as we can record media and as long as we can share content online (through encrypted channels), there will be breaks in the system. Realistically, there will be a lot more. I think that the likelihood of devising bulletproof DRM is about as high as the likelihood of stopping spam."[2]

An anonymous respondent referred to Internet-based content consumed through "tethered" devices, commenting, "The success of iTunes and move of Netflix toward pay for Internet-fed films continue to reinforce a pay-for-consuming scenario." Another wrote, "The current trends and efforts by lobby groups are outrageous, but I believe they will prevail because most people still don't understand the question and will only wake up when it's too late. I foresee a time where *nothing* is free and things like potatoes, chickens, counselling, clichés, and Scrabble will not be able to be accessed without paying a toll at every point."

At least one respondent said hardware makers are going to see more profit if they support open information sharing. "Technological protection of intellectual property seems to make good business sense for copyright holders, particularly compared to the difficulties of enforcing these rights through slow and expensive justice systems," wrote **Lea Shaver**, A2K (Access to Knowledge) program coordinator at the Yale Information Society Project. "But ultimately consumer demand for openness will largely prevail over the effort to preserve predigital business models. There will always be a market for new tools to subvert DRM, and the pace of innovation is much faster than that of the legal system. Just as important, the many companies who stand to gain from greater content openness—such as makers of hardware and providers of indexing and remixing services—are increasingly going to organize to block legislation that puts the teeth in DRM."

NEW ECONOMIC MODELS SEEN AS LIKELY

A number of the respondents reflected some optimism that people living in a highly networked age will adjust to new ways of thinking about the exchange of goods and services, including what is now referred to as *intellectual property*.

Louis Houle, president of the Quebec chapter of the Internet Society, responded, "A new capitalism will rise with the Internet (only an infant now)." **Fred Baker**, Cisco Fellow and an architect of the Internet, noted, "The current attempts at DRM mostly curtail a growing business, and the business will eventually be allowed to grow."

Paul Greenberg, president of The 56 Group LC, commented, "The fact that Gen Z or whatever they are called at the moment will have grown up in a peer-to-peer-empowered environment by 2020 will be (and is) the harbinger of social change that, when it comes to copyright control, will break down the traditional barriers that protect intellectual property."

An anonymous respondent predicted, "Audiences will be monetized in ways other than the traditional pay for access. The competition to attract the audiences will force free distribution as the norm." And another noted, "Traditional 'content providers' are doomed. Where legacy, large-content sets exist, those businesses will be able to capitalize somewhat. Copyright law will change substantially and people will give away more entertainment-based content. Dollars will be made through sponsorship, advertising, endorsements, and product placement."

Ross Rader of Tucows Inc., an active ICANN participant, wrote, "The walls are starting to fall already and as more content providers come to understand how to make money without DRM, the less meaningful DRM will become in the market. DRM, like communism, sounds great in theory but is very difficult to implement with meaningful effect. It sounds attractive from the viewpoint of the content providers, but as the unanticipated consequences become more readily understood, implementations will become increasingly rare. As with communism, we can expect holdouts for the next 20 years, we may even see a DRM superpower emerge, but it is unlikely that this form of content management will truly be the standard of distribution upon which the entire economic framework is based."

John Abell, new-media project director for the Committee of Concerned Journalists, matter-of-factly commented, "Content owners will have to adapt to an á la carte world. Protecting content becomes less necessary when it is unbundled, easily accessible by legal means, and reasonably priced."

Paul Miller, technology evangelist for the U.K.-based company Talis, responded, "There is early evidence of a more pragmatic recognition that value is shifting. With a recasting of the value proposition with respect to content, it becomes less necessary to overcontrol the content itself, more useful to have that content widespread, and increasingly possible to recoup more revenue on value-added services built around the content and its community of use."

Payment will come in new ways, according to **Scott Smith**, a futurist and consultant who formerly worked with Yankee Group and Jupiter. "By 2020, costs will be recovered in other ways," he noted, "from subsidies built into device costs to live performance to embedded ads, but DRM-locked content will be in the minority for mass-market entertainment. Looser DRM systems designed to protect small producers may still be in place—a hybrid between Creative Commons and limited-play versions."

Clay Shirky, author of *Here Comes Everybody* and a professor at New York University, agreed that things will change, writing, "By 2020, alternative licensing regimes will have superseded the DRM rationale."

Tze-Meng Tan of Multimedia Development Corporation, an architect of the Malaysian Internet, responded, "In 2020 most content will be distributed 'free' or for very low cost but supported by advertising, which will be embedded in the content."

Jeff Jarvis, blogger at Buzzmachine.com, commented, "When audience and content can be metered and monetized, then it will be in the interest of copyright holders to have their content distributed as widely as possible, with the knowledge that this is how they will make money through advertising or through the expansion of their brands (that is, the reduction of their marketing costs)."

Thomas Quilty, president of BD Consulting, a firm that investigates software piracy among other high-tech crimes, predicted that, by 2020, "though content control is in place, competition in the form of royalty-free content competes with products with high usage fees. This competition forces the rights holders to lower their fees to be competitive. Additional changes to laws worldwide place restrictions on the length of time after creation of a work that fees can be changed, using a schedule where the fees are reduced and finally eliminated over time."

Josh Quittner, executive editor of *Fortune* magazine, formerly of *Time* magazine and *Business 2.0*, and a longtime technology writer, responded, "As a content producer, my heart (or, rather, stomach) would like to see some form of IP protection going forward, but my brain tells me copyright will pretty much go away. From a tech perspective, I could see this going in either direction. If online advertising fails as a way to monetize content, I could see a micropayments system evolve, and that could easily go hand in hand with ironclad DRM."

Peter Kim, a senior analyst with Forrester Research who specializes in e-strategy, suggested, "The advertising model which supports media will collapse; both sides of DRM must learn to coexist, because content must be circulated with ease to build audiences, and consumers alone cannot and will not subsidize the commercial model which incents artists to create."

Fernando Barrio, program leader for the Master of Arts in e-business regulation at London Metropolitan University, commented, "A totalitarian scenario as described would inevitably lead to the abandonment of certain technologies and the creation of a parallel culture DRM-free." **Jerry Michalski**, founder and president of Sociate, a technology consulting firm, wrote, "Open-content sources will flourish because open works are much more easily woven into the new fabric of commerce and society." **Rafik Dammak**, a software engineer for STMicroelectronics in Tunisia, commented, "New media will appear and raise new issues. Open-source and Creative Commons movements will be the only 'fort' against the DRM and IP monopoly."

John Jordan, a professor of communications at the University of Wisconsin–Milwaukee, wrote, "The money to be made in technologies comes when user-consumers feel free to play and experiment. If all content is governed by a set of complex laws, user-consumers will feel stifled and will engage less with these technologies. They may not protest; they simply will withdraw. Once that happens, companies will be forced to realize that the content they offer and seek to protect will not, in fact, sell itself—they must, instead, accommodate user-consumer desires and ease restrictions in order to see growth."

And **Hamish MacEwen**, consultant for Open ICT, New Zealand, noted, "Even now, in 2008, DRM is failing as monarchies and dictatorships have done and will always do. Vista, with its capitulation to the centre, is lame, and Open Source is growing. The core is driven by the edge where there is no centre that can hold. Even Sony, the maven of proprietary systems, is opening thanks to the crowbar of market demand and competition. Gatekeepers who seek to dole out abundance as if it were scarcity are doomed. It's more likely that we will be rewarded for anarchic distribution than punished. The long tail wags the dog."

MANY SAY "INFORMATION WANTS TO BE FREE"

Many respondents said people will continue to get what they want at the price they are willing to pay; sometimes they will pay with their attention, sometimes with money, and sometimes with the decision to ignore politically constructed mechanisms established to compensate the creators of content. "You cannot stop a tide with a spoon," responded **Giulio Prisco**, chief executive of Metafuturing Second Life, formerly of CERN. "Cracking technology will always be several steps ahead of DRM, and content will be redistributed on anonymous networks."

"Information will always want to be free," wrote **Fabrice Florin**, executive director of NewsTrust.net, "and repeated attempts by governments and media companies to impose a digital rights management system will remain largely unsuccessful." **Dan Lynch**, founder of CyberCash and Interop Company, now a board member of the Santa Fe Institute, commented, "Copyright is a dead duck in a digital world. The old regime based its power on high distribution costs. Those costs are going to zero. Bye-bye DRM."

Geoff Arnold, senior principal and software development engineer for Amazon.com, responded, "This is a classic 'arms race,' but in this case, technology is going to be decisive. Every individual will have access to sufficient computing power to simulate every relevant content consumption use case, and DRM won't be able to keep up."

Christine Satchell, senior researcher at the Institute for Creative Industries and Innovation at Queensland University of Technology, agreed, writing, "Users will always find a way to overcome barriers put up by those with sheer interest of generating capital, and industry will have to look at ways of aligning themselves with a new generation of savvy users."

And **James Jay Horning**, chief scientist for information systems security for SPARTA and a former director of research for a DRM technology company, wrote, "The content owners are largely going to be defeated in an absolutist stance. This will be primarily for social, rather than technological, reasons. DRM-Lite (easily broken) will prevail in the marketplace. This is an extension of iTunes, Kindle, and Unbox. Enough of the world will be willing to pay for their convenience and legality to keep the content companies in business, but anyone willing to expend some effort (e.g., most students) will be able to rip off the content for free, and the content companies will have given up trying to hunt down and exterminate them."

Adrian Schofield of Johannesburg, South Africa, a leader in the World Information Technology and Services Alliance, responded, "The developing world will assert its belief in the value of open source and open content, and the combination of technical and legal complexities will impede the movement to imposition of stricter copyright controls. These two factors will further polarize the multinational corporations and the liberation movements and the (surviving) trade unions."

One respondent said he wished he could choose to "*totally* disagree" with the scenario. **Richard Hall**, codirector of the Laboratory for Information Technology Evaluation at the Missouri University of Science and Technology, commented, "As long as network neutrality remains in place, there is no way that DRM will survive, not just because the technologies will always be hacked, but also because the philosophy behind it is so onerous and evil. All through our history, if we owned a physical device, the device did what we wanted. When I purchased a record, it played on any record player, and if I wanted to record it for my own use, of course I could. Once people truly come to understand the nature of DRM: (1) I don't actually own things that I purchase, and (2) I am punished (e.g., my media won't play on my own players) because someone else might commit a crime. They absolutely won't stand for it, and, once this philosophy is widely understood, the open Web will send it crumbling to pieces more and more, and politicians will have to work with the will of the people. One other issue is to keep in mind (though it's more abstract, and people may not respond) is that virtually all innovation occurs when one thing builds on another, and that is why the law has always held that intellectual property is not eternal like physical property."

Alexander Halavais, a professor and social informatics researcher at Quinnipiac University, wrote, "While I have little doubt that there will be strife and problems with the interpretation of copyright in 12 years, we'll be seeing support for access to knowledge and knowledge commons, particularly in the international context."

Christine Boese, researcher and analyst for Avenue A-Razorfish and Microsoft, commented, "The people who are intent on destroying the public commons with excessive digital rights management controls and strictures may win some battles, but they will lose the war, may have already lost it. They killed their own golden goose. Cultural forces are much stronger than corporate fascists, and whatever they seek now to block will simply arise from other providers in other sectors, even if it means a return to singing around campfires and pianos, or making homegrown media products. Here's a thought: Maybe as the digital rights management Nazis kill their golden goose, they will also force creatives beyond excessive postmodernist remixing as an aesthetic, and artists of all stripes will start to value 'originality' over 'derivation.'"

An anonymous respondent predicted, "A collective commons of creativity emerges on the global scene in response to and favored by the cross-generational world population whose quality of life is more directly and immediately shaped by overarching problems of the century—such as viable access to water, energy, and medicine—than by what is generally perceived to be intellectual-property fundamentalism." And another wrote, "Free alternative content will force major content producers to accept that their business model has changed. This will lead to less restrictions, not more."

Some Suggest Alternate Methods, Adaptations

There will still be some controls, but they will come under a different system, according to many survey participants.

An anonymous respondent wrote, "We have to find a whole new model of how content can be used in multiple places, and the creators can still get the credit/payment they deserve." Another commented, "This is an unfortunate trend that will continue unabated without fierce public awareness campaigns and government support for alternative distribution mechanisms based on new business models. The property metaphor has had its time, and it is time to reevaluate the value copyright has within society. Incentives need to be put in place for creators, but free access needs to be negotiated on a collective level. Fair use should be the default position; copyright only when mass profiteering means real revenue loss. Individual freedom to access and transmit cultural artifacts is more important than individual reward in the end. However, artists must still be paid, perhaps a licensing registry where profits are shared from pooled low level—connectivity-level funds."

Nicholas Carr, author of *The Big Switch: Rewiring the World, from Edison to Google*, wrote, "By 2020, there will likely be a monthly arts fee added to ISPs' Internet-access charges, and the resulting pool of money will be split among copyright holders depending on usage. The fee will give users unfettered access to most copyrighted works."

Susan Thomas of S2 Enterprises LLC agreed, adding, "Content control through copyright cannot prevail. What IS likely is that access to the Internet will be controlled, and Internet service providers will charge a toll at the onramps."

Some expected that added scaffolding of regulation will make IP law work better than it does now. "There will be multiple levels of copyrights, some with very few restrictions," wrote **David**

Moschella, global research director for Computer Sciences Corporation's Leading Edge Forum and a *Computerworld* columnist.

"New forms of cooperation will emerge which are less win/lose," predicted **Mary Ann Allison**, principal of The Allison Group. "Commons will become a standard."

"UGC, Creative Commons, and open source are too powerful to suggest that the strict standards and complete micropayment systems these scenarios describe will be universal standards," responded **Susan Mernit**, an independent consultant and former senior director for product development at Yahoo. "I think we will see parallel systems for content and copyright management—the 'integrated systems' that are walled gardens much like AOL was for an ISP [Internet service provider] in the '90s, and the 'open media/open source' distribution sites that are smaller, more fragmented and that represent the long tail. The popular wisdom of crowds will dictate what is most popular, and payment structures will vary."

"We're already seeing new models of shared, commons-type ownership," commented **Cameron Norman**, a professor at the University of Toronto. "It will continue because in too many cases the free ownership or shared products are simply better and more responsive. The ability for open-sourced products to respond as we get faster and faster in terms of turnaround in all sectors will continue, and the old ways of copyright only limit that."

Havi Hoffman, senior editor, product development, Yahoo, noted, "In a perpetual panopticon (superveillant society), most media consumption will be trackable. But an alternative economy of reputation and information intermediation could begin to develop in parallel to the money system, which even today is traumatized by the technology of total connectivity."

Jeremy Yuille, digital media coordinator at the Royal Melbourne Institute of Technology (RMIT) in Australia, formerly director at the Interaction Design Association, wrote, "This scenario describes what may begin to happen in mainstream media. Grassroots media will tend away from this locked-down model to things more in line with Creative Commons and open source...The idea of content and entertainment will change, and the 2020 person will spend more time enjoying participative media than they currently do."

Clement Chau, research manager for the Developmental Technologies Research Group at Tufts University, commented, "As the world begins to assimilate into a culture where creativity is collaborative and participatory, and where the lines between the audience and the creator are blurred, IP and authorship will be redefined. Rather than creators having the 'rights to own' intellectual property, audience will pay to have the 'rights to participate.'"

And an anonymous respondent wrote, "The nature of the Internet has changed from simple information sharing to person-to-person empowerment. The generations growing up with the Internet as a ubiquitous tool believe it is there to be used for their individual needs. While companies will try to use old paradigm legal strategies to control the online space, they will be no match for the Internet's creative and willful users."

Regulators Are Likely to Remain at Odds

Some people do not think the industries and political groups involved in digital rights management will be able to find enough common ground internationally to secure more complete control.

"Things will stay lumpy and unpredictable for the DRM world," wrote **Susan Crawford**, an ICANN board member and visiting professor at Yale Law School. "I see two alternatives here. If network providers, law enforcement, and content companies continue along their present European path towards authentication, retention, surveillance, and control of every possible online communication, and if this route is adopted by the rest of the world, then—yes, DRM becomes perfect, perfectly charged for, perfectly controlled. But the world is a diverse and competitive place. Somewhere, somehow, there will be countries and network providers who just don't want to go along. There will even be competitors in providing DRM technologies who don't want to go along."

Robin Gunston, a consulting futurist for Mariri Consulting, wrote, "The only way this scenario can come about is if Asian countries agree to this accord, which I believe will take far longer than 2020." An anonymous survey participant wrote, "No chance. Too many legal entities in the world."

And **Hal Varian**, chief economist for Google, wrote, "Regardless of whether one thinks DRM is desirable or not, the coordination (in standards setting) and competition problems (inevitable due to zero marginal cost) are too great to overcome."

Many respondents say the system is likely to survive to 2020 as it is currently tiered.

"The world will be increasingly divided between creators of proprietary content and creators of open-source content: two worlds with different kinds of information ecologies," commented **Joan Connell**, the online editor for *The Nation* magazine, formerly an executive producer for MSNBC.com.

"The situation will be much like it is today and much like it was 100 years ago—major content producers will continue to find new ways to overprotect their investments, and consumers will continue to find ways to subvert these systems," noted **Alexis Turner**, webmaster at Greenwood Publishing Group in New York. "Cat and mouse are eternal."

And several respondents echoed what this anonymous respondent wrote: "What happens in China will dominate the future in DRM/copyright."

DRM and IP Law Have Support

There were a few respondents who expressed satisfaction with the current trends in digital rights management and IP law.

"You don't have to read Marx or Foucault (though it helps) to understand that, contrary to 1990s techno-utopianism, power tends to replicate itself, no matter how 'democratizing' or otherwise liberating a new technology may appear to be," wrote **Charles Ess**, a researcher on online culture and ethics at Drury University. "While there will be modestly successful resistance at the margins, most of us, most of the time, will find ourselves happy to drop 99 cents for a song from the iTunes store rather than fuss with copy protection workarounds."

Joanna Sharpe, senior marketing manager for Microsoft, commented, "DRM is important and critical in helping protect IP. New DRM tools that digitally protect copyright materials give attribution between content owners and producers and their work, so I don't believe using DRM is too restrictive. The arguments against DRM are weak, in my opinion. On the flip side, legislation that is too overbroad in shutting down all P2P [person-to-person] networks, and P2P innovation, doesn't make sense. P2P networks could be viable tools to educate and share information between groups, so it isn't in the public's best interest to shut down these technologies, just the exploitation of copyright infringement via networks. There has to be a balance between technology innovation and usage rights where people or companies are fairly compensated and technologies can advance to drive more open real-time communications online."

Buddy Scalera, vice president for interactive content for CommonHealth Qi, wrote, "We've been spoiled by free content that isn't really 'free.' This isn't limited to the Internet, it includes radio, television, and newspapers (so cheap as to almost be free). Content developers need to be paid for their work or else they will not be able to create new content. By 2020, people will grow accustomed to paying for subsidized content and subscriptions they value, but most people will pay micropayments for deeply relevant, vertical content."

William Winton, product manager of digital media for the 1105 Government Information Group, commented, "The Licensing Act of 1662 was greeted by many as the potential downfall of the free press. History proved this assumption wrong-indeed, English literature and art flourished in the Restoration period as never before. The seemingly eternal give-and-take between the creator, publisher, and public in regards to intellectual rights will not abate. Only a strong, fair, and effective system of digital content control will enable artistic expression to flourish while at the same time protecting the substantial investments that are required to enable such expression."

SOME AGREED WITH THE SCENARIO

Few of the respondents to this survey appear to be supporters of a perfected, global digital rights management system or universal law of intellectual property—the word *draconian* was used often by respondents in reference to the scenario presented. A significant majority either answered it cannot happen or said they wish it would not come about but think it likely.

"I agree it will happen; I don't agree it should," wrote an anonymous respondent. "Big media concerns will look out for their own narrow, short-term interests, and they have the lobbying power to have it their way. This will eventually increase the price of content while limiting its movement and usefulness. It should be noted that the creators of content and the owners of it are often two different entities. I don't believe many of the creators of content benefit from DRM." Another commented, "Technological convergence has also spurred the horizontal mergers of content providers, facilities-based network providers, and large IT/CE [information technology/computer engineering] into super media firms who will use their power and weight to ensure the digital genie does not get out of the bottle." And another wrote, "As corporations become increasingly powerful, including having the ability to buy not only politicians elected into legislatures but judges (as in John Grisham's book, *The Appeal*), we are likely to see greater control over so-called intellectual property."

A number of respondents mostly agreed that the 2020 scenario presented for discussion is accurate. "This is the 'Big Brother' trend we anticipate in 2020," commented **Janet D. Cohen**, blogger, futurist, and trend analyst. "This scenario is likely as the result of an increasing share of Internet access delivered via a smaller number of global wireless providers and partnerships (driven by threat of lawsuits) between these wireless providers and content producers," noted **Timothy McManus** of Nuance Communications. And **Seth**

Finkelstein, author of the Infothought blog, wrote, "Much of this is the case now!...Note my 'mostly agree' response doesn't indicate endorsement."

An anonymous survey participant commented, "I am disheartened by copyright controls as they currently exist and will continue to develop. The original intent of copyright was to provide a reasonable time period for the creator to receive exclusive benefit for his work, after which it was to be available in the public domain. Exclusive benefit in perpetuity is anathema to creativity and innovation. Until we realize this and the corporations relinquish control, we will have the problem of piracy." Another wrote, "Unfortunately, unless strong progressive reforms, particularly in the U.S., come into play soon, it appears that industry will continue to consolidate, control, and capitalize the marketplace of ideas."

Steve Goldstein, ICANN board member, formerly of the U.S. National Science Foundation, responded, "My main reason for agreeing is the increasingly oligarchic evolution of the service-provision marketplace. I would further predict that there will be cross-linking of content provider giants and Internet service provider giants and that they will find ways to milk every last 'currency unit' out of the unwitting and defenseless consumer. Governments will be strongly influenced by the business conglomerates and will not do much to protect consumers. (Just think of the outrageous rates charged by cable and phone company TV providers and wireless phone providers today—it will only get worse.)"

Catherine Fitzpatrick, a lecturer on humanitarian issues with the Open Society Institute, wrote, "Despite the strenuous efforts of the copyleft movement, no viable business model has emerged or will likely emerge to pay artists who create content in any other way but in selling copies of their content, which they must therefore copyright. Making the content free hinges on a philosophy that the state or philanthropy must pay all content creators, and that has many troubling ramifications for the freedom and viability of content creation. ISPs will simply find ways to bill for microchunks of content more expertly and efficiently, and, as more and more people monetize time online, billing micropayments will become normalized."

"With DMCA [Digital Millennium Copyright Act] becoming the baseline for WIPO [World Intellectual Property Organization] and WTO [World Trade Organization] treaties, this will be a hard trend to reverse, although the story may not be as bleak as postulated here," noted **Todd Spraggins**, strategic architect for Nortel Carrier Networks and president and chairman of the board of directors of the Communications Platform Trade Association.

Sam Ozay, e-learning and e-communication specialist at Postmodern (Asia/Pacific), responded, "DRM is bound to prevail. A capitalist economy will not allow the continuation of free access to material."

"I regret feeling that I must agree with this scenario," wrote **David Maher**, senior vice president for law and policy for the Public Interest Registry, the Internet top-level domain registry. "The content providers are overreaching. Their power is primarily due to their intense and effective lobbying of the U.S. Congress (and other legislatures worldwide), with content consumers unable to make their voices heard."

"Much as I would like to see openness and abundance triumph, I don't see any political will to overturn the Digital Millennium Copyright Act," wrote Internet sociologist and author **Howard Rheingold**, "to the contrary, other countries, most notably and recently Canada, are turning to similar legislation. Incumbent culture industries have the ears and pocketbooks of political leaders in the U.S.—witness how the U.S. has slipped from the inventors of the Internet to number 15 in broadband Internet access. There are plenty of hopeful signs—both iTunes and Amazon are stripping DRM from downloadable music because that is what music customers demand. Free Culture is a growing anti-enclosure movement. Digital technologies continue to enable infinite reproducibility. But at this point, only a highly caffeinated optimistic could make hopeful signs into a strong argument that the forces for enclosure might lose. Right now, the RIAA, MPAA [Recording Industry Association of America, Motion Picture Association of America], and other copyright abusers are winning."

The State of IP Law and Copyright as of Mid-2008

Most survey respondents expect that the "information-wants-to-be-free" ideal (first expressed by Stewart Brand at a 1984 hackers' conference) will continue to thrive in 2020, they expect cracking to continue, and they anticipate that political and economic systems will evolve to a more open system of property use and sharing by 2020. But at this time, governments and corporations in the global north are continuing to pursue methods to bolster the current, more closed approach, and the word *monetization* dominates the topic lists at Internet business conferences of all kinds.

In May 2008 the U.S. House of Representatives passed an antipiracy bill that proposed stiffer penalties for illegal copying and distribution of content and the creation of an "intellectual property czar" at the White House level. The bill, known as the Pro IP Act, passed on a vote of 410 to 11, and it was applauded by entertainment companies, drug makers, and unions. Many in the technology industry voiced their opposition.[3]

AT&T, the largest provider of connection in the United States, announced in 2008 that it has been in discussions with the U.S. entertainment industry about filtering unlicensed copyrighted material on its networks. The Recording Industry Association of America has been lobbying to get copyright filtering built into network access at the PC level, with the technology bundled with virus-scanning software. RIAA has made headlines by filing lawsuits against more than 20,000 individuals alleging violation of copyright law, and in October 2007 Jammie Thomas was ordered by a federal court to pay $220,000 to record companies—$9,250 for each of the 24 songs she was accused of buying and then sharing online. Music fans continue to share songs—a *Los Angeles Times* survey found that 69% of teens said they thought it was legal to copy a CD they purchased and give the copy to a friend.

One of the IP-support proposals in play globally in 2008 has been a "three-strikes-and-you're-out" law under which ISPs would monitor users for illegal downloads and terminate a customer's Internet access after the third such use. Legislation and/or industry agreements along these lines have been implemented or discussed in Australia, Canada, Japan, and the United Kingdom. A three-strikes agreement (also known as a "graduated response") was formalized in late 2007 in France. The government, owners of film and music rights, and ISPs joined forces to announce the move, and French President Nicolas Sarkozy said, as he endorsed the deal, "The Internet must not become a high-tech Far West, a lawless zone where outlaws can pillage works with abandon or, worse, trade them in total impunity." Under the French and U.K. proposals, broadband firms failing to enforce the rules could be prosecuted.

There was pushback in reaction to the three-strikes idea when it was proposed to the European Parliament (EP) in April 2008, where it was pronounced as counter to "civil liberties and human rights and the principles of proportionality, effectiveness, and dissuasiveness." Musician Bono testified to the EP, "I am firmly opposed to the position of some member states, whose repressive measures are dictated by industries that have been unable to change their business model to face necessities imposed by the information society. The cut of Internet access is a disproportionate measure regarding the objectives. It is a sanction with powerful effects, which could have profound repercussions in a society where access to the Internet is an imperative right for social inclusion."

As more people across the globe adopt the Internet, property owners who sell copies of their products see potential for bigger profits, and they are frustrated by the correspondingly high potential for unauthorized copying and distribution by consumers who are unwilling to pay the price they are charging for content. With 1.3 billion people using the Internet, the U.S. government figures U.S. businesses lost $250 billion last year in potential sales because of pirated goods. Most of the next billion people who will be getting online as the Internet is diffused across the world live in nations where U.S. and European laws do not prevail. Established content producers are not likely to achieve the optimal international profits on each song, film, or piece of software they create from expanded distribution of their products if things continue as they are right now. Digital rights management and enforcement of intellectual property expectations as currently legislated have been eroded more than ever in the past few years by several factors, including code cracking and peer-to-peer distribution technology and a lack of support for IP laws in Asia, where most of the illegal copying of Western-produced entertainment products is taking place.

Who owns ideas, artistic creations, inventions, and creations of the mind? Who is entitled to use them? Who is entitled to profit from them? The World Intellectual Property Organization (WIPO), one of 16 specialized agencies of the United Nations, was established in 1967 to try to help people across the globe resolve these questions. Treaties worked out through WIPO have significant influence, as do initiatives such as the World Trade Organization's Agreement on Trade-Related Aspects of Intellectual Property Rights (TRIPS Agreement), which was aimed at helping to bring global IP law into more alignment. There is, however, a rising concern that businesses and governments are leveraging these organizations to establish and support intellectual monopolies through a system of intellectual protectionism. An Electronic Frontier Foundation statement on WIPO reads, "All too often, governments are using international treaties negotiated through WIPO as well as other bilateral trade agreements to ratchet up IP rights at the behest of copyright holders."[4]

Some *Future of the Internet III* survey respondents called the escalating struggle between media companies and the people who copy and sell or give away those companies' property an "arms war." The following are two examples of battles in the war: In 2008 MediaDefender, a company paid by media corporations to monitor and slow the sharing of content on peer-to-peer networks like BitTorrent by planting fake files to slow file sharing, was hacked by a teenager who stole and then made public a series of internal e-mails that documented MediaDefender's P2P network sabotage methods. In 2007 Viacom sued Google, owner of YouTube, for "massive intentional copyright infringement" and sought more than $1 billion in damages, stating that Google avoided taking the proper steps to prevent the uploading and viewing of nearly 160,000 clips of video material owned by Viacom on YouTube, a video-sharing site. Google worked up a tool to help copyright owners manage their content, removed infringing videos, and structured language on the uploading sections of the site to encourage original material and warn against copyright violations. Google added $2.6 billion in advertising revenue to its bottom line in 2007; Viacom added $400 million. In April 2008 Viacom amended its complaint against Google to include a request for a jury trial. Google said it has adhered to the U.S. Digital Millennium Copyright Act, which offers protections to Web-based businesses in the case of users' postings of copyrighted materials.[5]

There has been incremental industry adjustment in regard to digital rights management. In 2007 several companies began to provide DRM-free versions of their content to consumers who pay a higher price, and more commercial artists were offering their work for free or at any price a consumer was willing to pay for it. By 2008, Amazon.com had come to an agreement with the top four music corporations to sell songs without DRM technology. Apple has come under some fire for creating its own DRM technology and refusing to license it to third parties, thus limiting those with iPods to downloading music only from the iTunes store and allowing Apple to profit from both iPod sales and music sales. After coming under fire for this, Apple has made some new deals with music producers, for instance, songs from EMI are now available DRM-free through iTunes. Many broadcast production companies have now begun posting streaming videos of their already-aired television programming online for free use on their own Web sites and on aggregating sites such as Hulu; viewers generally see advertising and "pay" with their attention.

Limited-use DRM is commonly used in online video and films. This is a factor that annoys many consumers who have been accustomed to buying and then "owning" a copy of content. An example is the system employed by Apple Movie Rentals. You download a film, and you have 30 days from that point in which you can start using it, and once you have started using it, you have 24 hours to finish. You can transfer it to another device, and watch it on a PC, iPhone, or Mac, but the clock is ticking.

There is some movement in business to monetize content without DRM. Microsoft won a patent in 2007 for a technique titled "stealthy audio watermarking" that would possibly track usage of files unobtrusively while providing IP owners with data about consumption patterns. This could possibly replace DRM, but it would not solve privacy issues tied to this sort of surveillance over content.

Meanwhile, consumers are finding that they often enjoy user-created content (UCC), also referred to as user-generated content (UGC), just as much as or more than commercially produced information and entertainment. This type of content is often posted with no claims to copyright or ownership. The Creative Commons, founded in 2001 by Stanford Law School professor Lawrence Lessig, offers a number of free licensing options to creators, allowing them to retain copyright while setting a level of distribution, for instance, allowing free use with no modification.[6] Many respondents to the *Future of the Internet III* survey voiced support for this model.

The vast amounts of free content being produced compete with entrenched industries, many of which are now pursuing litigation or, at least, are threatening producers of UCC who incorporate snatches of copyrighted content in the material they produce, sometimes creating new creations totally composed of mashups of copyrighted material. Some people expressed the fear that with the new ability to track and tag things digitally and store the knowledge in vast searchable databases, most every human idea will be licensed and claimed, allowing no freedom to build new ideas spun off from previous content. Sessions at recent conferences have included topic titles like "Is User-Generated Content the Future or the Last Gasp of the Past?" and "Will Copyright Kill Social Media?"

Concern that the 1998 Digital Millennium Copyright Act has caused an imbalance in U.S. law pertaining to fair use by the public of copyrighted material inspired a 2007 report by the Computer & Communications Industry Association (CCIA). It showed that companies that "benefitted from limitations on copyright holders' exclusive rights" represent $4.5 trillion in annual revenues for the United States, representing one-sixth of the gross domestic product and employing 17 million workers. CCIA president Ed Black said the Internet "depends on the ability to use content in a limited and nonlicensed manner...We must keep fair use as one of the cornerstones for creativity and innovation."[7] Representative Rich Boucher (D-Va.) is the author of a bill that would expand fair-use protections in the Internet age, saying it is "essential to our future economic growth and global competitiveness."[8]

Thus far in the Internet age, most businesses built around the pre-Internet, content-at-highest-profit-possible paradigm are finding it difficult to make the adjustment to a different approach to the economics of success. They were built on a foundation of content control in a time of information scarcity and are losing that control in a time of information abundance.

Sam Punnett, president of FAD Research, wrote in his *Future of the Internet III* survey response, "The problem lies with what is partly a hangover from traditional business models and partly due to consumer intoxication with access to 'free' media. Content owners and distributors are accustomed to dictating terms, and consumers (having now been exposed to extensive media choices with a 'free' price point) have become disenchanted with the value proposition of content owners...Consumers are becoming mesmerized by the sheer volume of choices (many of them free), and there is fragmentation of the audiences for any single platform. The value of media has diminished greatly. The outcome will be dogged by an inability of policy makers and media companies to craft viable regulatory-based alternatives. Innovation in the Internet world will continue faster than any coordinated effort that is likely to occur amongst rights holders, ISPs, and lawmakers. There are more tough times ahead for content owners. Successful innovation will determine the final outcome, and this is not something media companies are traditionally good at. The starting point for them must be to attend to audience habits, to increase value somehow, and to leverage the new characteristics of Internet-delivered media into a viable business model."

Hardware hegemony is the latest area of interest for those concerned about maintaining current forms of control. Zittrain's book, *The Future of the Internet—And How to Stop It*, is a warning that care must be exercised when taking a generative tool like the Internet and introducing nongenerative aspects that could take over and kill innovation.[9]

Zittrain noted that digital devices such as the iPhone are being built to be monitored, updated, and controlled by their makers; vendor updates can detect the hacks that people can now perform to "unlock" such devices to allow them to operate in an open format. These devices often provide more security, convenience, and easy updating, but using a "locked-down" device to access the Internet eliminates the openness of the system, eliminating privacy and anonymity. Zittrain also noted that people are beginning to spend most of their online time in a handful of aggregating destinations, such as Facebook and YouTube, that can be constricted by their own business plans or by the influence of outside forces. Tethered appliances and applications allow vendors to control content and limit use and to share people's private information with others, including outside vendors and government officials. As with every development in communications technology, there are positives and negatives to this evolution at the hardware and software levels.

Future of the Internet III respondent **Jill MacNeice**, information architect at the U.S. Library of Congress, offered up a tongue-in-cheek statement to illustrate how she thinks current structures might look in a future in which data can be tracked in heretofore unmatched ways: "By around 2020," she wrote, "I will have patented my device, now in development, that records human thought and evaluates it against a database of trademarked expressions. This device will be implanted in the brain and will keep an ongoing account of expression generation, with usage fees automatically deducted from bank accounts. Of course, this requires the development of significant infrastructure, but I'm working on that, too. I expect ongoing royalty payments will encourage the participation of infrastructure providers and content owners alike. Gaining subjects for participation could be a challenge, but I'm expecting government support by tying the project to defense and antiterrorism concerns. (And, by the way, I've gone on the record here with this idea, so potential violators, beware.)"

ADDITIONAL RESPONSES

Many other participants in the survey shared valuable input on the IP law and copyright scenario. Among them:

> "Unfortunately, stricter DRM is probably in our future at a high cost to the industry. To counter this, most of our future entertainment will be derived from free sources with no DRM like YouTube and the future online channels for games and entertainment. The 'old guard' will cling to out-of-date IP practices at the cost of their audience." —**Gerard LaFond**, founder and chief strategy officer, red TANGENT, and cofounder of Persuasive Games; works with leading brand companies such as Sony, Ubisoft, and Kraft

> "This scenario does not take into account the potential increase of alternative (and not necessarily legitimate) methods of accessing copyrighted material. For example, it ignores the potential increase of sophistication and ubiquity of cracking technologies, as well as the potential number of social networking and virtual worlds, which could make monitoring the exchange of such copyrighted material much harder than implied." —**Fadi Salem**, research associate, Dubai School of Government; research focuses on e-government and development in the Middle East and North Africa

> "Totally agree. It takes time and money to do quality work. Creators need to make a living, or they won't create." —**Trisha Creekmore**, interactive executive producer, Discovery Channel Interactive

> "The cost of implementing such a system across the Internet is astronomical, not to mention getting cooperation and buy-in. Moreover, Internet tech-savvy users will quickly adapt and defeat such a system. Their goal will be to maintain the free flow of information and expression across the Internet." —**Mack B. Rhoades Jr.**, Web services product manager, Michael Baker Corp.

"Kevin Kelly writes that the Internet is a giant copying machine. 'Value' will reside in what is not easily copied. DRM is just an attempt to extend a dying paradigm." —**Dick Davies**, partner, Project Management and Control, Inc.; past president of the Association of Information Technology Professionals

"Governments will find a way to protect the entertainment industry through copyright protection as an important part of the economy." —**Lawrence Swiader**, chief information officer, U.S. Holocaust Memorial Museum

"A crisis in rights management for cumulative work will encourage all-you-can-eat licensing, combined with machine-readable code to control artists' demand and attribution; sampling will be the base of redistribution the licensing fees. The key question will be to identify works that have been reused in spite of their authors' explicit refusal, and the contradiction between the approaches of two creators: Can one prevent the other to use her work?" —**Bertil Hatt,** researcher of Internet and social services, innovation valuation; employed by France Telecom and Orange (information technology and services industry) while completing PhD

"In an economy where creators of content can't be sure of compensation, creative work will diminish in quality and quantity. To avoid this, companies and lawmakers will work to protect creators through technology and law." —**Robert H. Rich**, PhD, strategic planning and evaluation, American Chemical Society

"While a new legal economic regime is highly desirable, it is highly improbable that real-world economic fear and greed would disrupt legal institutional inertia." —**J. W. Huston**, president of Huston Consultancy and futurist

"Nothing in the still-short history of the Web suggests that copyright holders and old-school distributors will innovate faster than passionate consumers. The tide's already shifted." —**Matt Gallivan**, senior research analyst, National Public Radio

"DRM will continue to develop, but the hackers will continue to keep pace, and the casual attitude of most users will overcome centralized controls on most electronic intellectual property." —**Jim Lucas**, Web manager, CACI, a provider of national security, defense, and intelligence-related solutions in the interests of the United States

"We are moving toward an open-access information society where most information will be easily and inexpensively available to end users. I suspect that an alternative business model will be developed to compensate information producers based upon either advertising fees, royalties paid through subscriptions, equipment surcharges, or minimal user fees (or, perhaps, a combination of these compensation mechanisms)." —**Gary Kreps**, chair of the department of communication, George Mason University; formerly founding chief of the health communication and informatics branch of the National Cancer Institute

"With the rise and expansion of an open-source or conditional use sector of IP creators, more direct artist-to-consumer relationships that cut out distributors and encourage payment as an element of community participation, and increased use of interactive massively-multiplayer-online-games-style experiential entertainment that cannot easily be copied, strict DRM controls will be found to be ineffective and easily subverted. An ASCAP [American Society of Computers, Authors, and Publishers] fee setup will be built into the price of electronics for legacy IP holders through an agreement between IP holders, government, and hardware manufacturers. The latter will gladly agree to increase sales and deflect lawsuits." —**Ted M. Coopman**, lecturer, San Jose State University

"This will only happen if external events—terrorism, etc.,—lead national governments to adopt the kind of authoritarian/fascist governance model necessary to establish/sustain absolute monopoly control over both media and telecom. After creeping up to that point for several years, the U.S. will follow the usual decennial pattern and resume liberalization of access to network inputs c. 2014." —**Tom Vest**, IP network architect and consultant, RIPE NCC Science Group, Office of the Chief Scientist; consultant for the Internet Society, and for OECD Economics and Statistics Division

"IP regimes are likely to get more complex, rather than less. We'll continue to see fairly rapid circumvention of any DRM scheme by hackers, and a teeming P2P black market in digital files. Expect to see much more file sharing go into the 'darknet' using encrypted, trusted social-network-based sharing. E.g., AllPeers." —**Anthony Townsend**, research director for the Technology Horizons Program of the Institute for the Future, providing long-range forecasts on technology; he is also a cofounder of NYCwireless

"DRM will be dead long before 2020, as will most of the current music- and film-producing companies unless they significantly change their business model. With every new technology, the entrenched companies claimed they were under major threat and something must be done to stop audio cassettes, VCRs, burnable CDs and DVDs, etc. Yet ironically, many of the new technologies actually provided more profit and more methods of distribution than before. The bottom line is that people will find a way to obtain the content they want at a price (even if it's zero) that they feel is reasonable. The companies who can figure out ways to monetize other aspects of their content will prevail." —**Scott Brenner**, technologist, Web developer, consultant for clients ranging from Fortune 100 companies to small nonprofits

"Two things will happen. Copyright protection technology will be only one of them. The other change will be new globally accepted copyright laws that will be significantly different from current copyright laws. On the one hand, these laws will partially comprehend the kinds of changes necessary to open up exchanges of information and to open channels of free expression. On the other hand, they will also have special clauses and enforcement procedures to protect large commercial interests, especially mass-media-and-entertainment corporations. These same laws will promote news as entertainment and stifle professional journalism and real news media." —**Benjamin M. Ben-Baruch**, senior market intelligence consultant and applied sociologist for Aquent, working at General Motors Corporation

"Content controls assume that all relevant sources of content development can be monitored and kept in line. Even today, that is proving to be an impossible task. The Wikinomics of the marketplace will not abate, in my view, but accelerate. The technology industry and media companies will adapt to this reality; however, *Minority Report*–style command and control tactics will not be their most successful strategy. Rather, sharing and collaboration will foster the emergence of newer models that will allow content owners, Internet service providers, and others to develop and garner revenue from content in both traditional and new ways." —**Barry K. Chudakov**, principal, the Chudakov Company, a marketing and advertising strategies creative consultant who has worked with many major corporations, including Microsoft and Disney

"Digital rights will be so contested that the current rights systems will have crumbled—unless there is some better accommodation for use, reuse, and sharing. Micropayment schemes will fail because ownership will be so obscured as to make claims impossible to work out in a rational way." —**Paul Jones**, director of ibiblio.org at University of North Carolina–Chapel Hill; the original manager of SunSITE, one of the first Web sites in North America

"2020 isn't likely to be very different than it is today with respect to IPR." —**Anthony M. Rutkowski**, cofounder of the Internet Society and a founding trustee; longtime leader in International Telecommunication Union; vice president for regulatory affairs, VeriSign

"The current copyright system is bankrupt and cannot be reinvented or protected. The question, for me, is not how to reengineer intellectual property. The challenge for the 21st century will be to set up a peer-to-peer distributed payment network in which artists can be paid directly. We have to question the techno-libertarian agenda, which preaches that the only way to get rid of copyright is to give away code or content for free. This is a bad proposition for artists. We can do better and invent a technical solution in which creative producers do get paid, but outside of the legalistic license structure." —**Geert Lovink**, professor and expert on culture, sociology and the Internet; based in Amsterdam; author of *Dark Fiber* and *Uncanny Networks*; responsible for the Institute of Network Cultures

"The privatization of content, with oligopolistic corporate control, seems likely. However, this will continue to hinder access for the poor, whether in the U.S. or a third world. This is a critical element for the expansion or contraction of the current expansion of the world into 'haves and have-nots.' If education is the primary force for social mobility, as I believe, then this may be the most critical policy issue of at least the first half of this century. My preference is for a system providing everyone access to all information. However, that begs the question of how do the major contributors get a fair return for their effort." —**Ed Lyell**, professor of business and economics, Adams State College, Regis University, San Luis Valley Board of Educational Services; pioneer in issues regarding Internet and education

"It's simply too easy to subvert content-control software, and there's little indication that this condition will change any time soon. Moreover, the trend at present is for content companies to move *away* from strict controls; while this could certainly change by 2020, my sense of things is that they'll tend to find that looser controls are better for them overall." —**Jamais Cascio**, originator of Open the Future, also works with the Institute for the Future, Institute for Ethics and Emerging Technologies, and Worldchanging

"Once virtual private networks become ubiquitous, the only way of enforcing IP/DRM restrictions will be with draconian punishments, and I just don't think content producers are going to want to be associated with those." —**Fred Hapgood,** technology author and consultant; freelance writer in technology and science; in the 1990s, he took on the role of moderator of the Nanosystems Interest Group at MIT

"This battle of control and ultimate ownership over access to the 'goods' has gone on for as long as forgery, fraud, and counterfeiting have existed. Middleman technologies will continue coming along to anonymize transactions that are illegal. ISPs will always exist that don't buy into the concept of content control. Walls will continue being built, and new paths around them or under them dug out." —**Dian Schaffhauser**, writer and editor for CampusTechnology.com, THEJournal.com, *Redmond Magazine*, *Computerworld*, and Web Worker Daily; founder of Sourcingmag.com

"DRM will disappear by 2020. We could witness in 2007 how iTunes eliminated this option. It is a strategic decision to benefit from the global markets in a lucrative way. We need to admit that more copyright law implementation will limit the knowledge of those who can't afford it from the least developed economies. That's why there is a need to find a balanced solution to guarantee the rights of publishers and the users." —**Hanane Boujemi**, ICT researcher for DiploFoundation, working on educating people about Internet policy and Internet governance, Malta

"It's difficult to imagine technology becoming secure enough to make this scenario possible. Copyright enforcement has always relied on honesty and trust between producer and user, and that will remain the case." —**Neil McIntosh,** director of editorial development for guardian.co.uk, based in London; one of the first to report on the commercial possibilities of weblogs; one of Britain's first blogging journalists

"Much as content owners insist on DRM, much more dire priorities have taken precedence. Criminalizing copyright violations looks petty with so many other issues to deal with." —**Cliff Figallo**, social innovator and original member of the first online community—The WELL, now of AdaptLocal.org; expert in fitting and implementing social Web applications to groups

"My response is based almost solely on my own experience, but I imagine that we will move to a subscription model for most popular digital content, and the issues described in this question will be mostly fought at the 'premium' level of experience, like access to the newest 3-D holofilms. For 2-D A/V media, we'll all subscribe to a combination of XM, Rhapsody, subscription TV, and Netflix Online, where we pay a monthly fee for the content that we then use on any connected (and approved) device. There might be variations in service level and packages, and there will certainly be young people and reprobates who try to subvert the system. But on the whole, most of us will pay monthly. As an aside, most of what we receive by subscription will suck." —**Peter Eckart**, director of health information technology, Illinois Public Health Institute

"I see the Creative Commons approach as gaining validity as stricter copyright laws turn people toward more open options. The economic incentive to use a copyright disappears when people find fungible sources that are more open. This leads to shorter copyright and more price-conscious pay options." —**Thomas Vander Wal**, principal and senior consultant, InfoCloud Solutions, Inc.; coined the term *folksonomy*; expert on tagging, the social Web, and social information use and reuse

"A vibrant hacking community races against the tech and media companies to provide free access to digital content. Their open-source, collaborative model allows them to compete with even the largest tech and media companies, and pirating remains an issue." —**DJ Strouse**, international relations and computer science student, University of Southern California

"This prediction is a horror prediction. Recent legislative initiatives in Sweden and the U.K. point in the opposite direction—to liberalize copyright protections." —**Norbert Klein**, member of ICANN's GNSO Council and Internet Society leader who works with Open Institute Cambodia, a company whose primary focus is on information

"Unfortunately, this scenario may become reality without the intervention of more forward-thinking and creative souls who can offer a completely revamped business model. For instance, there is no reason a music artist needs a record label anymore—they can hire their own lawyers, publicists, marketing firms, etc., and cut out the middleman. They can offer their music for a fee over the Internet, and encode it so that they always can trace downloads through technology. But that would be the end of the music industry as we know it. And that takes a lot more creative thinking to make it work." —**Janie Graziani**, manager of new media and technology for the American Automobile Association

"Can I say 'maybe'? The automatic billing seems interesting, but encryption of Internet traffic would defeat this pretty quickly. There will be a larger division between traditionally produced IP and IP produced in nontraditional ways/open source." —**Ben Spigel**, master's student in the department of geography, Ohio State University; researches microgeographies of academic knowledge exchange

"The public will always be cleverer than the content industry. Content owners won't rely on control but rather on providing additional reasons (persuasion, physical purchase trophies, better quality) to access content legally. Also, there will always be places in the world where the idea of IP is less well established." —**Jeremy Swinfen Green**, Telecom Express, an interactive marketing company

"And, this may set aback the open-source movement." —**Hakikur Rahman**, chairman, SchoolNet Foundation; coordinator of

Sustainable Development Networking Programme in Bangladesh; active in Internet Society Board, South Asia Foundation

"[I mostly agree,] however, open-source software and shared material on a noncommercial basis (i.e., Creative Commons type) will be more widespread." —**Julian Hopkins**, social scientist and PhD candidate at Monash University, Malaysia

"The ability to subvert the system will remain a step ahead of legal, technological, and industrial controls. Entertainment IPRs will not die, but alternative revenue streams (such as music labels taking larger cuts of ticket sales and merchandising) will be the main focus of all parties wishing to profit from creative cultural productions, and the drive to enforce IPRs will be weakened." —**Jade Miller**, PhD student, Annenberg School for Communication at the University of Southern California, with a research focus on global flows of information and culture

"In terms of the 'policing' of the materials of media, legislation will have played the game of 'catch-up'...The identification of 'users' and 'clients' for ownership will represent the next turning point for protection on the Internet; this will probably be linked to user ID such as ID cards as well." —**Maz Hardey,** social analyst, blogger, "defender of new media" completing a doctorate funded by the Economic Social Research Council in the U.K., based at the University of York

"Technically, it will be possible [to have regulation]. There will be a flourishing community situating themselves in countries that have less strict copyright laws/enforcement supplying content that has been adapted to get round the hardware demands (or ways of hacking the hardware). If agencies want to enforce copyright, they're going to have to work out a better pricing model (and a way of enabling backup copies to be legitimately made and/or content to be passed on in the way it's currently possible to buy secondhand CDs, etc." —**Emma Duke-Williams**, lecturer in the School of Computing and researcher, University of Portsmouth, U.K.; education blogger

"I do not see either side winning or losing this battle—it's just the current mix of open materials (with licenses like Creative Commons variants), fully protected DRM media, and people working around these in creative/illegal ways depending on one's perspective. Note that the world situation in 2020 will also depend very much on the dominant economic powers at that stage and not necessarily on the current levels of influence, so the U.S. influence may be slightly less than it is at present." —**Micheál Ó Foghlú**, research director, Telecommunications Software & Systems Group, Waterford Institute of Technology, Ireland advisory committee; member of W3C; blogger

"I agree that through emerging technologies such as the Semantic Web and the already-present Web services model, components of content and software will be serviced through gateways that may or may not charge a fee. However, unless there is a singular centralized system or set of gateways to deliver sets of content or services, there will effectively be no difference between the scenario today and in 2020. The presence of open-source software and open copyright will be sustained as a parallel and possibly more effective system along with the protectionist parties." —**Amit Kelkar**, consultant and sociology researcher, Postmodern

"This will only be true if there are also draconian antidemocracy features embedded into the network. Strict content control is censorship. It can only work if there is 100% censorship." —**Robert J. Berger**, CTO for Cinch; expert on backbone networks, access networks, wireless networks, and innovative Web applications, a frequent speaker on next-generation networks

"Much of this is the case NOW! [For instance:] '…strict content controls are in place thanks to the efforts of legislatures, courts, the technology industry and media companies.' DMCA, DeCSS, AACS, HDMI-HDCP, etc. 'Those who use copyrighted materials are automatically billed by content owners' It's not automatic. But one-click ordering is nearly so. '…and Internet service providers automatically notify authorities when they identify clients who try to subvert this system.' Check out the court case of Viacom suing YouTube. 'Protesters rarely prevail when they make claims that this interferes with free speech and stifles innovation.' See any of several DMCA court cases. Note my 'Mostly Agree' response doesn't indicate endorsement." —**Seth Finkelstein**, anticensorship activist and programmer, author of the Infothought blog and an EFF Pioneer Award winner

"I have taught copyright at a law school and feel I have to be more 'optimistic' than this. A portion of the content will be under 'strict content controls,' but a huge amount will be 'user generated' outside the content-control system—including some of the movies, sound recordings, books, magazines, photographs, paintings, et al." —**Rollie Cole**, director of technology policy, Sagamore Institute for Policy Research, a nonprofit, nonpartisan think tank headquartered in Indianapolis, IN

"By 2020, DRM will have passed for the majority of cases, and the argument will have been won with the activists wondering what happened. The IP owners will continue to make more money than ever." —**Sam Smith**, Web interface developer, University of Manchester, U.K.

"Simple DRM may serve to support some level of price discrimination if the user interface is not too awful, but strict enforcement of content control will be met with civil disobedience, as is true today." —**Brough Turner**, chief technology officer and cofounder of NMS Communications; oversees evolution of technology and product architectures

"Whilst I do agree that copyright protection technology will play an ever-greater role, I do think that the efforts of big-name performers, public service broadcasters, and even individuals creating user-generated content will ensure that content control measures will not 'dominate' entirely." —**Victoria Nash**, PhD, Oxford Internet Institute, University of Oxford, U.K.; formerly a research fellow at the Institute of Public Policy Research

"Without some protection for the creators of intellectual property, innovation will suffer." —**Don Heath**, Internet pioneer; former president and CEO of the Internet Society; member of U.S. State Department Advisory Committee on International Communication and Information Policy

"I expect to see some big changes in the industry business model that cope with the actual infringements through an optimization of the value equation. There is no need to pay $20 for a music CD, and the distribution will be dramatically improved by the usage of Internet-based technologies." —**Sebastian Ricciardi**, associate with Jauregui & Associates, a law firm in Buenos Aires; leader in the Argentina chapter of the Internet Society, formerly of ICANN's At-Large Advisory Committee

"Here in the U.S., we have a long history that tells us that you can't force the majority of Americans to do something they don't want to do, with Prohibition and the current War on (some) Drugs being the prime examples. It's quite clear that college kids have no sympathy for the RIAA's whines, and it's just a matter of time before DRM goes away. Besides, DRM flies in the face of the clear meaning of the copyright clause in the Constitution, which includes the keyword 'limited' to describe the length and scope of protection." —**John Levine**, founder of Taughannock Networks; a leader of the Internet Research Task Force's Anti-Spam Research Group and the Coalition Against Unsolicited Commercial Email

"I agree with this scenario more because I hope that this is what the future will bring rather than a certainty that this will be so. The innovative process is not dependent on theft, nor is attribution an enemy of free speech." —**Hinda Feige Greenberg**, PhD, director of the information center for the Robert Wood Johnson Foundation, dedicated to improving health care for Americans

"This is really more an issue of politics than technology. Copyright needs to evolve, but entrenched issues may stymie that. There are related constraints with issues such as microcredits and authentication, but those are pretty straightforward, especially compared to the politics." —**Greg Laudeman**, utilization catalyst and facilitator, community technology specialist, Georgia Tech Enterprise Innovation Institute

"Hard work needs to be done in each individual country and collaboratively to make this possible all over the globe. It is possible in many developed countries where use of the Internet is very common and e-governance is fully functional. For the countries that are still in the process to know more about the importance of Internet and only have heard the terms like *e-governance*, what I would like to suggest is that these countries organize high-level training programs for politicians, bureaucrats, diplomats, etc., the key persons of the country. If they could just invest 2 hours in this training course, then they can plan, decide in line with the Internet/Internet governance.

What is happening now is so many projects are planned, decided, and implemented without taking the ICTs in mind." —**Sudip Aryal**, president, Nepal Rural Information Technology Development Society

"Research indicates that young people expect Internet content to be free and not encumbered by DRM. As young people become voters, wage earners, and business/government leaders, innovative business models and changes in IP law will encourage open and free, rather than closed and monetized, content." —**Michael Edson**, director for Web and new-media strategy, Smithsonian Institution

"There will be (1) a black market analogous to the current DVD black market, but probably using nonphysical content. (2) A group will create its own media world modeled on Linux openness that deliberately refuses to copyright their content." —**Bruce Turner**, director of planning services for a U.S. regional transportation commission; retired from U.S. military

"While in 2020, copyright protection technology will dominate, there will still be a thriving underground P2P scene. Also, I don't think an automatic "use-then-get-billed" model will flourish." —**Aaron Schmidt,** Walking Paper Consulting, a blogger who writes about libraries, technology, and usability

"I agree with this scenario as far as it pertains to big companies. But more and more individuals will apply a Creative Commons sensibility to their creative and intellectual work, creating a robust, free sphere of new music, articles, documentaries, and other work. It will temper the efforts of larger enterprises." —**Jan Schaffer**, executive director of J-Lab, the Institute for Interactive Journalism; former business editor and a Pulitzer Prize winner for the *Philadelphia Inquirer*

"Sticky ads and data mining to target effectively will make DRM enforcement meaningless. People will like that they only get ads that apply to them and their lifestyle and needs, and the ads will take care of the need to enforce DRM." —**Garland T. McCoy**, founder, Technology Policy Institute, a think tank focused on the economics of innovation; formerly senior vice president at the Progress and Freedom Foundation

"Free trade notwithstanding, some protections will be guaranteed to protect intellectual capital. Alternatively, however, most content will be shared openly with no charge, except for selected categories like movies, some books, etc." —**John Murphy**, director, USA.gov technologies, General Services Administration of the U.S. government

"The digital rights management technology has not been shown to be particularly effective in preventing unauthorized sharing. The social shift in cultural norms required to move an entire population to see 'sharing' of intellectual property as the same as sharing of physical property will take longer than the 12 years you are projecting." —**Jill O'Neill**, director of planning and communication, National Federation of Abstracting and Information Services; author of the Infotoday blog; based in the Philadelphia area

"I doubt that the legislatures, courts, technology industry, and media companies will EVER agree on an automatic copyright system—although it sure would be nice. I also doubt that, even if they get a scheme in place, that ALL the producers would participate in. Look at CCC [Copyright Clearance Center]—it doesn't cover everything. There will always be companies that think they can control things better in-house." —**Judith Siess**, president of Information Bridges International, Inc. and publisher and editor of the One-Person Library newsletter, author and blogger

"Content will always be king, and the content providers have to make money in order to continue to gather and produce quality content. I don't know how we could avoid this." —**Tiffany Shackelford**, consultant who works with clients such as Phase 2 Technology, Stateline.org, Foneshow, WebbMedia, and Daily Me

"Workarounds will make this control effort moot, and new services and models will have been implanted, albeit sometimes in the margins of the law. Centralizers (governments, big corporations) will continue their pace, but users will be invincible even if they have to break the law." —**Alejandro Pisanty**, director of computer services at the Universidad Nacional Autónoma de México; active in leadership with Internet Society, ICANN, and the Internet Governance Forum

"Content will be generally free, open sourced, and available to all." —**Tim Terpstra**, cofounder of Teletekst Is Dood and cultural (media)

entrepreneur, editor, researcher, developer for The Generator/TodaysArt Festival, The Hague, The Netherlands

"Hackers will always be smarter than content creators. Legislation of the Internet will be understood to be futile, resulting in more and more 'walled gardens' of communities of practice. Those above ground will be heavily regulated as described [in the scenario]. Those underground will be 'anything goes' with a brutality and creativity that we can only imagine today." —**Kathryn Greenhill**, emerging technologies specialist, Murdoch University

"DRM is an arms race between standards creators and circumventors. I don't see this changing in the future. It's more likely that eventually the pro-DRM lobbies will be forced to develop different business models that don't lean so heavily on rights management for their profitability." —**Dave Coustan**, author of the Extraface blog and an independent consultant who formerly wrote EarthLink's official blog, Earthling, and Stuffo, a subbrand of HowStuffWorks.com

"There is little doubt that this is the direction towards which we are going, but I wonder whether this would not create a reaction in the medium term that we cannot predict today. I wonder, for instance, whether musicians would not end up in relying more on live performances for their income, and so attach less value to their IP rights, changing some of the parameters of the equation. I do not think that this scenario has a high probability, but I would not exclude it. Therefore, I mostly agree (but with qualification)." —**Roberto Gaetano**, ICANN board member; also responsible for SW development for International Atomic Energy Agency; an active participant in the ICANN policy-making process

"Since necessity is the father of invention, those who decide to break (or bend) whatever laws are enacted will figure out a way to both pirate information and do it anonymously. People will always want…and figure out how to get something for nothing." —**Dan Larson**, president and CEO of PKD Foundation, a nonprofit organization working for patient advocacy and education

"The seemingly endless space reserved for virtually free discussion and sharing of information and entertainment on the Internet makes this an economic impossibility over time. It is more likely that technology will advance to the point where making illegal copies is prevented." —**Bill Warren**, vice president of government relations, Walt Disney World; founding editor of the *Orlando Business Journal*

"There will be no way of controlling copyrighted content. Music and film business will focus on products that can be experienced and/or high-quality physical objects, trying to surpass the downloadable media." —**Tiago Casagrande**, works with social communications and new technologies for verbeat

"By 2020, copyrights will be more open. They will allow people fair use of the materials in question. People will learn in the years leading up to 2020 how to work within a more open system and society. Government, courts, and industries will have to take the time leading to 2020 to learn to adapt to the nature of the Internet and how it has shaped the global market. Those who cannot adapt will fail at some point because the cost of fighting so many legal battles is not sustainable for them and will push their consumers away." —**David Newberger**, founder of Blackdot Ventures

"The peer-to-peer file sharing is only part of a social change that is empowering. The transformation of who are trusted sources; the idea that 'the crowd' can make a difference (see Barack Obama campaign for one) and the fact that Gen Z or whatever they are called at the moment will have grown up in a peer-to-peer empowered environment by 2020 will be (and is) the harbinger of social change that, when it comes to copyright control, will break down the traditional barriers that protect intellectual property. There is already a change trickling through the corporate world where companies like P&G and Samsung are willing (somewhat) to share their previously well-protected IP with their customers or social networks that are useful to them (see Innocentive, for example) because they see the value in it for business. As insane as the music industry has been in their overboard attempts to prosecute it, they will all come to accommodate digital and copy rights with deals that will loosen the IP/DRM protection but still be reasonable enough, given the social changes. No question, the property holders deserve something for their property, but the laws need to change because the peer-to-peer trust and communications/

collaboration have changed the way institutions behave. The laws will be relaxed but still equitable for the property holder." —**Paul Greenberg**, president, The 56 Group LLC; BPT Partners LLC; MyCRMCareer.com, social media companies

"For another few years, I expect that the major content owners (i.e., labels, studios, etc.) will manage to lobby for significant political support for DRM schemes and will be able to enforce them, too. However, by 2020, massive backlash by both consumers and content creators (i.e., artists) will have forced said content owners to settle for DRM-free methods of distribution. The market will reject DRM schemes, smaller players will have created and established business models that do not only tolerate sharing of their content but encourage it. These business models will thrive thanks to what today is often labeled as 'piracy,' and they will embrace the social sharing habits that so many consumers seem to naturally adapt." —**Peter Bihr**, freelance consultant on Web strategies, communities, blogging and social media; based in Berlin, Germany

"There will probably be counter technologies that will weaken those mentioned efforts." —**Erkan Saka**, lecturer in media and communications systems at Istanbul Bilgi University

"I do not see such drastic change from now regarding to this matter. Both camps will continue to fight, and the level of piracy should remain high. It is an illusion to believe that technical means only will block the piracy and also because many technical resources are also being committed to break the protection technology. It may be possible that some content creators are using piracy like techniques as a tool of marketing." —**João Miguel Rocha Filho**, director, DataOne, a provider of software for connecting to Linux; based in Brazil

"In this case, we have a question of both sides being 'right.' The arguments put forward on both sides have reasonable justification, which leads me to believe that 2020 will see the relatively peaceful coexistence of protected and unprotected content on the Web and in other situations. In a society of abundance such as we have now, there are many sources of funding the creation and dissemination of knowledge and artistic expression given away free of charge out of a sense of altruism or just for the hell of it. At the same time, fully commercial activities (the 'maximalists') will continue to operate (as is just in an open, pluralistic society) collecting royalties. At present, I am studying the question of how to be sure that the Web is organized in such a way that users will find locators taking them to both commercial and noncommercial offerings of information/knowledge/artistic expression, where they can make use of that which is free and decide on which of the commercially available material is worth acquiring or obtaining through other means. For example, the first-world universities and school systems individually purchase subscriptions to services furnishing online access to thousands of scholarly journals, to be used by students and faculty. Here in the second or third worlds, governmental institutions at the national or regional levels do this and give access to academic institutions. So even costly protected knowledge gets into the right hands (sometimes) without cost. Coexistence of the two systems is not only just but workable." —**Fredric M. Litto**, consultant for Pearson Education Global e-Learning, president, Brazil Distance Learning Association

"The emergence of new ideas that happens when network linking is open will override the self-interest of those who would wall the gardens. Content in walled gardens will wither." —**Judy Breck**, blogger at GoldenSwamp.com—aimed at "watching the global golden age of learning emerge from the open Internet"

"The problems currently found on the Internet regarding DRM will continue. In 2020 access to content outside the U.S. via satellite and advanced networks will provide a 'workaround' for those seeking to subvert the system." —**Don Kasprzak**, chief executive officer of Panaround.com, a Web-solutions design company; former system engineer at Apple Computer

"I strongly agree that there are those who want to use copyright protection to control content, but I do believe all the people using the Internet will stage an online rebellion that will knock these digital dictators flat." —**Leonard Witt**, associate professor in communication, Kennesaw State University, Georgia; research interest is citizen journalism and user-generated content; author of weblog PJNet.org

"As an artist, the protection of an artistic element is important. As the writers' strike now publicly exemplifies, digital rights are expanding, yet laws are not in place to protect the creators of the music and film contributors." —**Janice Stevenor Dale**, president, JSDA, Inc., The Design University, interior design

"I think DRM is a little bit similar to controlling in huge enterprises (business science): It is an idea of the last century and not really effective. Organizations in the digital world will sooner or later realize that it would be much better to state their point of view clearly and to show trust in people." —**Oliver Quiring**, PhD, Institute for Communication Science and Media Research, Ludwig-Maximilians-Universität (LMU), Munich, Germany

"The combination of the technological tide plus competition between nations will gradually suppress copyright. Copyright will become a quaint legal oddity from the past." —**Alexis Chontos**, webmaster, the Art Institute of Pittsburgh

"I see this scenario happening, although at some point, it will be unfortunate if fair use is left out of the picture. I hope publishers relax a bit and let access to their materials be the focus. Copyright needs to be available for the author rather than favoring the content owners/publishers, as it does today, and as your scenario projects it to be. As a librarian, I hope that fair use of copyrighted materials will be still be allowed and encouraged, even if through specific channels, methods, or locations. On the other hand, if a library is billed through a micropayment system each time a patron used a copyrighted material, it will (hopefully) be much cheaper than the current charges for site licenses on information databases and copyrighted material." —**Teresa Hartman**, associate professor and head of education, University of Nebraska Medical Center

"I'm somewhat in the middle on this one. Artists need to be recognized and compensated for their efforts. Those who wantonly abuse copyright laws should be prosecuted. There does, however, need to be a middle ground allowing public access to such content. I don't see how the DRM movement stifles competition. If artists aren't rewarded for their work, why bother to compete?" —**Mike Samson**, interactive media writer and producer

"Alas, I mostly have to agree as the large corporations who 'produce' or at least distribute said content have the clout to force these things through the courts. However, we have a whole generation of users who've always used digital data in a free manner. Thus these new 'adults' may, in the name of 'creative freedom' possibly find another solution which will enable users, producers, and new creatives to coexist in a manner that pays the bills and allows a certain amount of creative flexibility. The big problem of stifling new ideas, which nearly always build upon old, will otherwise cause mass court congestion, demand significant intrusion in privacy (which is being infringed upon in the name of national security), and cause large amounts of the population to gain a criminal record." —**Robert Eller**, Concept Omega, a media marketing and communication company

"Large corporations that exert total control over large amounts of intellectual property will decline substantially. Independent content producers (artists, musicians, writers, etc.) will take advantage of low-cost, self-publishing opportunities and distribute their own content either as individuals or as small companies representing few artists. By avoiding major corporations, they will reap a larger percentage of profits and will therefore rely on loyal fan bases that have been drawn to them by the wide distribution of unrestricted content." —**Jamie Richard Wilson**, journalist and freelance Web developer

"As a practical matter, the copyright authoritarians will never be able to catch up to users of media. More likely than making the Internet a permanent sting for file sharers is an arrangement where users pay a small flat fee as part of their ISP/mobile access bills that goes into a file-sharing fund, which compensates artists proportionally for those works that are being exchanged on P2P networks." —**Jacob Kramer-Duffield**, blogger and student, University of North Carolina School of Information and Library Science

"Alternative scenario: Some sort of Creative Commons–inspired copyright laws reformulation will occur, allowing for nonprofitable (attribution only) content sharing." —**Luis Santos**, Universidade do Minho–Braga, Portugal

"...Think like an engineer: Solve the problem and enable growth. At Davos, I saw this contrast in approaches to the environment by Al Gore and the movement (who favor carbon taxes and prohibitions on behavior) and by the Google Foundation (which favors investment and invention to affect the market, reducing the cost of electricity to less than coal, and solving environmental and economic problems together). In a postscarcity media economy, you won't be able to make money anymore telling the public what it cannot do; you will make it only by helping them do what they want to do. We may well pay for content—witness iTunes. We may distribute it for free. We may give our attention to your advertising. We may give you valuable behavioral data. But you have to open up to enable us to do that. Otherwise, you give yourself a gigantic cost of marketing and customer support when everyone else around you is free." —**Jeff Jarvis**, top blogger at Buzzmachine.com; professor at City University of New York Graduate School of Journalism; contributor and consultant to the *Guardian*; advisor to start-ups

"As a content creator, I like to get paid for my work. However, much of what I write, in the end, has always been pro bono, and the Internet has been a boon to writers disseminating their work. What I see, though, is that the increased corporatization of the Net is leading us in the direction of increased copyright protection, which is good for commercial materials, but which also threatens to subvert the areas of the Net where noncommercial material proliferates. This is becoming obvious already on YouTube and will continue in this direction as big business figures out how to wring profit from the public spaces of the Internet." —**Dennis Baron**, professor of English and linguistics, University of Illinois, runs the Web of Language site and researches the technologies of communication

"My expectation is that the 'control' over copyrighted works [exercised by] commercial entities, such as publishers the recording industry and video producers will diminish, not increase, and that the move toward 'free' access will expand. I don't think that the issue will be resolved or clear by 2020. It will still be messy and contentious, with lots of litigation still going on." —**Timothy F. Richards**, library director, University of Michigan-Dearborn

"DRMs will die. They are a huge invasion of personal privacy. They don't allow format shifting among individuals who have legitimately paid for copyrighted material. They don't allow for fair/dealing or use. It becomes a nightmare for the visually impaired. They make it impossible for libraries, museums, and archives to preserve our heritage. DRMs will be gone." —**Robert Tiessen**, head of access services, University of Calgary Library

"Companies that fight a war against their customers only pave the way for their successors. Already, services that offer legal downloads of DRM-free music are gaining favor over those that require DRM. In addition, there has been an explosion of not only new (noncentralized) distribution channels for music but, more importantly, the supply of music, as expenses for every factor of music creation have decreased. Just as BMI [Broadcast Music, Inc.] succeeded versus ASCAP with a 'lower quality' of music, modern-day industry associations will not be able to stand against the self-organizing, self-spreading, self-popularizing mass of modern-day and future content creation." —**Jay Neely**, social strategist in the process of founding News Armada, a Boston-based company working to advance Internet-based news and commentary and community online

"I think we are definitely moving in this direction, and there is a lot of power behind this movement. That being said, the amount of user-generated content will likely make DRM less constricting than this scenario describes. More and more people are developing free content for the Internet, and I think this trend will continue despite any types of DRM that may be in place. People have found their voice on the Internet, they have things to say, and many are very willing to share their content without any compensation. A few of these content creators will have large audiences that will help to make legislation and DRM less powerful." —**Ilima Kane**, program manager, Klein Buendel Inc., a Colorado-based health education firm

"DRM is an arms race, and while those on the side of command and control have been winning battles, they cannot win the war. Web 2.0 is all about participation and sharing. New monetization mechanisms will have to be developed to support cultural creatives, as everyone—not just the 'stars'—becomes a prospective producer of digital content." —**Joe McCarthy**, principal instigator, MyStrands; formerly principal scientist at Nokia Research Center, Palo Alto, CA; he also has worked at Intel Research, Accenture Technology Labs, and Nokia

"I hope by 2020 media has figured out that currently copyrighted content, especially audio and video files, are consumables and should be given away free in order to attract viewers to participate—i.e., actually go to the concert for a more complete experience. We're already seeing the beginning of the end for DRM—3–4 major music publishing companies have recently dropped DRM. More are sure to follow." —**David Lee King**, digital branch and services manager for the Topeka and Shawnee County (KS) Public Library

"Any sort of DRM system depends heavily on the user community's general agreement as to the value of the content. The UGC trend and increasing value and use of 'amateur' content could partially or completely undermine any DRM system. The real differentiation will come with a yet-to-evolve level of credible arbiters who can help identify, and establish the value of, content worth purchasing or renting." —**Bob May**, founder and chief creative officer, Lucid Media, a Georgia-based company that provides Internet content as a driver in consumer branding

"DRM is being challenged already as consumers protest the restrictive nature of its setup. Consumers will limit the efforts to expand DRM." —**Naomi L. Lacy**, assistant professor, research division, University of Nebraska Medical Center

"I hope that DRM, as we know it, fails. I think that the media corporations will need to adapt and adopt. They will need to adapt to meet consumer demands and adopt new policies and, perhaps, even a new-media standard. File sharing and copying will always be with us. I think the media companies wish we would never have progressed beyond the days of vinyl records. The freedom of creativity will prevail over the restrictions of greedy corporations." —**Mark Fennell**, senior Web engineer, Athens Regional Medical Center, Athens, GA

"I listened to Lawrence Lessig on *TedTalks* this week talk about what has ALREADY happened with the sharing of digital information. Already, artists and content producers easily circumvent any laws. Like the record industry with CDs, owners of content will need to innovate and find a way to share appropriately or prepare to close up shop. Also, the marketplace is now global and if one country, such as the U.S., attempts to increase restrictions, content will simply be published online through Chinese unregulated means." —**Theresa Maddix**, satisfaction research analyst, ForeSee Results

"As technology permits society to become more Orwellian in its control and use of information, I believe this will unfortunately come to pass. And as it does, the privacy of information users will also diminish accordingly." —**Lisa Carr**, director of strategy, Targetbase Interactive, health care strategist and writer

"I agree that, as this issue evolves, the lawyers and other copyright holders will prevail—to the detriment of promoting and spreading their brand. A better alternative would be to allow for use, as long as users are not earning money from others' work/talent, and that users are not defaming content providers." —**Mike Driehorst**, messaging strategist; leads social media for Hanson Inc., an interactive communications and video production company in Ohio

"I hate the thought of everything being copyrighted, but money talks in government. I believe the media companies and others will push (pay) to get this kind of control in place." —**Susan Frede**, vice president for research and panel management, TNS, a global market research company

"Copyright owners will realize the promotional ability of people using their work. A whole new type of licensing will be developed where items can be used, if copyright holder gets a percentage of ownership of the work." —**Chris Myers**, webmaster at the University of Michigan

"Policing these rules would be nearly impossible." —**Don Ranly**, PhD, professor emeritus, University of Missouri School of Journalism

"By 2020, a new tech-savvy generation of executives will be running content companies. They will have learned from past examples such as what's going on in today's music business. There has to be a balance between reasonable content protection and ultimately making the content easy for customers to consume and share." —**Brian T. Nakamoto**, cofounder of MrJoy, Inc. and product-line manager for Everyone.net (a leading provider of outsourced e-mail solutions for individuals and companies around the world)

"Content controls will solidify their grip on mainstream content, but with significant caveats. It seems unlikely that large organizations like studios will regain the profits or clout they had in the past. And as DRM gets tighter, creators favoring DRM-free distribution models will also become more mainstream. The two worlds—DRM and DRM free, studio and indie—will continue to coexist, but the former's clout will ebb and the latter's will rise." —**Ivor Tossell**, technology columnist/journalist for the *Toronto Globe and Mail*, known as "the blogging journalist" and a social observer

"Advocates of freedom of information and privacy hopefully will fight against strict content control. We count more on self-regulation instruments than on control by legislation." —**Jutta Croll**, managing director, Stiftung Digitale Chancen—Digital Opportunities Foundation, Berlin, Germany, promoting and supporting access and equal opportunities for all online

"Property rights in information will be transient, and service-based revenue models will predominate." —**Ed Steinmueller**, professor, science and technology policy research, University of Sussex; researches industrial structure of high-technology industries, coevolution of technology

"International legal agreements across many nations are protracted and complex affairs. The trend, if anything, is toward freer exchanges and less regulatory control between countries." —**Tim Grafton**, market research director for UMR Research Ltd., a market research company based in New Zealand

"The media industry will eventually have to rethink copyright as the sole method of protecting and extracting value from content. Value-added options (such as higher sound quality in audio) may be a part of that equation. Other options such as tagging of content as a way of identifying the originating source will also play a role during the transition and after." —**Jasmine Sante**, Sante Strategies, independent Web strategy consultant in the Washington, DC, area

"This is possible only if we make microtransactions commonplace online and through cells—microtaxation goes with this as well." —**Jennifer Jarratt**, principal, Leading Futurists LLC; works with formalized methodologies to assess and interpret potential futures

"Reality and hope are two different things. In reality, for the bulk of the world, there will be legally mandated controls for the content 'owned' which will extend everywhere the courts and tracking can reach, for the content. However, users will be able to create their own content, and some platforms will allow/encourage this 'free' content to flourish on their delivery systems. The money will flow, then, from related advertising revenues and search (and other future revenue generating avenues to be developed). Like the speedy line at the airport screening, money will = convenience; free = inconvenient but adequate." —**Nikki Waters**, product manager, Internet Services Group, Kaiser Permanente, a medical HMO

"In 2020 DRM remains opposed only by a few. Libraries and academics remain marginal in policy discussions. Fair use has morphed into a focus on DIY content (examples from 2008: Creative Commons–licensed content, FreeSound)." —**Bryan Alexander**, director of research National Institute for Technology and Liberal Education, blogger, expert on computer-mediated pedagogy; based in Ripton, VT

"This would make 'Big Brother' a reality! I can see a person inadvertently using documents and then receiving a huge bill for the copyrights." —**Jana Vanderslice**, counselor and technoethics researcher

"Although it is still unclear which business model will support the further development of the Internet, one of the drivers will be the extended role of DRM and IP. This will be especially true for specific areas of content. Others may be more of the 'open source' and free space." —**Wim van de Donk**, professor of public administration and chairman of the Scientific Council for Government Policy in The Netherlands (WRR)

"I think this unfortunate trend will lead to a much larger pay-for-use segment of the Internet; free sites will be rarer, and users will be charged low rates for just visiting a site. Charges will be by stealth; maybe there will be a usage bill at the end of the month that is automatically deducted from a person's account." —**Ruth Martin**, National Chengchi University, Taiwan

"Information wants to be free. Copyright holders will continue to battle with bloggers and file sharers who wish to expand freedom to access information and knowledge via the Internet. I believe copyright will still exist and be honored in 2020, but new types of commerce and new commercial models will develop that will enable copyright holders to make money while users will gain freer access to music and other forms of entertainment." —**Richard Silverstein**, blogger, specializing in the U.S. and Middle East politics

"It's proven difficult to actually 'control' anything on the Internet. As an occasional creator of content, I recognize the importance of providing due compensation to copyright holders. I'm just not optimistic that whatever successes we have in setting up a system that deals fairly with content owners, as well as buyers who wish to use that content for personal use (not for profit), that piracy and other as-yet-unidentified forms of abuse won't still exist and continue to be termed a major problem." —**Jim Wiljanen**, president, Evans/Greenwood LLC, MI

"This may be more hopeful than predictive—but just as Internet 'filters' for adult content are easily circumvented by those who are supposed to benefit from them, and get in the way more often than they help, relying on DRM to enforce intellectual property is a flawed approach from both a technology and a legal perspective. Ultimately, large content owners and producers will be forced to recognize that the 'digital restrictions' approach to copyright and IP management—relying on technology to enforce legal standards—is costly, futile, and misguided." —**John Eckman**, practice director, Next Generation Internet, Optaros, Inc., a professional services firm offering strategy, design, development, and consulting services tied to open-source software

"This will probably be pretty true in the OECD countries, but I cannot see it happening in most developing countries by then." —**Joseph Straubhaar**, professor, University of Texas–Austin; research interests include international communication and cultural theory, information sciences, and the digital divide

"Probably. Inevitably. Sucks. Although I think artists/content owners should get something, the whole thing just reeks of red tape and greed." —**Virginia Bisek**, Web content developer and writer

"Copyright protection will diminish rather than increase as grassroots organizations and individuals increasingly mobilize to resist and even undermine the efforts of commercial interests to 'protect their assets.' If big business continues to place restrictions on their content, folks will move increasingly to those arenas where 'user-generated' content is being created and freely distributed, using such means as provided by the Creative Commons licenses to get around copyright restrictions. DRM will not likely be going away, but I don't think it will have the stranglehold on content that is being predicted here." —**Kerry Anderson**, library consultant for the government of Alberta, Canada

"The DRM/IP battle will be fought long into the 21st century, if not beyond. Digital content is too hard to police, and the alternatives are going to be too restrictive to implement. Incremental cooperation and implementation will be the scenario, but 'automatic notification' to authorities isn't right around the corner." —**Woody Degan**, chief executive officer and operations director, Memphis Sound Entertainment; Consumer First Consulting, IT consulting

"I hope that copyright evolves to a more open format—and that artists, not middlemen, reap the fruits of their labor. But, copyright needs to evolve to Creative Commons." —**Beth Gallaway**, Information Goddess Consulting, a Web 2.0 consultancy

"There will be a revolution of sorts before we get to better copyright controls. As the general population becomes dissatisfied with having to purchase content multiple times to use on different devices, this will push a new form of copyright management around usage. It should not matter if I want to access content on an iPod, computer, TV, etc. This would bring down the per-usage cost but could provide more revenue for the music and film business…This also makes it easier for the digital public to not buy formats multiple times (DVD, CD, digital file, etc.). Or, perhaps better explained by how if you buy it on LP, 8-track, again on cassette, and CD, and then pay again for digital, the producer will face a backlash as consumers are tired of paying for the same content on a new form of media." —**Chris Miller**, senior vice president, digital operations and new business for Element 79, an advertising agency

"Corporations and copyright holders have too much to lose; they'll do their utmost to protect their property." —**Lynn Blumenstein**, senior editor, *Library Hotline*, *Reed Business Information*

"DRM and other forms of copyright protection technologies will fail in the short and long term. The open-source concept will be the victor." —**Daniel Fisher**, lecturer

"I think the current trend is toward less, rather than more, strictures with regard to DRM. This, of course, will be predicated on individuals and companies being able to obtain an income from their work." —**Steven Hausman**, president, HausmanTech Consulting

"I think content ownership will be difficult to define given the current (2008) trend of user-generated contents. By 2020, the majority of content producers will not be the music and film business of 2007; rather, they will be from Gen Y-ers, Gen C-ers, etc." —**Philip Lu**, vice president and manager of research analysis, Wells Fargo Bank Internet Services Group; formerly a senior analyst at Gartner and a senior manager at Schwab

"There has to be effective methods for protecting intellectual property rights; however, those who attempt to maintain high historical rates based on expensive means of distribution will be marginalized as the public finds reasonably priced alternatives." —**David F. Salisbury**, associate director for science and research communications, Vanderbilt University; formerly science and technology reporter for *Christian Science Monitor*

"In 2020 copyright protection will continue to be an issue, and clients will continue to find ways to subvert the system." —**Christopher Brown**, strategist and managing editor of new media for the U.S. television program *America's Most Wanted* on FOX

"The largest barrier to this scenario that I see is the ability of the different stakeholders to agree on standards that work across different platforms." —**Gretchen Pruett**, assistant director, Georgetown Public Library

"This is, without a doubt, the direction we are heading in. However, the great strength of cross-cultural and transnational grassroots opposition from young consumers and young media producers convinces me that as this generation moves into positions of power, there will be an end to this direction of progress and, perhaps, indeed a liberalizing backlash. This should be the case by 2020." —**Francis J. L. Osborn**, futurist and activist, philosophy department, University of Wales Lampeter (formerly St. David's University College)

"Quite the opposite occurs—in a wiki world, public authorship dominates and only skilled content producers will be able to charge for their copyright title. Conversely, voluntary payment usage fees or donations will become the cultural norm—somewhat like tipping in the hospitality business." —**Cambria Ravenhill**, manager of national channel planning at TELUS Communications, IT sales executive with expertise in the IT hardware, telecom and wireless space

"The more companies try to control content, the less control they will have. The more control they are willing to give up, the more they will be able to accomplish what they need." —**Jonathan Dube**, president of Online News Association, director of digital media at CBC News, the Canadian Broadcasting Corporation, publisher of CyberJournalist.net

"Whilst the embedding of information in digital files will grow, so that authors can be attributed, the actual content itself will continue to increase exponentially. This increase in content will lead to much more diverse consumption habits, with consumers increasingly choosing homemade content over that 'officially' provided, and attempts to enforce projection of an increasingly shrinking market share will only serve to make that market share disappear even faster." —**Richard Osborne**, Web manager for the School of Education & Lifelong Learning, University of Exeter; research focus is e-learning

"There are strong moves towards this outcome now. It only makes sense that intellectual property rights will be maintained by technology, particularly when the status quo is inadequate." —**Richard Fowler**, auditor specialist, Northrop Grumman

"I am not really sure how this will play out, especially since the whole 'Net neutrality' thing is still up in the air and undecided. But I do think it will be very difficult to copyright everything. There will always be hackers to circumvent the 'rules.'" —**Rachel Kachur**, behavioral researcher, U.S. Centers for Disease Control & Prevention

"DRM is proven, by 2020, to protect the corporations at the expense of the creators. The creators of intellectual property will reach a synergy with the users of said property and will gain their protection through a true supply/demand, one-to-one relationship with their consumers. Legislators and the courts will be largely excluded from the operation of a free, global network." —**Eric Kreider**, director of Web services, the University of Akron, OH

"Regarding innovation, fledgling authors and artists might thrive, if warranted, as they make some money on their work. I trust the 2020 scenario enables those who create the materials that are automatically copyrighted to allow those materials to be shared without billing." —**Thomas Lenzo**, business and technology consultant, Thomas Lenzo Consulting

"I am with the 'protesters who rarely prevail' on this: There will be even less so-called freedom for anyone without the capital to control resources, and certainly, freedom of speech will be severely restricted as well." —**Alex Don**, linguist and educator

"The overall trend in all aspects of life is increasing government control and regulation. This will be no exception." —**John Sewart**, professor at College of San Mateo

"The Creative Commons model is a beautiful one where the creators can find more of a middle ground for sharing or charging for their work. I also hope for a model where people are supporting for their creations. I hope we can come to a place where people who create content are paid for their work and people who enjoy work will be happy to pay and support work. If everyone can get behind a micropayment model to encourage and support quality, everyone wins." —**Tiffany Shlain**, founder of the Webby Awards, cofounder of the International Academy of Digital Arts and Sciences; named one of *Newsweek*'s "Women Shaping the 21st Century"

"How depressing! Unfortunately this looks all too likely. The idea of being billed micro (I hope!) sums almost every time one accesses the Internet, and somehow keeping track of this in order to deal with probable mistakes, is both daunting and damaging to the (relatively) free and open use of the Web." —**Roderick White**, editor, *Admap* magazine, World Advertising Research Center

"Perhaps I am optimistic when I strongly disagree. The public simply will not stand for any form of DRM, which is already being abandoned by most major labels. It has become clear that technology has rendered it impossible for any company or industry to control the production and distribution of music. For better or worse, the Internet and inexpensive recording software have turned that control over to musicians and music lovers. More and more people expect free content from the Web. Search is free. Most news and entertainment are free. Indeed, nearly all Web content is free. Much to the dismay of the RIAA, people do not feel as if they are stealing when they copy music from someone else. It is nothing like slipping a CD under your coat and slinking out of a store without paying. In fact, nobody misses the music you just copied at all. Providing advertising-supported free music is the next logical step in the evolution of the Internet. For me, this is not just a theory. You will soon see it in action through an initiative I will launch in early 2009." —**Peter W. Van Ness**, president, Van Ness Group, a Web-development company; founded Personal Computer Solutions in 1983; cofounded StockPlan, Inc. and MyStockOptions.com

"I am very sad to agree with this statement, but attempts from organizations like the Electronic Frontier Foundation, libraries, and other groups have so far proven to be mostly fruitless in the face of billionaire companies and megaconglomerates. The more that content control abounds, the less likely that libraries will be able to offer material to our users, and even worse, we will be unable to preserve that content for future generations." —**Sarah Houghton-Jan**, consultant for the Infopeople Project; digital futures manager, San Jose Public Library; author of the Librarian in Black technology blog

"This will most likely be the case. The government has sided with property owners in the past and will continue to do so as a result of the lobbyists representing the industry. Unless something massive happens, people will not rally around this cause because they don't feel that they are directly affected. People don't feel that creative property is worth fighting for." —**Brian Hare**, Barco Law Library

"Unfortunately, I'll have to mostly agree with this one, but I think it will be less in the form of Orwellian surveillance and unbreakable DRM (which has proven to be extremely breakable) and more due to a convenience-of-access and price-leveling phenomenon which will bring digital content (including games, movies, music, and text) simply and affordably to most consumers. There will continue to be some 'piracy,' but working within the system will be the much simpler, and relatively painless, choice used by the majority." —**Jason Stoddard**, managing partner/strategy at Centric/Agency of Change, an interactive strategies company; he is also a popular speaker on social media and virtual worlds

"Laws will abound and lawless mercenaries will find ways around." —**Robert Grant**, chief executive officer, VoyaCare, Inc., a medical connector company

"Copyright protection and content control will only work when content is made affordable on a mass level. Otherwise there will always be a reason for subversion." —**Dan Weingrod**, vice president for digital operations, Cronin and Co.; oversees creative online initiatives for integrated marketing communications company

"Strict content controls will continue to be eroded and shown to be far more a hindrance to the profits of the content creators for many common media. Creators will increasingly use distributors to just distribute rather than completely control their creations. Profits will be generated by making content available ubiquitously and charging small amounts for the convenience and quality of the delivery of nearly any IP the user wants. In short, they will give away the product but charge a little for the delivery and make profit in volume. The delivery can be priced by the item or subscribed to in tiered levels up to being unlimited. If obtaining the IP at a small cost is easy and inexpensive enough, the motivation to steal it by bypassing the profit-generating distribution channels will be reduced so much that, for the potential pirate, it will simply be worth it to just subscribe at a nominal fee and save the hassle and risk of stealing the content. Many IP-based businesses have found that offering content by subscription offers considerable advantages, including a steady revenue stream and predictable operating costs not dependent upon the latest blockbuster to boost sales between sags." —**Shawn Kelly Apochromantic**, configurations manager, designer, technologist, futurist, General Atomics, and volunteer builder in Second Life

"All contents will be DRM free, very soon. People will pay for specific content if they find value in usages like access anywhere, on any device. I will be happy to pay $1 to view a high-definition, 3-D movie in 2020. Why DRM it if I don't have time to view it 20 times?" —**Louis Naugès**, president, Revevol, an enterprise 2.0 company with offices in France, Spain, the U.K., and the U.S.; a founder of Microcost, an IT services and hardware company based in France

"I think that there will be a revolution in copyright protection. There will be a stratification of what we call the Web. By 2020, we will have the ability to have open public information and more premium info, similar to basic cable and premium channels with the purchase of tiers of information." —**James Gorman**, principal, Working Technology Partners, a company offering technology solutions to businesses

"In the first-world nations that have adopted such enforceable and predictable controls, this is what the average consumer/user's or citizen's experience will be—but content continues to seek its own freest and least encumbered state—especially entertainment content like music, movies, and video games. There are significant numbers of hackers, data pirates, thieves, and content redistributors located in nation-states that are sympathetic to, or at least condone, such activities. There is a continuing market for pirated content and a healthy black market in China, Russia, India, Africa, the Middle East, and Latin America—outside the control of the U.S., EU, UN, etc." —**Sean Steele**, CEO and senior security consultant, infoLock Technologies; leads corporate business strategy for insider-threat management consulting services and solutions company

"A quick tour of the physical storage media for software/data in the last 30 years illustrates that this is highly unlikely. It also fights against trends in user-generated content. Finally, perhaps most importantly, the provenance data (lifecycle of data) needed is just not being captured, stored, transported, and interpreted effectively. For example, look at Phil Agre's "Living Data" in *Wired* magazine 11/1994—he was echoing similar ideas from the '60s and '70s, and we are no closer today." —**Duane Degler**, user-centered designer and strategist for

Design for Context, writer and editor for IPGems, focused on knowledge management, semantic integration and performance improvement

"There is no way to control and monitor global content. The problem is that there seems to be no way to make money from providing content." —**Nancy W. Bauer**, chief executive officer and editor-in-chief, WomenMatter, Inc.

"I mostly agree for the type of communications that you describe in this scenario. However, there are other modes of communication—research findings, for example—where there may be DRM protections to ensure proper attribution and credit but not necessarily for the purpose of generating revenue. Similarly, it will be interesting to see what new forms of expression evolve and how the older systems of attribution, credit, recognition, and so on are manifested and whether a broad economic model can be built that is analogous to open source but applicable to more diverse forms of creative expression." —**Amy Friedlander**, director of programs for the Council on Library and Information Resources, a nonprofit that services research and higher education

"I just can't see everyone getting their act together and staying ahead of the technology curve enough for such a seamless sounding setup to prevail." —**Douglas Schulz**, managing editor for online publishing, America's Health Insurance Plans; formerly a Web team director for a biotech industry organization and manager of Internet and Web services at the Council of Better Business Bureaus

"So much content will be available that those who protect it too strongly will risk that their content will not be consumed." —**Kathryn K. Goldfarb**, president, KG Communications, an independent consultancy

"While I would like to see this happen—that content providers prevail in protecting their rights and are able to see an income stream—I don't see sufficient examples that the software manufacturers and search engines are working to make this happen. Certainly, if one looks at the copyright issue surrounding Google's indexing of the world's libraries; Google's, Ditto's, and Alexa's indexing of all the JPEG files on the Web; and most recently, Microsoft's Photosynth, I fail to see adequate protection to maintain use rights. I do hope Useplus.org succeeds and

Adobe CS4 incorporates a system to reference use rights in the IPTC data. Having it automated place deposits seems too good to be true." —**Robert Visser**, owner, PageRank-SEO, a search engine optimization and Web marketing company in the Washington, DC, area

"DRM is already dying a painful death. This idea is great in theory but next to impossible to achieve in reality." —**Jay Buys**, vice president for digital development, Fleishman Hillard, an international marketing and communications company

"Concepts of open source and copyleft are based on respect rather than financial gain. Perhaps what is needed is a clearer divide between those in profit-making and those in liberality mode. In the end, the properly informed customer can choose sides." —**Elizabeth Cleary**, student, National College of Art & Design, Dublin, Ireland

"The ad-driven, open-source, everything-free aggregation model of content has become the accepted norm. Attempts to control or change that model have failed so far. But if the cannibalization of content continues, it will mean less original content and more replicated, unoriginal content with people accepting 10% of knowledge as 100% knowledge and understanding." —**Michael Castengera**, senior lecturer at the University of Georgia's Grady College and president of Media Strategies and Tactics Inc., a media consulting firm

"It will be essentially impossible to enforce national laws over the Net. Applications such as Second Life will alter the individual's relationship to government. As well, sharing amongst individuals, regardless of nationality, will continue to increase, thereby increasing the sharing of content. The old concepts of copyright are in their last days." —**Richard Hammond**, knowledge management team leader, United States Environmental Protection Agency; knowledge management expert currently examining Semantic Web and RFID

"I think that some forms of creativity would be stifled if there were no controls over copyright. Why would an artist or writer bother creating for a living without controls over their own work? This is for copyrighted material only, not everything on the Web." —**Patti Nelson**, a webmaster who works on U.S. government sites

> "'Billing' will not be according to actual access of copyrighted material but according to modeling of access behaviors. The charges will be so minimal that most people will not care about them. Let's say I spend $100 per month for various forms of Internet access. My charges for copyrighted access will be on the order of $1 if I am a moderate user. Owners of copyrighted material will be compensated out of the pool of collected fees according to the prevalence of their copyrighted material on the Internet. The compensation schedule will not be linear, but will be biased to give the small holder of copyrighted material a slight advantage. The schedule will discourage the massive buying of copyrights." —**Dixon Hutchinson**, software engineer

> "The horse has already bolted, and enforcing DRM is one way to really get up the noses of your fans and customers. Companies have to learn to trust their customers who, on the whole, spend more on content than get for free. If companies continue pursuing DRM, it will just make them look even more corporate and nasty than they do already. And anyway, the hackers will always find ways round this." —**Helen Keegan**, founder of Beep Marketing, a self-employed consultant and a judge for the Webby Awards

> "I disagree with the statement characterizing the intent of DRM efforts: They are intended to aid the reproduction of capital and not ensure creator control." —**David Hakken**, Indiana University School of Informatics and a professor of anthropology who studies social change and the use of automated information and communication technologies

ANONYMOUS COMMENTS

The following section includes a selection of contributions to the discussion of the future of the Internet and IP law and copyright protection from survey participants who chose to remain anonymous. Among them:

> "Content won't get created without an economic incentive to do so. However, users of content will not put up with severely restrictive copy protection schemes. A happy medium will be found, pleasing everyone other than those with the mantra that content must be free. There is no such thing as 'free.'"

"Information wants to be free."

"I would not be surprised that the professional intellectual property community continues to attempt to lock down its intellectual property. The solution is the Creative Commons movement. While the number of people who formally participate in the Creative Commons appears to be limited, a huge number of people create content that is informally a part of the Creative Commons. This base of material that is out there, free for anyone to use, is growing. There is WikiCommons with an excellent media resource. There is Flickr with its collection of Creative Commons photos. There are blogs and Web sites where the content may not formally be CC, but it is free to distribute, discuss, and debate. The IP community may continue to try to lock down its content, but there is an alternative that they can't lock us out of—and if they alienate the masses sufficiently, it will drive people toward the Commons."

"Digital rights management (DRM) technology doesn't work. While legal controls will, no doubt, be in place, DRM has already failed and seems unlikely to be resuscitated."

"While content owners are entitled to compensation, the scenario reflects an extreme in enforcing those rights. There must be a better balance."

"Backlash is already under way. We'll still see a mix, and people will be a lot smarter about avoiding onerous DRM."

"Yes and no—content owners will find there's more cash to be made by allowing mashups, sharing, and through other revenue sources than just direct billing. I disagree with 'Those who use copyrighted materials are automatically billed by content owners, and Internet service providers automatically notify authorities when they identify clients who try to subvert this system.' Relationships will be more blurred."

"The key issue is different media have different needs. For instance, musicians make much of their money touring, so giving their music away may actually increase their income. DRM for music may make no sense long-term, while for movies, giving away movies eats into the income source. What I cannot determine is whether we'll continue to see a one-size-fits-all DRM or whether we'll see variations by industry. If we do, then I think the scenario is far more complex than discussed here."

"DRM as we know it today will collapse. The stable mode will be where the artist and performers are paid in accordance with final audience size."

"Newer content providers will eschew DRM so as to reach more people. Hackers will screw DRM systems, too."

"The purpose of copyright was to lure creative people into the open by ensuring a decent return on their creativity, as an alternative to keeping their methods or creations secret. What you describe is almost the opposite, a perversion of what patents and copyrights were for. People will resist and evade when 'intellectual property' is used as a bludgeon for the sake of giant corporations, but they will see justice in individuals being compensated for their creativity."

"The trend is clearly against these controls, and the law can't keep up with technology. The focus will be on incentives for people to buy legitimately."

"Most business models will assume free content sponsored by advertising or some other means of generating income. Some things will be tightly protected, but fundamentally, this debate will become moot."

"Content value will crash, as public discourse becomes more valuable; copyright will be superceded by a larger intellectual property designation, which will not be available for art-related content."

"This would be like taking away existing rights: no simple matter. People are just now getting more and more used to 'free-for-all' programs and widgets, developed collaboratively; they'll extend that frame of mind to other works as well."

"The public's reaction to the RIAA's shenanigans regarding its strict Internet piracy punishments, which are already seen as ridiculous by most, will cause a revolution of sorts in the way intellectual property of all kinds is transferred."

"The complexities of enforcing copyright will likely exceed the value of the content it is trying to protect."

"Although I fear that the answer in 2020 will be mostly agree, my fervent wish is that strict content controls will not come to pass. Recent developments in Canada (see the Fair Copyright for Canada) show that there is some hope for protesters' views to be heard. Also I note that the term *protesters* in the question is used pejoratively, so the question can be seen as being biased."

"The amount of user-generated content will overwhelm efforts to control content."

"People will resist any infrastructure that is put in place to enforce this, and as quick as companies are to create encrypted mechanisms for content control, hackers are as quick to create workarounds. I believe, instead, that new models for payment will come into play that will subvert the old mechanisms."

"Improvements in advertising will allow content creators to make revenue while allowing free access to content."

"Large film projects costing hundreds of millions may never be financed in the future without DRM; however, smaller film projects may still be feasible. Consumers are likely to want a more on-demand service rather than the broadcast nature of today's media offerings, so flexible delivery/control is required; however, DRM will need to be fast changing to overcome piracy and DRM avoidance."

"As more young people move into adulthood, their assumption of free access to music/entertainment will become the norm and begin to be reinforced in legislation and policy."

"The degree of monitoring required to prevent breaches of copyright protection technology creates too many potential privacy hazards to be tolerated in most places. In particular, it would require ISPs to be able to decode encrypted transmissions, at which point personal communications, financial transactions, and every other form of online activity is no longer private."

"There will be better tracking of those using, but it will be tough to identify those who subvert the system."

"File sharing will actually increase the amount of content that will circulate without any copyright protection."

"Never! Even some of the people who are asked to protect IP, such as IT support staff at universities and lawyers, breach IP restrictions for private use because they think it's stupid. Our IT support department has a secret MP3 server, and my lawyer friend specializing in IP law rips DVDs all the time. Nobody but some weird bureaucrats actually believes that the old Hollywood IP regime will stand much longer."

"We've already seen iTunes and Amazon move to less DRM in response to the economics of consumer demand. New business models will be emerging that don't rely on the exploitation of IP and copyright."

"I have to agree with the scenario, unfortunately, but it will imply a strong, almost nondemocratic control of everything citizens will do on Internet via the Internet providers who won't have a choice but to cooperate with authorities."

"Things will become more 'open' as people realise the value of knowledge increases when shared."

"Corporate interests [will make this happen]."

"They are trying to do this, but it seems to be failing. If this does happen, the likely pathway is to install the various controls in the name of 'national security' and then extend them to a vast number of other areas, including IP control."

"The bad guys win."

"Too true! The record companies are the ones forcing these laws through, and we (the consumers) are going to suffer from their greed!"

"Not only will large companies take advantage of this automatic billing, but so will individuals, whether they be in an equivalent of YouTube or Second Life or wherever people are producing new things. The big shift here will be people will have copy protection on more things. Current social trends around creation will be moved to similar models as the big companies—so standardized. Multiple currencies will have automatic converters."

"Content will proliferate through networks and copyright will be nonexistent."

"Available technology will be able to subvert available DRM techniques for the foreseeable future. This will require an ongoing rewriting of copyright and related laws."

"In relation to this subject, one interesting notion is molecular rights management, from Jamais Cascio: http://openthefuture.com/2007/09/molecular_rights_management_1.html."

"We have to admit human nature. We can't have a good copy protection mechanism, like you can't stop viruses, spam, and even spyware."

"This is forcing an out-of-date business model onto a new technology, but it depends on whether someone can come up with something more compelling. I suspect most will come from subscription services...Alternatively, there will be an agent function: I pay for what I access, but again, the processing is done by the middleman. But there will be a battle for access to the customer...And that will skew everything. As for free speech and innovation: I don't buy this one! Web 2.0 makes it easier for individuals to carve out a space and communicate to limited or global audiences. It will get easier. The threat is whether people actually use it."

"Copyright enactment for products that can be copied easily like music or movies is doomed to fail. A copyright tax is more likely."

"Copy protection schemes will be mostly defeated by market demand and technology. Consumers want to move their music and video to their tablets, PCs, TV without hindrance. As well, there will be an explosion of user-generated video and music content—even full-length feature movies."

"Unfortunately, this will likely come to pass unless there is an unlikely turnover in law."

"The public revolt, in the form of 'voting with your dollars,' will force companies to back down, and will result in changes in laws and in legislatures, too, if the rascals have to be thrown out."

"Much of this drive is coming from American companies, and resistance to the invasion of privacy represented by these technologies is widespread and deep. The impending change of political alignment in the U.S. may have a positive outcome on this development."

"This model is too unbalanced to be stable for very long. The question becomes simply, 'Who will be destroyed in the battle?' as in, 'Will individual careers be cut short?' or, 'Will whole nations falter because of their commitment to such a one-sided deal?'"

"DRM will dominate certain sectors, but user-created content will dominate overall."

"Copyright protection will apply to a large amount of content, but will not dominate as derivations and expansion of 'open-source' and virtual communities grow."

"Let's hope this world, which some are intent on seeing become a reality, is overtaken by an uprising of advocates for consumer choice. I would guess that by 2010 or 2012, there are enough cases of consumer DRM disasters (not being able to transfer the content for which they paid to a new device when the old one becomes obsolete) that policy makers step in. In addition, business models will also demonstrate that DRM is not a market accelerator but a content market inhibitor."

"A much more likely scenario is managed chaos in which the most egregious violators are prosecuted, but the majority of small-scale, relatively innocuous (read: nonprofit) copyright violations continue unabated."

"Yes, such a system will exist, and may even function well. Piracy in some form or another also will exist, and it will be a pain in the neck. This battle will take place across the globe and will never be won by one side or another."

"This simply would not be tolerated. Brilliant minds will always collaborate to circumvent such technologies."

"Don't see how this type of control can be implemented given the shifting technology landscape."

"The interests of the users of material are diffuse while the interests of the owners are concentrated. This asymmetry ensures the owners will prevail, but less smoothly than the [survey scenario] suggests."

"Wishful thinking—the thieves are going to win this one, the rest of us will lose."

"If this prediction does come through, then I see a techie rebellion that could be damaging to the network itself. Bottom line is that there will always be ways to get around DRM, and the hacker community will continue to work ways around DRM and IT systems."

"Companies hoping for this scenario will vanish, or at least shrink dramatically. New business models will evolve, and the major players may not be the same old, same old."

"Due to developing countries being so far behind in technology, in value placed on copyright protection, etc., it will take more than 12 years to reach this state."

"DMCA clones have failed in the EU and Canada due to popular resistance after observing the effects of the U.S. law. Legal business models that make media consumption ubiquitous, cheap, and fashionable will succeed. When most consumers use legit services, the die-hard pirates will be easier to prosecute. Hollywood and especially the music industries will have to adapt to lower profits, but they will survive. Some DRM may linger on closed, specialized systems like videogame consoles. Finally, there will be robust laws passed to protect remixes, mashups, and fan-created media."

"If the entertainment industry were to ever become this restrictive, a grassroots production system would develop that would bypass such excessive safe guards. One of the early indicators that Microsoft's Vista operating system would struggle, in my opinion, was the outcry on many technology blogs about its excessive use of DRM in the design of the OS."

"There's going to be a big battle coming, and there's no clear sign of who's going to win yet—but consumers have a powerful voice, when they wish to use it, and the generation that's growing is also growing awful sick of being told how they can use items they paid for."

"This is one area where I expect open source and the sheer numbers of uncooperatives to win."

"Open source is becoming comfortable for more people, and will become increasingly so. People are unhappy with DRM and are working to avoid it, and content providers will have to accept this and work with the increased open access."

"People will stop using systems that enforce restrictive DRM. The current trend (Amazon MP3 downloads, for example) points towards an overall loosening of DRM and allowing consumers to use their goods in more reasonable ways. This will especially be important to consumers as they come to have more and more different machines (iPods, computers, phones) on which they want to use the same files."

"Unfortunately, I agree. I wish instead that Amazon and iTunes would partner with public and academic libraries so that library patrons could use iPods (and Apple computers) and Kindles with their public library cards. Amazon and Apple should also work with subscription database vendors like Ebsco and Gale to make browsing databases easy on iPods and Kindles."

"Movement is the other way, and rightly so."

"When most of the users of the Internet are below the poverty line, does anyone really think Disney is going to be able to bill someone in Bangladesh every time an image of Mickey Mouse is displayed?"

"It is really hard to imagine such things would be happening 20 years later in countries like China."

"Companies such as Apple have trained us to sip. Of course, this will also kill libraries as soon as e-book technology hits its prime."

"Money talks, and the rights of consumers have routinely been overlooked in favor of those claiming copyright."

"Copyright protection was invented for print. It is going the way of the dinosaur."

"This assumption does not take sufficient account of the growth of the open-source movement and people's willingness to share materials they themselves control. Commercial copyright owners ARE likely to do what they can to continue to assert ownership and protect their intellectual property; but content will represent an increasingly smaller percentage of overall creative content. Evidence: lifting of DRM controls on some commercial sites, increasing number of open-content journals succeeding, maturation of Creative Commons."

"The amount of money invested in protecting existing business models is bigger than the amounts invested in developing alternatives. The IP lobby is very well funded and protects established business models."

"This seems to be the overall direction of intellectual property in general, even more so vis-à-vis the Net. While the peer-to-peer networks will continue to play a role for quite sometime, they will slowly disappear yet remain key players in the developing world, where infringements are harder to enforce."

"We will see more legislation, but enforcement will be challenging, especially in countries that do not adhere to international copyright agreements."

"The 2020 scenario sounds like a plausible solution to the problem, but difficult to implement."

"Information wants to be free—money will be made from ads—regulation of this type will hurt everyone."

"The ability to verify plagiarism will be instantaneous with all published works accessible digitally. Software to scan the Web will be used regularly by anyone who publishes anything as a natural part of doing business."

"Business will drive content protection. Government will be reluctant to interfere. ISPs will not be able to see violations and will not be a part of the content value chain."

"We are already seeing young artists bypassing firms and their expensive advertising agents and going directly to the consumer via the Internet. The huge profits for many people will probably begin to diminish."

"People will always find a way to keep the Internet free. In that light, there will be protections on intellectual property, but not exactly restrictions, as long as people continue to expand their use of the Internet for sharing ideas."

"Technology will not rule the intellectual property/digital rights management. Creative Commons will become more common, duration of copyright shorter (from 70 years now to 20 years)."

"Intellectual property concept will change."

"This push for a more 'totalitarian' control of IP will ultimately have a negative effect. I can see DRM acceptable for people when it is not intrusive and invisible. This means that those scenarios listed would not happen because people would have accepted it…I also don't see ISPs ever becoming the gatekeepers. This requires them to monitor all traffic at all times and makes THEM liable. Who would allow themselves to become liable to a trillion-dollar industry?"

"For every protection, there will be challengers to break it. Creators don't control the content, publishers and corporations do. Creators will always have problems with the restriction."

"Technology and copyright need to meet somewhere, and technological perpetuation of the existing system is unlikely. Encryption is going to make the enforcement scenario here very unlikely."

"DRM seems to be dying a deserved death, replaced by some form of Creative Commons licensing."

"This may be true in the United States, Australia, and Europe, but much less likely in other parts of the world."

"The ad model is in. The *New York Times* is taking the lead. Other newspapers will either follow their lead or lose revenue."

"We're already well on our way to this happening."

"The Creative Commons movement will be significant."

"Things will go on much as today with a balance between copyright protection and libertarianism on the Net."

"Technology used to enforce IP protections is so easily and often circumvented that it becomes impossible to enforce on a worldwide scale. Companies also begin to realize that it is far more cost-effective to support open standards and platforms in software and that freely available content simply generates more buzz about the artists' other material. Pay-access to services is likely to be more viable than access to data/content."

"We will see a growing open-content movement and models of content control that are more directly overseen by the actual creators of the content."

"If the above scenario dominates, intelligent, progressive, free people will not be using the offered devices in the 'approved' way, or true creativity will be stifled and distributed almost exclusively underground."

"A hybrid of some free access with more options for a fee seems more likely than complete control."

"Despite the efforts of many, the old adage that 'money talks' still holds true—especially in politics. As long as this is the case, media companies will get the laws and regulations they desire."

"Your fully operational scenario actually is pretty grim. I can't see it actually happening because too many people want too many different things. That's all for the best."

"Two systems will develop. One will be controlled by copyright protection technology. The other will be self-generated. Open-source software and Creative Commons licensing are examples."

"The copyright model won't be the primary way to deal with intellectual property."

"Users will not tolerate this approach and will demand greater ability to manipulate and use content as they desire. Change is coming…"

"This is so Orwellian, it frightens me. I imagine us monitoring things more closely (look at the U.K. and CCTV [closed-circuit security camera systems]) but find it difficult to see this sort of protectionism in place."

"Fair use for education is being restricted, and the balance is tipping in favor of copyright holders. This will greatly stifle creativity and innovation."

"The powerful recording and publishing industries will pressure Congress to support the interests of their businesses. Innovation will be confined to big businesses."

"Freedom, connection, and collaboration will rule."

"There will still be a lot of outlaws."

"The recent [2008 U.S. television] writers' strike reinforces what is at stake for businesses and creators. Money. The efforts to limit 'fair use' in the area of education will continue."

"Money is the very simple reason that copyright content will be more closely monitored."

"Eventually, notions of IP ownership will shift as ever more noninstitutionally created content floods the world. Individual 'amateur' creators will have less sense of ownership of their work and, indeed, will welcome its wide reuse."

"The DRM issue will be resolved as legislatures finally look at how digital media differs from others. Laws will change to reflect the reality of 21st century copyright. But things will get worse before they get better."

"There must be a way to stop the privateering that is going on, particularly in China. All the legislation outside China will not stop that."

"The protectionists will have a slight lead on this issue, but both sides will prevail."

"Compensation for intangible products will have to be achieved by some sort of uniform tax or other allocated charge."

"Copyright does not protect the authors, but only mercantile interests, and it damages the poorest countries/peoples."

"Agreement in this case does *not* signify…well, agreement. Unfortunately, I think this is how it *will* be."

"Methods for micropayments will increase availability to those who can afford it and much more adequately remunerate producers while liberating information to those who cannot afford to pay."

"Free speech has historically been messy and will continue to be so."

"If, at the same time, the 'greed factor' can be controlled so that there is a sense of equity both on the part of content providers and content users, this could be possible."

"This draconian view is one of two possible outcomes. If public outcry is sufficient and networked 'sales' mechanisms (think Radiohead's *In Rainbows*) show sufficient velocity, then we may see the opposite; a few content dinosaurs grasping at the last vestiges of the empire they once thought they controlled."

"The key issue is basic economics...The vast majority of information on the Internet is already free, and as copyright protections expire, the proportion will continue to grow. And, of course, there will always be those who don't want to pay—and as one of the founding fathers of science fiction wrote, what science can create, science can circumvent."

"I can't envision Internet service providers taking on the policing role and notifying authorities of violations."

"There will be lots of new licensing schemes available to make IP widely accessible, and many writers, artists, and other creators will distribute content freely for mashup purposes."

"Protesters, geeks will prevail."

"Open-source and social media have demonstrated the power and willingness of people to provide free content. The Creative Commons tagging will be more popular and CC terms need to be revised."

"Removing individual choice will force people into seeking other ways to share content."

"It is impossible to execute this scenario. Fastest Internet connection (most wanted) relies on peer-to-peer or mesh organization. Once something is published on the Internet, it becomes public property, for all practical purposes. There is no way to bill for use of this, no effective way that can keep people from circumventing any copy protection. Internet providers share in this, since they deal with the free-market wishes of their clients, who want their privacy and will only pay for 'first rights' to something, not residual payments. See Archives.org."

"We will move from an era of highly protected intellectual capital to one of shared and open access. It can't be any other way, and organizations will be less challenged to monetize the new way than they currently think."

"This may become a reality in Western countries but not worldwide. Cultural differences of copyright protection are too large. In academia, the 'open-access' movement will continue be strong."

"A new business model will emerge, one that is better adapted to the variety of media/data sources available in a new digital world."

"Years from now, children will sit at their grandparents' feet and hear tales of how wide-open content sharing was in the old days."

"Court systems, responding to the growing caseloads of unfounded litigious allegations, are becoming overwhelmed with time-consuming work that prevents them with dealing with the real problems of our society."

"Content creators will maintain ownership and rights for their content, but will find alternative ways to benefit from broader distribution rather than solely focusing on stricter controls."

"The Internet is too vast to be under such control."

"Some offerings of music without DRM and open-access archives offer hope that copyright protection technology will not dominate."

"The mass of material on the Internet is too great. Copyrighted information will be swamped by alternative sources of material."

"Ironically, China may become the best example of true free enterprise with free entry and exit."

"There are too many political entities involved to allow this [scenario] to be effective."

"The benefits of free information on our economy and the general well-being of society will greatly outweigh any rights to proprietary information. Companies that own copyrights will be forced to find a system to monetize their content while promoting its free flow through the Internet (i.e., product placement for embedded ads, etc.)."

"Creative individuals will continue to want to protect content and get paid for that which is used and enjoyed by others."

"I mostly agree with the scenario, with the caveat that people will retain some 'fair use' rights, for instance having the ability to transfer a song from computer to car or a movie from TV to TV."

"Two parallel streams will develop: (1) IP holders who wish to maintain full control of their IP will find ways to distribute the content at lower cost or for free, monetizing it via advertising or other methods that benefit them. (2) Open source and GNU [a project enabling the free and open software movement] will become the predominant source for content, spurred on by social network sites and a new generation believing knowledge should be owned by the human race, not commercial entities."

"Have you read *The Starfish and the Spider*? I suspect that although copyright protection will become a dominant theme, those pesky starfish will be hard to corral."

"This is the best scenario for copyright management agencies. Such a brave new world may exist, but there will be parallel worlds enjoying free access and/or no robot police."

"The advancement in technological development will continue apace as before, and the legislatures and courts around the world will continue to walk many steps behind, trying to catch up with their nation's technological development."

"It is difficult to think of a scenario that would see some of the extreme powers granted by the U.S. (and many other) governments being clawed back. Furthermore, the technology convergence has seemed to also spur the horizontal mergers of content providers, facilities-based network providers, and large IT/CE into supermedia firms, who will use their power and weight to ensure the digital genie does not get out of the bottle."

"The man always wants to protect the bottom line."

"At the same time we see corporations solidifying controls on digital content, I am encouraged that the general public is willing to share, as seen on YouTube and Flickr."

"The growth of high-quality free content will probably apply significant pressure against the value of protectionist scenarios."

"Those who produce the materials should have a right to protect their utilization."

"Attempts from organizations like the Electronic Frontier Foundation, libraries, and other groups have so far proven to be mostly fruitless in the face of billionaire companies and megaconglomerates."

"Even though these industries are lobbying and pushing hard, they will have to develop other business models. The younger generations don't want to pay and will not pay, and Internet service providers are currently unwilling to be the watchdog agencies."

"Content producers will market more directly to their clients without the need for the organizations that drive up the cost."

"'Information wants to be free' advocates have lost the head start they had from major corporations' slow adoption of the Internet. The Internet will continue to be more and more fully integrated into the business and intellectual property practices of corporations."

"The value of digital goods and the cost of their distribution will not be high enough to sustain such DRM-dominated world…DRM will always be broken as a technology due to its fundamental structure, i.e., your customer is the attacker. The role of DRM will be limited…The process will be painful due to the slow understanding of the changing market and economics by the media industry."

"The more strictly companies try to control their material, the more diligently people will work to pirate it; the more narrowly 'fair use' is defined by the courts, the more some people will feel it is actually their moral duty to subvert the copyright laws."

"Access to larger markets and micropayments represent a more efficient model."

"The extremely arcane legal system surrounding copyrights has a lot of cleaning up to do in the next 12 years!"

"In 2020 the dominance and influence of big media companies will be diminished…content that is free and grassroots will play an increasingly important role."

"The big businesses behind content production will win over the free spirit of many in the Internet world."

"ISPs would fight tooth and nail any requirement that they be held responsible for policing their networks."

"This will lead to user profiling and potential targeted marketing."

"If they insist on prosecution and legal punishment as their only course of action, they will eventually destroy their own market of consumers."

"Most people understand that information is power. With so many pulls on our attention, it's clear that those who give away content gain the ability wield significant influence over behavior."

"Content owners will offer affiliate programs through which end users will generate and split ad revenue when displaying copyrighted content. Everybody will become a content distributor and share the ad revenue with content owners."

"Ideas will continue to be the currency that drives economies. IP owners will always need to balance control with viral, word-of-mouth distribution and buzz. I don't think you're going to end up with the draconian controls you're talking about; creators (the IP owners) understand they need 'earned' distribution."

"The current system cannot be sustained because it does not have public support. In the future, consumers will pay a fee or tax for 'copyright' technologies, artists will make their money as they did in times past—from performances, not media sales."

"Copyright will only get looser, with maybe fee structures shifting to pay authors more up front. Copyright *tracking*, however, will improve."

"There has never, ever been any example of media that was not copied, stolen, or forged."

"Despite the power and money of the information giants, grassroots, bottom-up technologies and systems will prevail."

"People will want to be able to enjoy their music and movies using whatever platform they choose, without technological hurdles. The resulting economic force in the entertainment market will engender a new paradigm for handling copyright."

"Production and distribution costs are close to zero, there's no reason to give old-style media companies a slice of the action."

ENDNOTES

1. From "The Economy of Ideas: A Framework for Patents and Copyrights in the Digital Age (Everything You Know About Intellectual Property Is Wrong)," by John Perry Barlow, published in *Wired*, March 1994, available at http://www.wired.com/wired/archive/2.03/economy.ideas_pr.html.
2. See *The Future of the Internet—And How to Stop It*, by Jonathan Zittrain (New Haven, CT: Yale University Press, 2008), available at http://futureoftheinternet.org/.
3. See U.S. House of Representatives Resolution 4279, December 5, 2007, available at http://thomas.loc.gov/home/gpoxmlc110/h4279_ih.xml.
4. See *Privacy and Human Rights 2006 Executive Summary*, by Electronic Privacy Information Center and Privacy International, December 2007, available at http://www.privacyinternational.org/article.shtml?cmd%5B347%5D=x-347-559060.
5. See the Digital Millennium Copyright Act of 1998, published by the U.S. Copyright Office Summary, December 1998, available at http://www.copyright.gov/legislation/dmca.pdf.

6. See *Privacy and Human Rights 2006 Executive Summary*, by Electronic Privacy Information Center and Privacy International, December 2007, available at http://www.privacyinternational.org/article.shtml?cmd%5B347%5D=x-347-559060.
7. See *Fair Use in the U.S. Economy*, published by the Computer & Communications Industry Association, September 12, 2007, available at http://www.ccianet.org/artmanager/uploads/1/FairUseStudy-Sep12.pdf.
8. See the Creative Commons Web site, available at http://creativecommons.org/.
9. See *Fair Use in the U.S. Economy*, published by the Computer & Communications Industry Association, September 12, 2007, available at http://www.ccianet.org/artmanager/uploads/1/FairUseStudy-Sep12.pdf.

Scenario 4

The Evolution of Privacy, Identity, and Forgiveness

Prediction: *Transparency heightens individual integrity and forgiveness. In 2020 people are even more open to sharing personal information, opinions, and emotions than they are now. The public's notion of privacy has changed. People are generally comfortable exchanging the benefits of anonymity for the benefits they perceive in the data being shared by other people and organizations. As people's lives have become more transparent, they have become more responsible for their own actions and more forgiving of the sometimes unethical pasts of others. Being "outed" for some past indiscretion in a YouTube video or other pervasive media form no longer does as much damage as it did back in the first decade of the 21st century. Carefully investigated reputation corrections and clarifications are a popular daily feature of major media outlets' online sites.*

Expert Respondents' Reactions	
Mostly Agree	45%
Mostly Disagree	44%
Did Not Respond	11%

All Respondents' Reactions	
Mostly Agree	44%
Mostly Disagree	45%
Did Not Respond	10%

Note. Since results are based on a nonrandom sample, a margin of error cannot be computed. The "prediction" was composed to elicit responses and is not a formal forecast.

Respondents were presented with a brief set of information outlining the status quo of the issue 2007 that prefaced this scenario. It read:

> People openly share more intimate details of their lives online every day, and they are flocking to social networks and uploading and/or viewing homemade videos by the millions. Ubiquitous computing is diffusing into everyday life. Much of what goes on in daily life is more visible—more transparent—and personal data of every variety is being put on display, tracked, tagged, and added to databases. The number of mobile camera phones in use will top 1 billion in 2007; miniaturized surveillance cameras are simultaneously becoming extremely inexpensive, sophisticated, and pervasive; clothing is being designed with technology woven into the fabric; and it is expected that most surfaces can and will be used as two-way interfaces in the future.

OVERVIEW OF RESPONDENTS' REACTIONS: The comments supplied by respondents, who split their vote evenly, were widely varied. Some noted that transparency is an unstoppable force that has positive and negative impacts. Some said more transparency might somehow influence people to live lives in which integrity and forgiveness are more likely. Some said that there is just as much chance it will not have any positive influence, in fact, it makes everyone vulnerable, and bad things will happen because of it. They noted that the concept of "privacy" is changing, and many commented that it is becoming scarce. Some expressed the hope that privacy will be protected, and many expressed the fear that it is threatened by emerging innovations. Tracking and databasing of citizens and consumers are expected to be ubiquitous. Reputation

maintenance and repair will be required. Some people will have multiple digital identities; some people will withdraw from a world in which surveillance and exploitation is so easy.

The response to this scenario was evenly divided between mostly agree and mostly disagree. At least part of this reaction is due to the variety of issues the scenario encompasses; the multiple layers inspired a bounty of thoughtful insights that provide a wealth of telling detail about our times and our expectations for the times to come.

"Poignant personal stories about the erosion of privacy are already starting to creep into the media," wrote an anonymous respondent. "These include such simple things as the 'outing' by the street-level photography in Google Maps of a man who had been trying to conceal from his family the fact that he had not given up smoking. As similar news reports become more prevalent and focus on cases of more serious damage resulting from the assault on privacy, there will be a public backlash that starts to reverse the culture of careless disregard for personal privacy that exists at present. At the same time, I do expect people to become more forgiving of people whose indiscretions have been 'outed' by YouTube and similar phenomena. Given that reputation considerations depend strongly on nuance, and given how sloppy journalism often is at getting even the facts right, I do not think that published 'reputation corrections' will ever be an effective way to reverse reputational damage."

William Winton, product manager for digital media for 1105 Government Information Group, wrote, "To be certain, social mores change; human nature does not. By making every action public, we open ourselves up to scrutiny that, using more measured judgment, we might not desire, either as individuals or as a public. Humanity, perhaps, is not as evolved as our conceits would have us think. While there is private behavior that befits public scrutiny (there always is), there is a great deal that does not. To make everything 'transparent' is to lay bare our own shortcomings. Does this humanize us or make us ever more vulnerable to ill-considered attack, calumny, or worse? Will this discourage future potential leaders who may be fully qualified in every respect but feel restrained by past behavior that might come to light? Are we to be exposed as being 'all-too-human,' or taken to task? *Ecce homo*?"

A number of respondents noted a generational divide, among them **Alex Don**, linguist and educator, who wrote, "This is not a world in which I would be comfortable living. The younger generation however, having grown up with these cultural backdrops, will adapt fairly well to this type of scenario, or they will not be able to partake of their brave new world."

Jerry Michalski, founder and president of Sociate, wrote, "Gen Y has a new notion of privacy. The old 'never trust anyone over 30' will turn into 'never trust anyone who doesn't have embarrassing stuff online.'"

And **Lynn Blumenstein**, senior editor for *Library Hotline*, *Reed Business Information*, commented, "A significant minority…will opt out of the transparency scenario, which will remain the domain of the young."

It must be noted that a vast majority of the respondents to this survey are not of the "digital generation"; they are over 30 and thus may not have the same sensibilities in regard to this question as those who actively participate in emerging online communications forms of all types. Age differences are a probable influence on the quantitative results on this survey question.

Many respondents said the pendulum of people's trust in one another will swing from more to less.

"New innovations come in and sometimes become major tidal waves of change," explained **Walt Dickie**, executive vice president and chief technology officer for C&R Research. "But they tend to be overplayed and soon their internal contradictions and dysfunctional, overzealous applications become clear. Then there's a pullback, and the change is integrated more sensibly into the culture. Thesis/antithesis/synthesis, remember?"

Peter Kim, senior analyst for Forrester Research, responded, "Although society will seem more transparent, most people will guard many private aspects of their lives with great tenacity."

ICANN board member **Roberto Gaetano** said there will be a mixed future in regard to transparency. "We will probably have a distinction between 'public' people, who will be exposed more and more to openness and transparency, and will consider that a necessary condition for being a public person, and 'normal' people, who will have more the tendency to hide in anonymity," he wrote. "The pressure for transparency in public people will come from different pressures. For politicians, for instance, it will be considered a prerequisite for office. But the people who do not have the need for divulging personal information will develop even more fear than they have today that private information might be used by wrongdoers."

Roderick White, editor of *Admap* magazine, summed up the position of many respondents when he wrote, "Obviously, there are two possible views of how this will develop. At present, there is clearly a developing backlash against the exploitation by third parties (from insurers to recruiters to sexual predators to all-purpose criminals) of such transparency as already exists. Given the evident desire of a large proportion of humankind for 5 minutes of fame, it may well be that we do all come to wear our hearts on our home pages, but the potential downside is there, and it should only take a few major scandals to change this climate. I'd say the jury was out, and the prospects pretty evenly balanced."

"As author of *The Transparent Society*, I agree that this is the best of many difficult possibilities. The alternatives are far worse. We must adapt. In an open world, at least we'll be free," wrote futurist and writer **David Brin**.

Transparency Makes Everyone More Vulnerable, and Technology Will Not Change Human Nature

Many of the respondents who did not agree with the scenario took a dim view of the future framed by their present paradigm. **Marco Rivera**, an Internet specialist for Vistronix, Inc., an information-management firm, wrote, "Ubiquitous computing (UC) does not change human nature. While I'd like to believe that most people will use UC to create a more open and 'forgiving' society, there are always those who will use it to substantiate, defend, and evangelize their particular bias. UC will reenforce ancient hatreds and may even radicalize those who, in past times, would have been uncommitted and unconcerned."

Jim Horning, chief scientist for information systems security for SPARTA, Inc., a former fellow at Xerox's Palo Alto Research Center, wrote, "Yes, there will be a lot more information about a lot more people readily accessible to a lot more people, but inequality will continue, and those with the most power will have the greatest influence on what will receive widespread attention and what will quietly disappear from view. Character assassination will continue to be a blood sport, now carried out on a global scale. The division of society into mutually distrustful enclaves, each taking seriously only what appears in media it trusts will enhance neither integrity nor forgiveness."

Frank Thomas, a respondent who chose not to share his place of employment, wrote that the scenario does not take cultural differences under consideration. "In 2020 the majority of global Internet users will live in China, India, Indonesia, and other Asian countries with a completely different culture of shame and of identity," he responded. "The scenario also implies that the trend towards increased transparency will continue without limits. The massive identity frauds that become more and more common will make people more hesitant in publishing (real) individual information on the Internet. As people can play with multiple identities, a large overload of fake information mixed with genuine will limit the trend towards transparency. So, in 2020 there will be an Internet world with a heightened transparency, where fake and genuine information is mixed and another one with restricted transparency. Concerning forgiveness, this has nothing to do with technology but with cultural values." And an anonymous correspondent noted, "The U.S.-specific premise that privacy can be traded won't necessarily travel."

"Viciousness will prevail over civility, fraternity, and tolerance as a general rule, despite the build-up of pockets or groups ruled by these virtues," wrote **Alejandro Pisanty**, ICANN and Internet Society leader and director of computer services at Universidad Nacional Autónoma de México. "Software will be unable to stop deeper and more hard-hitting intrusions into intimacy and privacy, and these will continue to happen."

An anonymous survey participant responded, "I'd give the following scenario instead: The loss of privacy leads to more identity theft. The increased retention of personal data in government and corporate databases leads to more leaks and abuse of that information. Placing more personal information online will expose individuals to anonymous attacks from those who disagree with them."

John Jobst, an IT specialist for the U.S. Army Corps of Engineers, commented, "People are going to realize that their privacy is becoming nonexistent and resent the intrusions. Personal tabloid journalism will be so prevalent that reputation corrections and clarifications will be almost impossible to make. As more people try to hide in the corner to prevent the public spotlight from shining on them, forgiveness will shrink and intolerance will grow."

Mack Rhoades Jr., Web services product manager for Michael Baker Corp., projects that more people will feel the need to hide their identities. "People will be less open as more private sector or government intrusion occurs," he predicted. "Being 'outed' causes people to become less transparent and take more measures to hide or protect their identities."

Nancy W. Bauer, CEO and editor-in-chief of WomenMatter, Inc., noted, "People are learning the hard way that everything they say or show electronically will never disappear—and will never be forgiven. This is already the case. Nothing disappears."

Paul Jones, director of ibiblio.org at the University of North Carolina–Chapel Hill, commented, "We all yearn for the idea of the village or the small town until we feel how they work to stifle individuality. Transparency will be painful and asymmetric. So yes, more sharing and more knowing, but forgiving? The small-town accommodation might be made, but not without costs and sanctions."

Benjamin Ben-Baruch, senior market intelligence consultant and applied sociologist for Aquent, wrote, "Privacy will become increasingly compromised and increasingly important. People will pay a premium for services that limit practicable access to so-called 'public' information about them, and an underground will be created where people can try to hide from being surveilled and recorded. Organized crime will attempt to forge identities, mask identities, corrupt data about individuals, and sabotage databases of private information. Increasingly, there will be a gap between those who are protected from surveillance and from having private information exposed and those who lack privacy."

Several respondents noted that high-profile people are likely to continue to be the most exposed.

"The pendulum will not swing back to forgiveness by 2020," wrote **James Gorman**, principal of Working Technology Partners. "It will take more time. By 2020, the pervasive media will have taken down a presidential candidate and other high-ranking officials and the older (born prior to 1965) American public will have to mostly pass on until there is forgiveness described herein."

Brad Templeton, chairman of the Electronic Frontier Foundation, responded, "I disagree that the public will become that much more forgiving. Worse, there will be sins defined in the future that most people are not aware are sins today, and the records of those sins will come back to haunt the future as better AI-enabled search technology finds them."

Catherine A. Fitzpatrick, an expert on humanitarian issues with the Open Society Institute and Physicians for Human Rights, commented, "Far from leading people to become more human and more forgiving, the 'always-on' exposure of the Internet and aggressive data scraping by the IT industry will lead to more and more forms of escaping responsibility through subverting identity and the use of hacking and anonymous avatars and such, and will also lead people to become more and more conformist and tribalist and fearful of the opinion of the mob online. The new media will become more and more intrusive and aggressive, more and more unforgiving, and there will be a backlash by the rich, the famous, and the criminal to find ever-new ways of hiding or confusing this aggressive new power. The noise of a million confidences blaring all the time will drown out the meaning."

Social media researcher **danah boyd** called the survey's scenario "wonderful science fiction but dreadful social science prediction," writing, "There are two populations that most users want to avoid at all costs: those who hold power over them (parents, teachers, bosses, governments, etc.) and those who want to prey on them (corporations, marketing firms, bullies, etc.). We are going to see a lot of chaos around privacy in the next 13 years, yet I don't think that we will have equilibrium by then. Realistically, the only comfort we will reach will be over embarrassing material. I think that we'll be far less embarrassed by our pasts once everyone's are out there in some form or another. My prediction is that we will find ways of using content to talk at different levels, just as writers have in the past and just as Chinese activists do now. Much of the 'private' content will be produced in a way that is publicly palatable and can be read at multiple levels by those who are closer to the individual. Already, this is what teens are doing with their SNSes (while they are also trying to restrict access using whatever means are available)."

An anonymous respondent noted that global politics will influence outcomes, writing, "Ecological problems, conflict over fuel, food, and water supplies will nurture tribalism and extreme politics. The Web will facilitate this factionalism."

And **Nick Dearden**, campaigns manager for Amnesty International, wrote, "There is a rapidly expanding trend for the Internet to be used by governments and companies to exert control over what individuals can and cannot say, and the ways in which they can use the Internet. In more repressive countries, anonymity and privacy are the key ingredients in creating an Internet useful in the battle for expanding rights and social change. As the desire and ability to control the Internet spreads, privacy is likely to become more important in more countries."

Transparency, Along With Its Associated Positives and Negatives, Is an Unstoppable Force

The respondents who mostly agreed with the scenario were those who expected that a future that is more transparent will be one in which people are more likely to cut each other some slack. "Web 2.0 is all about transparency," wrote **Gerard LaFond**, founder of red TANGENT, a marketing agency. "When we hit that tipping point, where there are more people online participating in social networks and sharing personal information, then privacy no longer matters. This is a scary proposition, but it's already happening. The good news is this creates all new social mores and fosters a new order of morality." An anonymous respondent referred to a popular Hollywood film about a man whose every moment—unbeknownst to him—is a reality TV program, commenting, "*The Truman Show* will become as daily life to all. We'll get on with living instead of protecting everyone's collective a$$." And another anonymous participant responded, "Those who maintain a wall will always be suspect."

Jeff Jarvis, Buzzmachine.com blogger and professor at City University of New York Graduate School of Journalism, said the issue is not privacy; it is about control. "The digital generation realizes that one cannot make connections with people without giving up something of oneself—you can't meet skiers until you reveal that you ski," he explained. "We will enter a time of mutually assured humiliation; we all live in glass houses. That will be positive for tolerance and understanding, but—even more important—I believe that young people will not lose touch with their friends as my generation did and that realization of permanence in relationships could—or should—lead to more care in those relationships."

"In 2020 privacy will have merged as a best friend issue, where you tell the world what previous generations told their very best friends," wrote **Stan Felder**, CEO of Felder Communications, a marketing company.

Clement Chau, manager for the Developmental Technologies Research Group at Tufts University, commented, "Transparency in people's identity will bring people together closer in 2020. Rather than struggling between public disclosure and privacy, people will leverage the power of the Internet and other social networking media to form their own identities. People will assume that you know who they are and who they want to be. We will fully understand that we all have different 'selves' that we affiliate with different social-cultural groups. As a result, action will be valued much more than first impressions."

Mary Ann Allison, principal of The Allison Group, noted, "The past becomes less important in a society which is now and future oriented. Repressive control continues to diminish, not always for normative reasons...but also for practical reasons." **Virginia Bisek**, Web content developer and writer, celebrates the idea of transparency, writing, "Anonymity has provided a safe haven for cowards and ignorants. Although this reeks of loss of privacy, the good outweighs the bad. Yes, people will pause before shouting or doing something stupid. We can only dream."

Some respondents shared the expectation that repeated "outings" of people's previous indiscretions will make their errors seem less egregious. "When we *all* have skeletons in our cupboards, having a skeleton in your cupboard won't matter," wrote one anonymous respondent. "Time dulls all outrages," wrote **John Jordan**, an associate professor of communications at the University of Wisconsin–Milwaukee. "A newly minted teacher applying for a job at a children's school may find it difficult to explain away Flickr photos and YouTube videos of wild partying from just a few months ago. But an older teacher who has a wilder side exposed from 20 years ago likely will not have the same difficulty explaining away a 'youthful indiscretion.' What will certainly be true is that, given the number of such pictures and videos available, they will not seem as shocking. Something else will have come along to satisfy our shock quotient."

Nikki Waters, manager of the Internet services group for Kaiser Permanente, responded, "By 2020, the dark secrets that used to (perhaps rightfully) be things you should be ashamed of will now be 'OK' because people will be desensitized." And **Hank Dearden**, director of business development for Digital Industry, Inc., noted, "People won't care about past indiscretions mostly due to fatigue, which is, I guess, a form of acceptance."

Joe McCarthy, of MyStrands, formerly principal scientist at Nokia Research Center in Palo Alto, wrote, "David Brin was prescient in his book *The Transparent Society*. As recent Pew Internet surveys have revealed, younger people are leading the way in revealing more about themselves in various digital spaces. There will be an uncomfortable period—just like now we are finally (I hope!) getting past the stigma of anyone having smoked marijuana. I hope that radical transparency does usher in an era of radical accountability, compassion, and understanding."

A number of people agreed but qualified the agreement. "The bar of acceptable behavior will be set higher and we will be more tolerant," wrote **Ted Coopman**, a communications technology lecturer at San Jose State University. "However, I think that outing extreme deviancy for public figures will still grab attention and ruin people. Look for more libel suits and therefore more care in what people accuse others of."

Some respondents cautioned that transparency cuts both ways. "The opportunity to find more and more people who share our interests and appreciate our points of view encourages us as individuals to be more open about who we are," commented **Kent Kirschner**, media specialist for Neighborhood America, an online community-building company. "This will continue to evolve and open up as today's activities become ubiquitous. Simultaneously, we will see a rise in predatory behavior."

Nicholas Carr, author of the Rough Type blog and *The Big Switch*, observed, "This scenario is a great example of wishful thinking. By 2020, the Internet will have enabled the monitoring and manipulation of people by businesses and governments on a scale never before imaginable. Most people will have happily traded their privacy —consciously or unconsciously—for consumer benefits such as increased convenience and lower prices. As a result, the line between marketing and manipulation will have largely disappeared." An anonymous respondent was thinking along the same lines, writing, "'Give Me Convenience or Give Me Death,' to cite the title of a Dead Kennedys album. Everyone's concerned about privacy, but it's just too darn easy to buy things online, even though you know 'they' are tracking everything you do." Another anonymous respondent commented, "Information is used much more often to try to sell things and to control people than to understand or forgive." And another wrote, "The sheer volume of images and info about people is what will keep most of our lives private—there simply won't be enough time for someone to review your life!"

Some people who expressed views against the scenario's likelihood pointed out cultural differences across the globe as a reason, but at least one respondent saw the blending of global mores coming as a result of an expansion of familiarity and transparency. "We may find a massive amount of change as our societies integrate a general base view and allow for niche attitudes and ways of life," responded **Robert Eller** of Concept Omega, a marketing company. "Already we see this reality in larger Western cities where people play their daily public role and, due to a greater amount of anonymity, are also able to live 'their' lifestyle viewpoint with little risk of desocialization. In the U.S., it was virtually a stoning offence if you were to be divorced/be gay/be female/be black, etc., as a candidate for president; German chancellor Gerhard Schröder had his fourth wife...and did anybody give a hoot? Nope."

Privacy Will Be Both Protected and Threatened Through Innovations

Some respondents projected that systems will be adapted to afford at least some privacy.

One respondent proposed the following 2020 scenario: "Most individual data cannot be accessed unless by explicitly authorized relatives. Thanks to semantic coding, almost any information can be accessed, but the main process by computers is done to prevent people from deducting the information they are not supposed to have. More generally, privacy is enforced by the fact that excessive access to confidential data can be revealed," predicted **Bertil Hatt**, a researcher of Internet and social services and innovation valuation for France Telecom and Orange.

Duane Degler, a designer and strategist for Design for Context and writer and editor of IPGems, which is focused on Semantic Web integration, agreed. "Increasing individual-level tolerance has been a trend in modern societies, and is likely to continue as the novelty of this format of data sharing wears off," he wrote. "...It is probably not major media that will guard reputations, but background Semantic Web services and pervasive agents that individuals can control."

Peter Bihr, a freelance consultant on Web strategies based in Berlin, wrote, "Social networking sites will, by 2020, long have incorporated strong mechanisms for privacy control by their users. As an exception, there might be social networks with strong incentives to really openly share personal data. These networks will be used by a large number of people, partly for financial reasons (free of use; vouchers or other financial rewards), partly due to lack of understanding of the effects (low education)."

Thomas Quilty, president of BD Consulting and Investigations, responded with this 2020 scenario: "As technology makes the collection of information easier—at times, without the consent of individuals—laws are passed worldwide to protect the rights of an individual as to whether data collected even anonymously can be used or shared with others. Personal AI-presence programs that represent the individual constantly search databases—even private databases—containing information related to its owner for information in violation of the owner's privacy profile settings. Data in government or private databases, if found to be wrong or illegally collected, are disputed automatically without the individual's intervention."

Tom Vest, IP network architect and consultant for RIPE NCC Science Group, predicted there will be some moves aimed at reputation blurring. "More people will opt for greater affectation (celebrity-style image management on a microscale) or obfuscation, e.g., using bots to generate personal 'info-chaff' to obscure actual online and offline behavior."

Hal Varian, chief economist for Google, responded, "The key issue with privacy is trust: Will the organization to which I transmit private information use it in my interest? I am optimistic that we can move in this important direction."

And **Jim Kohlenberger**, director of Voice on the Net Coalition and a former White House specialist on telecom policy, pointed out this idea of a likely reality in the next few years: "Complicating this vision is that by 2020, storage is so cheap that a person's entire life can be recorded in video, audio, converted to text, and searched. Someone else's life recording, in which you may interact and be a part, could become posted without your consent. Thus, new privacy protections would, nonetheless, be put in place along to prevent digital defamation."

Our Concepts of "Privacy" Are Changing

Many respondents agreed that perceptions of privacy will change due to the changing communications landscape. "There is no way back towards the same old notion of privacy," wrote an anonymous respondent.

"The same way having a tattoo today is no longer a barrier to career growth or social access, the standard for what is considered the 'norm' will continue to change," predicted **Bryan Trogdon**, president of First Semantic, a company that leverages the Semantic Web. "The benefits of instant, autonomous social feedback (what movie to watch, where to vacation, which chair to buy) based on shared personal preferences will far outweigh the cost."

John Eckman, a director with Next Generation Internet, Optaros, Inc., wrote, "Our collective notions of privacy (there are many notions of privacy, not one notion of privacy, even today) will evolve—we will come to have a broader understanding of what it means to have a public record of statements going back to youth. I'm not certain, though, that this will result in more integrity or more forgiveness. I guess that the context of everyone having such a visible record will make any one individual's statements less impactful, but so far we have seen this tending towards more judgmental and discriminatory behavior, not more forgiveness."

Blogger **Richard Silverstein** responded, "While I agree that notions of privacy, rectitude, and sin will evolve over time in a freer direction. I don't think people will be more willing to sacrifice what they view as essential elements of privacy. This will still be a realm in which people will see a virtue in protecting the most personal and intimate facts about themselves and their lives."

Ivor Tossell, technology columnist and journalist for the *Toronto Globe and Mail*, wrote, "YouTube 'outings' will indeed become more commonplace and accepted, as will evidence of putative politicians' lewd and offensive senses of humor as 20-somethings. But one of the lessons of the Web thus far is that name-and-shame sites (remember www.dontdatehimgirl.com?) have had limited traction, despite their salacious premises. It seems more likely that a privacy-aware generation will instead take active ownership over its online identities and, instead of becoming comfortable sharing intimate information, move decisively to manage (and often limit) what the world sees, to its own advantage."

Peter Eckart, director of technology at the Illinois Public Health Institute responded, "It's more likely that people give in to having their personal information bought and sold in the marketplace, and kids grow up—and the culture changes—to not having understood the value of privacy at all, so they don't miss it. 2020 will see the latter stages of a culture war, fought by older folks (I'll be 58 that year) trying to hold on to what privacy is still left, and younger folks—distracted by the media marketplace—wondering what all the fuss is about."

Scott Smith, principal at Changeist LLC, and others projected a divide, with more people consciously populating one of the extreme ends on a scale that goes from total transparency toward total privacy. "What seems more likely is a growing division between those who don't mind transparency and operate out in the open light of day—warts, broadcasted SSNs [Social Security numbers], and all—and those who choose to avoid disclosure of any kind," Smith wrote. "The benefits of open disclosure/transparency will decrease as more people flood the open market with predominantly useless private information—constant location and status updates, multiple 'cosmetically retouched' life stories, and vast amounts of visual pollution from their personal lives."

Tracking Will Be Leveraged More, Surveillance Ubiquitous, Privacy Scarce

While most respondents concentrated on the aspects of the scenario tied to forgiveness and trust, others addressed the ways in which data about individuals' lives will be collected and used. **Sean Steele**, CEO and senior security consultant for infoLock Technologies, presented the following 2020 scenario: "Ubiquitous surveillance will allow those who are willing—or those unlucky enough to be forced—to place some or all of their lives online in real time for others' entertainment (á la *1984*, *The Truman Show*, *Max Headroom*, and/or *Running Man*). Pervasive one-way surveillance by government and law enforcement will exist in all major cities and nations, as it will online, and GPS tracking of persons, vehicles, goods, and possessions, etc., will be commonplace and easily accessed for those willing to pay for it. Narrowcast advertising will be used in virtually every public area and retail space, and ads will be customized, personalized, audible only to the individual and only while in proximity to the good/service being sold. Spot promotions will target impulse buying habits like never before. Mobile devices and/or RFID tagging will continuously communicate via short-range radio (e.g., Bluetooth) with corporate marketing databases, and marketers will cross-feed and share data in order to provide rich, up-to-the-minute, 'three-dimensional' profiles of consumers."

Havi Hoffman, senior editor for product development at Yahoo and blogger, wrote, "The volume and ubiquity of personal information, clicktrails, personal media, etc., will desensitize us. A superabundance of transparency will lose its ability to shock. Maybe there will be software-driven, real-time reputation insurance service, offering monitoring and repair to dinged reputations. This could be as ordinary as auto insurance or mortgage insurance is today, and as automated as the nightly backups performed by most online businesses. I don't agree that this will make us any kinder, gentler, or more open in our dealings with each other. I do believe the next generations will take a different view of public and private/much as our take on social mores and self-expression has changed radically when compared to the time when our grandparents came of age."

"Many people are not aware of the loss of privacy and freedom when they put all their data on the Internet," wrote **João Miguel Rocha Filho**, director of DataOne, a provider of software for connecting to Linux based in Brazil. "Not only other people are doing use of this data but also business enterprises, security agencies, and all sort of government bodies. Also people are not aware that their info will drive others to access it—health (or lack of it), familiar life, financial life, political life, etc. The technologies in use now are very helpful to people, but in time, without control, they may well be dangerous tools."

Jonathan Dube, president of the Online News Association and director of digital media at CBC News, publisher of CyberJournalist.net, commented, "By 2020, people will by and large have lost the notion of privacy because so little information will remain private." **Tiffany Shlain**, founder of the Webby Awards, wrote, "Privacy is an illusion." And **John Murphy**, director of USA.gov technologies for the General Services Administration, added, "The openness of the Internet guarantees that this will happen, whether one approves or not."

Leonard Witt, author of the PJNet.org blog and an associate professor in communication at Kennesaw State University in Georgia, responded, "My fear is how a repressive government might use everything they know about all of us—and in this open world, they will know everything."

On the other hand, **Judy Breck**, blogger at GoldenSwamp.com, wrote, "This transparency will deliver devastating blows to tyrants and terrorists who rely on working in the dark."

And **Josh Quittner**, executive editor of *Fortune* magazine, wrote that he expects privacy will be exchanged soon when it is decided that complete transparency is required for safety. "Total transparency for total security! Sounds Orwellian. Is Orwellian. Sadly, it'll be our response to the next major terrorist event in the U.S. (and then elsewhere)."

Some Expect People to Withdraw

A number of respondents said as people begin to see how their personal information is being databased and used, they will begin to back away and become more careful about public displays of private materials. *Backlash* was a word used in many responses.

"The backlash against social networking's incursions into personal privacy is already beginning," commented **Milton Mueller**, a professor at Syracuse University and expert on Internet governance and technology policy. "People will adjust their behavior to be more careful about the possible future uses and abuses of information about themselves. There will be more data, and more things done online, so there will be much more to keep track of and to hide."

Richard Osborne, Web manager for the University of Exeter, wrote, "I suspect more of a backlash as unscrupulous and manipulative people start to understand just how much power they can hold over others using freely available online information. Perhaps a couple of nasty cases will lead to a shift in public perception and changes in the law."

Susan Mernit, an independent consultant and former executive with Yahoo and America Online, predicted, "By 2020, we will have a backlash against openness and privacy and have a series of private networks that individuals can use with greater anonymity—they will be premium, secure channels. Rather than forgiveness, society will negatively rate a larger number of people and a backlash against transparency will occur—the New Privacy of smaller and more elite networks will rule."

Fernando Barrio, senior lecturer in business law and program leader for the Master of Arts degree in e-business regulation at London Metropolitan University, wrote, "The current privacy-unfriendly scenario, with the problems it creates for individuals, will backlash, and people will try to keep more data out of the public domain. For that data in the public domain, not forgiveness but trivialization and gossiping will be the norm."

"By 2020, there will be a backlash on personal information sharing," wrote **Nathan Botts**, community informatics research associate at the Kay Center for E-Health Research, working on the National Health Information Network. "As the technical ability to analyze, process, and aggregate shared personal information becomes more prevalent, people will be forced to seek increased measures of security and privacy in regards to personal information sharing so as to mitigate the prying of employers, insurers, and government-based organizations. Personal information sharing will become a more intimate transaction."

Peter Van Ness, president of the Van Ness Group, a Web-development company, responded, "Today, everyone on social networks has a small taste of world fame every day. By 2020, there will be a backlash against sharing personal information as we find we really don't like how this world fame tastes at all. The pendulum is about to swing the other way. I see it already. Most employers now check MySpace and Facebook as part of their hiring due diligence. This so-called 'sharing' will morph into personal marketing. People will create an Internet persona much as we all now have a public face. Transparency will vanish, and, unless we fiercely protect our privacy, it will vanish too."

Scott Brenner, a Web developer and consultant for Fortune 500 companies, noted, "There will be major data breaches and other negative aspects of all this 'openness' that will cause some people to push back. Schools, employers, potential romantic partners, neighbors, etc., will routinely obtain personal information on others (and use it for both good and evil purposes) that would have been nearly impossible to uncover in the latter part of the 20th century."

Tiffany Shackelford, a consultant for clients such as Phase 2 Technology, WebbMedia, and Daily Me, commented, "I think the oversharing in the social networking community may blow up in our faces." **Aaron Schmidt**, a blogger who runs Walking Paper Consulting, agreed with the scenario but wrote, "The culture of transparency might well peak before 2020; a significant backlash could already be in place." And **Jay Neely**, a social strategist and founder of News Armada, a Boston company working to advance Internet-based news and commentary and community online, predicted, "A gross misuse and exposure of personal info that people had made public but thought was too obscure to ever gain attention later in their life will occur within the next 10 years. The negative backlash to this will more likely result in a frenzy of new privacy legislation than an enlightenment of humanity."

Many respondents indicated that all of this will cause people to want to drop out of sight, off "the grid." **Chris Miller**, senior vice president for Element 79, an advertising agency, wrote that he sees three factors at play in his mostly agree answer to the proposed scenario: "(1) Lack of privacy will force people (who don't want public outing) to live their lives more openly and not commit the 'indiscretions of the past'—if anyone could tell if anyone was lying, people wouldn't lie. (2) There will most likely be a few high-profile murders, kidnapping, etc., based on someone monitoring another individuals' information. This will at first create a privacy backlash but will push for more openness. (3) Coming off of number 2 and a bit of 'who's watching the watchmen?' there will be a small part of the population who continues to live off the grid to an even greater extent. They will not trust the new notion of privacy. This will at first be people who have 'dropped out' but then will continue with their children, who are born off the grid and stay out of the openness of society."

Bruce Turner, director of planning services for a U.S. regional transportation commission, commented, "A significant group will drop out of ID technology and deliberately attempt sabotage."

Reputation Repair Will Be Commonplace

There was a high level of agreement on the growth of the reputation-maintenance business. "In 2020 your online identity will be more important than your physical one," wrote **Mark Youman**, principal at ICF International, a Washington, DC, consulting company.

A number of respondents expected to see people of privilege and power managing to rise above the exposure likely for the lower classes. "A high level of transparency (through profiles, user ratings, feedback, and other mechanisms) will be necessary for doing business by 2020—you simply won't be invited to the table if you don't provide that type of information," predicted **Jason Stoddard**, managing partner for strategy at Centric/Agency of Change. "Of course, gaming the system will be the new 'search optimization' of the day, but 'found media' will typically correct any gamed records. The highest social status may indeed be the people who are truly invisible, unknowable, and opted out of the system, since this will imply that they have large amounts of money and power."

Patti Nelson, a webmaster who works on U.S. government sites, wrote, "This has started; reputation cleanup services are already in business. Interesting though that this type of transparency might encourage people to behave better. It's as though people are creating a global Big Brother by choice." **Mark Eberman**, Web director at a direct marketing company, wrote "So far, we've seen this decreasing privacy resulting in less tolerance, not more. The result will likely be a class of prudes who can afford privacy, pushing their values downhill rather than the masses pushing their more relaxed values up. History-scrubbing will be a growth industry."

Jim Lucas, Web manager for CACI, a security solutions company, wrote, "As technology makes privacy more of a challenge, an industry will develop and prosper that erases selected evidence of current or past indiscretions."

Matt Gallivan, senior research analyst for National Public Radio, commented, "I see there being two main options in the future: (1) People shut themselves off to the interactive world and, as a result, lose the massive value and utility that sharing offers, or (2) people accept that utility and value and, in so doing, learn that everyone in this age—not just politicians and celebrities—has to work to maintain a carefully calibrated public-facing image. I don't imagine many people will choose option one."

Several responded that it is not possible to completely rehabilitate a damaged reputation. "I do not really believe that reputation corrections are really functional," wrote **Oliver Quiring**, a professor at the Institute for Communication Science and Media Research at Ludwig-Maximilians-Universität in Munich. "It is much easier to destroy than to build up reputation." **Brian Dunbar**, Internet services manager with NASA, commented, "'Truth' will become a quaint 20th-century idea. Whatever gets the most hits and most blog/Myspace/media coverage will be accepted as fact."

The portion of the scenario indicating that media organizations might publish reputation updates was mostly ignored and sometimes denigrated. "I got a big laugh out of 'Carefully investigated reputation corrections and clarifications are a popular daily feature of major media outlets' online sites,'" wrote Infothought blogger **Seth Finkelstein**. "This combined 'Carefully investigated,' 'popular,' and 'major media' in one sentence and wanted it taken seriously. I think the reality is going to be more like 'Sleazy reputation hit-pieces are a popular daily feature of tabloid media outlets' online sites' (like they have been as long as such media has existed—i.e., 'yellow journalism')."

An anonymous respondent commented, "'Major media' becomes an oxymoron, as they continue to shrink; Google/YouTube (and others) will become the 'major media.'"

And **Hamish MacEwen**, a consultant with Open ICT in New Zealand, wrote, "'Major media outlets?' You must be joking. Fragmentation and decentralization, combined with aggregation and collaboration, will remove those legacy institutions and supplement them with a bewildering range of sources and opinions."

Digital Identities Can and Will Be Multiple

A few respondents pointed out the complexities of "privacy" in a digital present and future in which people sometimes have more than one "self." "Digital duplicity will become a high art," wrote **Greg Laudeman**, a technology specialist at Georgia Tech Enterprise Innovation Institute.

An anonymous respondent noted, "More and more people will create multiple personas. The volume of information may provide cover, and privacy software will keep up with transparency." And another commented, "Driving forces compete not to shape public opinion but to create multiple 'sets' of personal identities that, in turn, shape perceptions of desirable membership in controllable groupings—while teens seek out restricted online communities that offer not anonymity but 'sameness.'"

Anthony Townsend, research director for the Technology Horizons Program for the Institute for the Future, responded, "Expect a whole new layer of social infrastructure for reputation and identity management to be layered on top of this. Social networks will proliferate, as will the ability to maintain multiple, sometimes conflicting identities and trails across each one. In the end, who will be the arbiter of what's true and what isn't?"

Barry Chudakov, principal of the Chudakov Company, a marketing and advertising agency, commented, "New digital identities deconstruct our singular notions of self, just as our 'life on the screen' obliterates the proscenium arch of literate theater. It is more than Pollyannaish to think that transparency heightens integrity and forgiveness; this ignores the growing dynamic of self and other merging, of copies and originals replicating each other. At stake here is our sense of self that grew up feeding on the alphabet and its linear outcroppings. I believe the more likely scenario is that we will realize that we must manage our digital identities, much as a corporation manages its messages and relations with the media. Further, as our lives become more transparent, we will regard privacy much as Rousseau regarded nature once the industrial revolution threatened it. The rarity of privacy will only be slightly affected by reputation corrections and clarifications, because these will be seen to be as yet another identity foray, another option in the malleable sense of self which will define each of us."

And **Luis Santos**, of the Universidade do Minho–Braga, Portugal, wrote, "We do not need to go forth a decade to anticipate a much more complex (hypercomplex, as Henrik Qvortrup calls it) social environment. People will most certainly adopt more flexible identities and more public facets of those identities, and that will not produce enhanced transparency; quite the opposite. Still, transparency in that particular sense is not a very desirable goal in itself—it rhymes with conformity, and that runs against the pillars of knowledge appropriation and development."

Our Definitions of Identity and Privacy Are Rapidly Changing in a More Transparent World

No previous communication technology has had the impact on privacy and identity that the Internet has already achieved; it packs a wallop. Between now and 2020, that landscape will continue to be populated with additional exciting possibilities and frightening threats.

In June 2008 privacy experts told the U.S. Senate Committee on Homeland Security and Government Affairs that the 1974 Privacy Act, the law governing how U.S. agencies handle information about people, is antiquated. It does not address data mining or the way corporations and governments in the United States are contracting with private data brokers.[1]

The U.S. Government Accountability Office (GAO) released a report calling for Congress to establish new privacy rules and issued yet another report recommending that the White House Office of Management and Budget appoint a permanent "chief privacy officer" to oversee the issue government-wide. The GAO report noted that privacy protections are not being addressed appropriately.[2]

Individuals who have invested themselves in work and play online generally establish one or more personal identities that are on display at varying levels of exposure, from one-to-one to one-to some to one-to-all. These identities are often initiated by individuals' construction of user profiles on social networks such as LinkedIn, Facebook, and MySpace.

Outsiders' perceptions of one's reputation also come into play in establishing online identities; they are built from the reactions of others in various ways, from ratings systems to news accounts, to others' responses to individuals' participation in online discussions. Most people have multiple online personal accounts they have built at various times, and some find they are difficult to track and keep up to date. Some people build fictitious accounts; some lie about certain personal information in their accounts or leave out unflattering details.

Internet-invested corporations have seen the value in online identity control and have begun to compete in this realm. Organizers such as OpenID.net (to allow Internet users to log onto many different Web sites using a single digital identity) and FriendFeed (to allow users to aggregate the personal content they have on many sites—Flickr, Twitter, YouTube, blogs, etc.—and see and share a stream of that data that friends can subscribe to) have been introduced in an effort to assist people in assembling what TechCrunch blogger Michael Arrington refers to as the "Centralized Me."

A group of corporations and individual volunteers concerned about the future of identity joined to start a massive effort named the Data Portability Project, which really got underway starting in 2008. Among the participating organizations are Yahoo, Google, Facebook, Dow Jones, and the British Broadcasting Corporation (BBC). There is hope that open standards and protocols can be leveraged to allow individuals to be able to move, share, and control their identity across various online services, tools, and networks.[3] Of course, centralizing data makes it simple for the individual to use but also simplifies government snooping, identity theft, and other crimes, so security remains the big question.

The London-based human rights group Privacy International has stated that Google's plans for the future of search (individualized, targeted search through the use of vast amounts of users' personal data) are "hostile to privacy."[4] There is a compelling fear of placing all the details of everyone's lives under the control of one or two entities; it does appear to many people as if Microsoft and Google are in a battle to see who controls the most information.

People have responded to the availability of Web 2.0 social networks, an overwhelming voluntary outpouring of vast amounts of personal data online on blogs and sites such as YouTube, Flickr, MySpace, and Facebook, where dozens of millions of people from all over the globe are sharing written, audio, video, and artistic content. Marketers, security companies, government spy organizations, and criminals have the same easy access to this data as one's friends and relatives do.

More companies are being founded on the basis of people's willingness to exchange their privacy for a service they consider to be of value. Pudding Media is a company that proposes to provide free PC-based phone calls in the United States or Canada to any users willing to allow their conversations to be searched by marketers who will respond by serving up contextual advertising.[5] Among the privacy worries this service raises is the threat to the person on the other end of the conversation who may not want to participate in a searchable communication.

A January 2008 Pew Internet study, "Information Searches That Solve Problems," found that 26% of those who used the Internet for research were concerned they might accidentally reveal private or sensitive information to others.[6] A December 2007 Pew study, "Digital Footprints," found that 60% of Internet users are not concerned about the amount of information available about them online; 38% said they have taken steps to limit the amount of online information available about them, just 3% said they regularly do searches to monitor their online presence, and 74% said they have checked once or twice. "Digital Footprints" found that many of the people who share information online are unaware of the implications tied to exposing their information. Many do not know that search-engine firms archive users' search terms, their computer's address, and the unique identifier for their Web browser for several months.[7]

A report prepared by the Center for Democracy and Technology in August 2007, *Search Privacy Practices: A Work in Progress*, found that search firms were moving toward "a new competitive marketplace for privacy," actively addressing their approaches to the issue, but most still do retain search data for at least a brief period because it is valuable on many levels.[8]

Privacy International and the Electronic Privacy Information Center reported in their December 2007 executive summary on the state of privacy in more than 75 countries that "the proliferation of new technologies and surveillance applications in the public and private sectors...was balanced by more regulatory activity and greater international norm-setting—many countries have established data-protection laws and oversight bodies." It reported that by 2006 or 2007, Bulgaria, Germany, India, Switzerland, and Taiwan all adopted federal open-government laws, and all but three European Union countries had data protection laws. Nongovernmental organizations and civil society groups have been leading successful campaigns to support privacy. The European Court of Human Rights is playing an increasingly important role in the interpretation of privacy rights in cases about issues tied to workplace privacy, identity documents, and communications surveillance. Many countries working toward privacy protection are located in Africa, Latin America, and the Middle East.[9]

U.S. presidential candidate Barack Obama's use of the Internet to combat rumors during the 2008 campaign is one way the medium has been leveraged to retain or regain trust and maintain accurate identity. Obama and his advisers decided to fight the online rumor mill with his own online rumor aggregator, http://www.fightthesmears.com. He encouraged supporters to hunt down rumors, share them on the site, and correct mistaken information. Acts to benefit the common good online and new initiations of trust relationships appear to be taking place in Internet-driven locales such as Facebook and Wikipedia, where people work to extend knowledge of themselves and the world.

Information-privacy expert Daniel Solove wrote in his 2007 book, *The Future of Reputation*, about the way the Internet is transforming how we communicate. "An entire generation is growing up in a very different world," he wrote, "one where people will accumulate detailed records beginning with childhood that will stay with them for life wherever they go...The Internet is bringing back the scarlet letter in digital form an indelible record of people's past misdeeds."[10]

Researcher Sam Bowles of the Santa Fe Institute argued in a June 20, 2008, *Science* article that there is experimental evidence that people often act against their own personal self-interest to serve the common good. He said that those who assess potential futures in the political, economic, and social sense should figure that altruism can sometimes be more valued than economic gain. He added that as the world becomes more interconnected adding new challenges, this altruism must be harnessed, "The moral nature of humans must also be recognized, cultivated, and empowered," he wrote in the essay.[11]

On the other hand, global criminal traffic and leveraging of Internet tools is at an all-time high. Terror networks operate online; it has been reported that religious extremists that use violent methods operate more than 6,000 sites for recruiting and promoting negative agendas. Governments and other entities are building biometric databases of fingerprints, palm prints, and facial scans. Millions of surveillance cameras are being added worldwide annually; Britain has more than 4 million, and France is aiming for a million.

Networked computers allow phone companies, GPS providers, and automated toll systems to know where you are, search companies like Yahoo and Google to know what you are looking at, news providers to know what you read, and even your local grocery chain to know what you eat.

The Internet is no longer a network of "dumb" packet-switching conduits; because of the need to satisfy desires for security, search, and economics, it has become application and user aware. Identity is being embedded in the network layer through initiatives aimed at solving the problems of crime, spam, hacking, and the updating of the Internet to allow IP numbers to be assigned to billions of objects—not just PCs. Many changes will arrive with the gradual adoption, now under way, of IPv6.

The Semantic Web—what World Wide Web innovator Tim Berners-Lee likes to refer to as Web 3.0—is expected to bring changes in individuals' abilities to control their personal information and outsiders' abilities to access it.[12]

In addition, hardware, software, and services companies are beginning to move consumers into a realm in which they are using tethered devices and inputting all of their creative content and personal information into what is known as "the cloud," a term derived from the concept of people's digital information floating in an accessible Internet-based space instead of residing on individual computers.

Both tethered devices and cloud computing force users into a paradigm in which information can be accessed and controlled by an outside source—the provider of the cloud space, hardware, and/or software. For instance, any personal computer equipped with an automated update feature can be used to conduct surveillance on a user. A targeted update from the maker of the operating system or any software provider whose product is installed on it can possibly allow searching and sharing of any document on the PC and can even turn on the machine's video camera or microphone unbeknownst to its owner. Mobile phones can also be programmed at a distance to allow them to act as microphones, even when they are powered down by the user.

Microsoft is developing and has applied for a patent for software to allow remote monitoring by employers of employees' productivity, physical well-being, and competence through sensors that read various inputs through their use of their computers.

Adware companies such as Gator Corp. and Claria Corp. market "free" software applications such as screensavers and games that they bundle with tracking programs that install simultaneously with the desired software and stealthily datamine personal information without users' permission. NebuAd and Phorm, Inc. have business plans to work directly with ISPs to contract for permission to scan customers' Web surfing habits and deliver targeted advertising. Public interest groups are asking the U.S. Congress and the U.K. Parliament to assess privacy protections in light of such surveillance initiatives by industry.

While all of this is complex, it does not even begin to scratch the surface of possibilities likely in the future if more people begin to construct multiple alternate identities in virtual worlds similar to but more user-friendly than Second Life and World of Warcraft.

Additional Responses

Many other participants in the survey shared valuable input on the future of the Internet and privacy, identity, and forgiveness. Among them:

> "There is a learning curve in this rather new phenomenon, and those who are excessive in what they share are sometimes suffering the consequences. In the future, while sharing personal information, opinions, and emotions will likely increase, it will be done with far greater discrimination and the material shared will be 'managed' more carefully by the person contributing it. People learn!" —**Don Heath**, Internet pioneer; former president and CEO of the Internet Society; member of U.S. State Department Advisory Committee on International Communication and Information Policy

> "People will discover that the anonymity of 2007 was the free love of the 1960s, and sadly many will live their lives with the consequences." —**Robert Grant**, chief executive officer, VoyaCare, Inc.

> "One can only hope that people will get a clue and become more circumspect about disclosing personal information. The EU's privacy protection laws are a step toward assisting them and preventing abuses. On the other hand, many social networks will become more Borg-like. So I am less than certain about my disagreement with the premise." —**Steve Goldstein**, ICANN board member, retired from National Science Foundation, where his job in the 1990s was to help diffuse the Internet globally

> "My big fear is that homogeneity prevails in 2020 with people talking about the importance of individuality as they continue to sink into a morass of mediocrity." —**Michael Castengera**, senior lecturer at the University of Georgia's Grady College and president of Media Strategies and Tactics Inc., a media consulting firm

> "Sooner than later, Web 2.0 will become old-fashioned. We are starting to see problems associated with private information shared on networks. Those problems will become larger and more frequent in the future." —**Sebastian Ricciardi**, associate with Jauregui & Associates, a law firm in Buenos Aires; leader in the

Argentina chapter of the Internet Society, formerly of ICANN's At-Large Advisory Committee

"I live in a small town where everyone knows everyone else. This does not lead me to believe that knowing more about people makes you forgive them." —**John Levine**, founder of Taughannock Networks; a leader of the Internet Research Task Force's Anti-Spam Research Group and the Coalition Against Unsolicited Commercial Email

"The type of transparency currently in vogue has nothing to do with either integrity or forgiveness—it is more closely aligned with exhibitionism. There is no reason to believe that it will evolve into a vehicle for responsible actions." —**Hinda Feige Greenberg**, PhD, director of the information center for the Robert Wood Johnson Foundation, dedicated to improving health care for Americans

"Reciprocal access to personal information will change the nature of human relations, increase the depth of social networks, and promote interpersonal cooperation and coordination." —**Gary Kreps**, chair of the department of communication, George Mason University; formerly founding chief of the health communication and informatics branch of the National Cancer Institute

"I doubt the primary assumptions regarding widely distributed benefits from transparency, especially with the rise of exploitative/manipulative promotion enabled through data mining." —**Oscar Gandy**, author, activist, retired emeritus professor of communication, University of Pennsylvania

"My students are already completely comfortable with living a public-private life online. They believe that nothing is private—they have no expectations of a quiet unknown thought. Nor do they think that the ability to speak anonymously is particularly valuable. In fact, they're suspicious of anonymity. They think you are required to carry a driver's license wherever you go—on foot—in America. So I'm confident that 'sharing' will become even more the norm. I'm not as confident (particularly in the political realm) that the past actions of public figures will be forgiven. The rest of us will be able to muddle along just fine, but political candidates will have to live completely examined lives from grade school on. They wanted to be loved." —**Susan Crawford**, founder of OneWebDay, celebrated each September 22; ICANN board member; associate professor, Cardozo Law School; visiting professor, Yale Law School

"Transparency might be the unexpectedly beneficial side effect of losing privacy. The time to do something about protecting individual communications and data about individual behavior from snooping by the state or your ex-spouse, your irate neighbor, the person you cut off in traffic today, targeted ads, or spam was about 10 years ago. At the same time, the same Net of media-veillance, data-veillance, electronic breadcrumb collecting, and voluntary hyperdisclosure that has forever obsoleted 20th-century notions of privacy also makes it more difficult for public figures and institutions to hide their gaffes, indiscretions, and malfeasance. The critical question is whether a sufficient number of people will be reputation literate enough to use online reputation information effectively or to erase bad information that stains their own reputation. It's not just the truth that has become nakedly visible—along with the heretofore hidden true information has come misinformation and disinformation. Wikipedia controversies have revealed the evolution of social mechanisms in the Wikipedia community to address the reputation-damaging possibilities of letting anyone publish; it remains to be seen how effective these evolving social mechanisms will be. Will it be easy to slander people—or easy to see through such attempts?" —**Howard Rheingold**, Internet sociologist and author; one of the first to illuminate virtual communities; author of *Virtual Reality*, *Smart Mobs*, and *Virtual Community*

"Overall, technology tends to increase privacy, not reduce it." —**David Moschella**, global research director for the Computer Sciences Corporation's Leading Edge Forum; *Computerworld* columnist

"Mostly agree, however, the Internet needs to find ways to both forget information and for information to be corrected (or put into context: a new right of reply)." —**Adam Peake**, executive research fellow and telecommunications policy analyst, the Center for Global Communications (GLOCOM); studies the intersection of public policy and the Internet

"I remember teaching in Tokyo in 1986 and how the Japanese handled living so close together, with paper-like walls. They could hear everything in the apartments next to them, yet admitted to 'hearing nothing.' Same would be said of offices without walls, crowded trains, etc. People learn to 'not care' or not listen, or to be careful to not talk about or judge what they do hear. Ideally, that will be what happens with our loss of privacy (whether self-disclosure or video/audio surveillance)." —**Ed Lyell**, professor of business and economics, Adams State College, Regis University, San Luis Valley Board of Educational Services; pioneer in issues regarding Internet and education

"This is an optimistic scenario, but one that is at least plausible. The 'shock then shrug' progression seems to appear time and again, across a wide variety of issues; I see no reason why it wouldn't be replicated with online media. I would qualify this scenario, however, by noting that there would be much discussion about what kinds of 'unethical' pasts would and wouldn't qualify for social forgiveness. Licentiousness would probably get a pass, for example, while cruelty probably would not. It's likely, but not certain, that this will skew generationally, with younger people being more willing to shrug and forgive than older people." —**Jamais Cascio**, originator of Open the Future, also works with the Institute for the Future, Institute for Ethics and Emerging Technologies, and Worldchanging

"Generally, increased transparency will heighten individual integrity and forgiveness—to a point. So long as the individual has some feeling of being 'part of the crowd,' transparency will continue to drive the behaviors described. However, when transparency shines a spotlight on an individual in the crowd, the individual will often retreat from the experience. This element of human psychology will require service providers to tread a fine line with their privacy vs. access policies and ensure some level of anonymity in the crowd without singling out the individual too expressly. The scenario described requires a high degree of conforming normative behavior. I'm not so sure that a community the size of the Internet has a sufficiently narrow band of behaviors in the community required to set forth a standard of behavior and accountability that could be acceptable to enough people to become commonplace. There may be pockets of behavior similar to what is described, but the Internet community is too fragmented and individual to truly drive the behaviors described." —**Ross Rader**, director of retail services, Tucows Inc.; works with ICANN in the Registrars Constituency, part of the Generic Names Supporting Organization (GNSO)

"Unfortunately, the scenario assumes a global level of morality that the human race shows little sign of achieving. Improvements in identity theft prevention will reduce the level of personal crime inflicted online, but this will not change the lack of personal integrity." —**Adrian Schofield**, manager of the applied research unit, Johannesburg Centre for Software Engineering, South Africa; leader in the World Information Technology and Services Alliance

"I agree with the first sentence, that in 2020 people will be more open to sharing personal information, but I disagree with the rest of the scenario. I don't think sharing personal information will be done due to some perceived public benefit, nor will people become more responsible for their actions or more forgiving. Rather, I think we'll see a sort of bloodsport made of discovering indiscretions. Perhaps *bloodsport* is too strong a word. Maybe a better way to put it is that we'll see a new form of what 'reality TV' already provides." —**Steve Jones**, professor of communication and associate dean of liberal arts and sciences, University of Illinois–Chicago, cofounder of Association of Internet Researchers

"While I agree that people are likely to continue to freely share personal information online, and that people might be willing to exchange anonymity for perceived benefits, this prediction misses the primary concern with such a scenario: the harms that users do not perceive. Many share data with companies such as Facebook because they perceive a benefit: improved social interaction. What many users fail to understand completely is the manner in which a company such as Facebook collects, aggregates, and uses that personal data. Even if the public's notion of privacy has changed, the potential harms have not. I also disagree with the notion that transparency will bring a new sense of responsibility and forgiveness. Rather, the fact that so much of my life will be indexable and searchable via powerful online tools such as Google/Facebook/Web 2.0, users will be more timid, afraid of their past, and less likely to share certain pieces of information online. We also might see a rise in an attempt to obfuscate one's online identity: creating accounts with misinformation in order to confuse/confound those who want to create a user profile or appropriate the data. Finally, while being 'outed' by YouTube might become less of a concern, I doubt media outlets will seek to investigate reputational corrections—it is the drama of damaged reputations that attract viewers, not the restitching of a torn one." —**Michael Zimmer**, PhD and resident fellow at the Information Society Project at Yale Law School; research includes social and cultural dimensions of new information technologies

"Yes…*if* appropriate checks and balances are clearly seen to be adopted and adhered to by those services holding data on our behalf. We might increasingly trust our peers, but our trust in data-holding corporations is falling ever more rapidly. Moves such as the Creative Commons/Talis Open Data Licensing initiative, Google's Open Social, the Attention Trust, etc., are all (small) steps in the right direction." —**Paul Miller**, technology evangelist on the senior management team at Talis, a company delivering human-centric Web applications, based in Birmingham, U.K.; a blogger for ZDNet

"This was more 50–50. If anything, I think that people will muddle through, learn from their mistakes—i.e., what harms others and themselves—and begin to develop an ethics, including an ethics of information privacy, that will work somewhat against the current tendency among young people to bare all in (virtual) public. Indeed, there is some research that shows that the more people are engaged in online venues such as Facebook, etc., the more they are concerned about privacy, not less—going against earlier assumptions that young people are somewhat naive and/or careless on this point. So, yes, I think people will become more responsible, but as that responsibility becomes more widespread, I don't think it will correlate with greater forgiveness for ethical lapses. Such lapses are more forgivable if there is no well-established practice or ethos that makes us aware of the possible harms, prevailing best practices of using a technology, etc. But once those are in place, then there is less reason to forgive people who thereby should know better." —**Charles Ess**, a professor of philosophy and religion and researcher on online culture and ethics, Drury University, Springfield, MO, and active leader of the Association of Internet Researchers

"Why people share intimate details online is a mystery to me, but they clearly want to do it. I disagree that people are becoming more responsible or more forgiving. There will be increasing damage to reputations." —**David W. Maher**, senior vice president for law and policy, the Public Interest Registry, the Internet top-level domain registry; formerly vice president of public policy for the Internet Society

"As far as transparency is concerned, Internet and tools like YouTube and the Daily Motion will have the main role in correcting the stereotypes generated by the traditional media about sensitive issues like terrorism, Islam, and the notion of doctrine. When it comes to privacy, everything has a price, and the price we to pay for we 'Internauts' is to be more flexible about privacy issues and admit the networks are not secured, but can we survive without Internet?" —**Hanane Boujemi**, ICT researcher for DiploFoundation, working on educating people about Internet policy and Internet governance, Malta

"We are moving away from the Fundamentalist influence, so the stigma of 'unethical pasts' may be minimized." —**Janet D. Cohen**, blogger, futurist and trend analyst

"People use the Net to resolve uncertainty and to search for the truth as major media has been taken over by desperate corporations. Individuals feel less vulnerable to identity threats and would rather be authentic online." —**Cliff Figallo**, social innovator and original member of the first online community—The WELL, now of AdaptLocal.org; expert in fitting and implementing social Web applications to groups

"As long as individuality prevails...Two phenomena are acting simultaneously. Internet will continue to offer exposure to anyone who wants some, and will emphasize the need for some individuals to abandon the Internet if it interferes too much with their private life. Which trend will prevail? Being in love with everyone, or knowing that everyone is (not) in love with you (-;)" —**Louis Houle**, president of the Internet Society's Quebec, Canada, chapter

"There will indeed be more transparency than there is now, but human nature will not have been changed. People will simply get better at choosing what to display." —**Fred Baker**, Cisco Systems Fellow, former Internet Society (ISOC) chairman of the board; Internet Engineering Task Force (IETF) leader; an architect of the Internet

"It is obvious that privacy is changing in meaning. Online profiles have more information (pictures, data, videos, etc.) and things will only get deeper—with trust playing a critical role in demystifying why most people prefer to remain anonymous online. Just as new residents watch their backs for a while in a new estate before opening up (and being accepted by other residents), online communities will become more open. One of the most disturbing problems today is online crime, which is usually masked by hidden identities. With increasing comfort to move beyond pervasive anonymity, various online communities will be able to move closer to identifying the source of crime-prone online information." —**Gbenga Sesan**, Internet for development consultant, Paradigm Initiative, Nigeria; his work is tied to the use of ICTs in socioeconomic transformation, focusing on underserved groups

"There will always be a need to keep certain personal information private (medical, legal, for example) and people will fight to have the right to keep this type of information private. However, as more people discover that revealing more information allows them to keep connected to their globally distributed friends in an increasingly online world, they will sacrifice a little of what is considered private information (what I like to eat, watch, read, etc.)." —**Tze-Meng Tan**, Multimedia Development Corporation (MDeC), Malaysia, and director at OpenSOS; Internet pioneer as a member of the team that started Malaysia's largest ISP, TMnet

"The Boomer generation will have to fade from social dominance (corporate and political hierarchy) before such openness, especially forgiveness, is achievable." —**Todd Spraggins**, strategic architect, Nortel Carrier Networks; president and chairman of the board of directors of the Communications Platforms Trade Association

"We are indeed moving toward a 'Transparent Society' (as per David Brin), and that's likely to be a good thing overall—provided, that is, that the transparency goes both ways, both outward from government/business and inward in equal measure." —**Mike Treder**, director, Center for Responsible Nanotechnology; expert on the implications of emerging technologies; a research fellow with the Institute for Ethics and Emerging Technologies

"Although we are seeing an increase of people using user-generated content technologies in order to share their views, emotions, photos, data, and video with friends, there is yet a tremendous reluctance of the connected population to share information mainly due to the uncertainty of what the companies owning those technologies will do with their personal information, and also because of the fact that they know they could be prey of identity theft schemes. Internet users are getting more educated every day about the potential threats of sharing their own personal information. I do not agree with the statement that 'the public's notion of privacy has changed.' On the contrary, the use of the Internet has made people, companies, governments, and policy makers much more aware on the importance to protect privacy and confidential information through numerous mechanisms." —**Cristos Velasco**, director general of North American Consumer Project on E-Commerce; Mexico-based attorney and active in the Internet Governance Forum and Internet Society

"I agree for the most part with the premise, but only if information authentication is also in widespread use. While everything may be out in the open, reputation maintenance and remediation needs to be safeguarded from malicious or unintentional errors. Thus, the ability for information to be disseminated which is falsified or, less heinous, unintentionally incorrect has to be brought under control." —**Rich Miller**, chief executive officer, Replicate Technologies, Inc.; an Internet pioneer with ARPAnet who implemented some of the first computer messaging and conferencing systems

"The future is more complex than this set of questions suggests. There will be more geographic/political diversity, and some societies will be more open, others will be more controlled." —**Norbert Klein**, member of ICANN's GNSO Council and Internet Society leader who works with Open Institute Cambodia, a company whose primary focus is on information

"Privacy concerns will still weigh in quite heavily…Some population segments will be fine with disclosing more in order to get their needs met, such as online dating and people actively looking to connect with others they don't know. They give more info to get more info in this context. I predict, by and large, people will be open to sharing more information in general, just by the sheer volume of more people online in the next 12 years, but not necessarily personal information nor personally identifiable information online. More opinions and emotions, yes, due to the ubiquity of online communications tools and social media sites and a larger WWW population using the Internet, but not full transparency and individual forgiveness about what is shared online. Not everyone wants to be Net famous…on online video sites like YouTube. Some do, I predict most won't want the visibility." —**Joanna Sharpe**, senior marketing manager, Microsoft

"Being outed for some indiscretion will still be punitive. Moreover, transparency will work both ways, as it will enhance the monitoring of public officials and working of institutions by private citizens." —**Bernardo Huberman**, senior fellow and director of the Social Computing Lab, HP Laboratories; consulting professor in the department of applied physics at Stanford University

"Reputation and image management is a big part of the public relations industry and it won't go away. Once an indiscretion is outed on the Internet, it will still be tough to get rid of it. Just because it becomes more common to see/read/hear about another's faux pas online, doesn't make it more acceptable or less embarrassing. Case in point: Britney Spears. Is there anyone left on Earth who doesn't think she's out of control? Just because the media crams coverage of her down our throats every day, doesn't make her poor choices any more acceptable." —**Janie Graziani**, manager of new media and technology for the American Automobile Association

"We will just get more used to sharing data, even when we don't particularly want to, and will learn to modify our behavior in this light. So the outing of past indiscretions will become more commonplace—and therefore less shocking—and so we will forgive them easier." —**Jeremy Swinfen Green**, Telecom Express, an interactive marketing company

"Certainly, conceptions of privacy will change, but increased tolerance for difference? Perhaps slightly, but not much." —**Alexander Halavais**, professor and social informatics researcher, Quinnipiac University; explores the ways in which social computing influences society

"This sounds pretty much on the money to me. Only thing I'd add is that the public's notion of identity has become more sophisticated. The citizens of 2020 seamlessly integrate online, digital, physical, role-based, and private aspects of their identity into something their parents will probably have real trouble understanding. This, of course, has precedents—e.g., the rise of teenage consumerism postwar, or the invention of the nuclear family brought on by the industrial revolution." —**Jeremy Yuille**, digital media coordinator at RMIT Communication Design and program manager at ACID in Melbourne, Australia; previously director at IXDA, the Interaction Design Association

"I agree and have published many papers on this exact topic. If everyone is outed for a past indiscretion, the indiscretion loses its scandal factor." —**Christine Satchell**, PhD, senior researcher, Institute for Creative Industries and Innovation, Queensland University of Technology

"People will be more tolerant of a general lack of privacy and public exposure, and expectations of privacy will shrivel. However, I think being 'outed' will be just as damaging, if not more so, as to be captured on film doing something damaging will show serious naiveté and a lack of sophistication about the world—perhaps even more damaging than drunken antics themselves." —**Jade Miller**, PhD student, Annenberg School for Communication at the University of Southern California, with a research focus on global flows of information and culture

"This scenario is partly 'right.' Yes, there will be a greater degree(s) of transparency, BUT in 2020 there will still be those populations who, whilst are immersed within new social media (NSM), remain at a distance from such technology. The generation that was born in the 1980s was the first to grow up with technology. The level of transparency in this scenario is something that resonates best with their social sensibilities. However, we (and 'them') are yet to be able to have a handle on the impact of having a digital identity that is out there and permanently available. Yes there is a generation of young people that are 'used' to being on YouTube, Facebook, etc., but the culmination of these types of information portals and long tail effect of such information pasts and biographies will still be up for debate in 2020 and not as clear-cut as the scenario above makes out." —**Maz Hardey**, social analyst, blogger, "defender of new media" completing a doctorate funded by the Economic Social Research Council in the U.K., based at the University of York

"In some respects, people will be more open, and in other respects, they may not be—this is not a clear-cut thing." —**Iddo Genuth**, founder and chief editor of *The Future of Things*, a science and technology e-magazine

"Already I see younger people who are happier to share more information publically than I am, and I see that trend continuing. Any other major cultural shift could impact this the other way of course." —**Micheál Ó Foghlú**, research director, Telecommunications Software & Systems Group, Waterford Institute of Technology, Ireland advisory committee; member of W3C; blogger

"I think, in the same way Bill Clinton made past drug use a nonissue, that having documented moments of embarrassment will be normal—embarrassing, by definition, but not fatal to most careers." —**Clay Shirky**, consultant and professor in the Interactive Telecommunications Program at New York University; an expert on the social and economic effects of Internet technologies

"I mostly agree, but privacy requirements will not disappear and will be catered to by anonymous and nym [pseudonum] networks based on open-source strong crypto." —**Giulio Prisco**, chief executive officer of Metafuturing Second Life; formerly department head at the European Satellite Centre, analyst at European Space Agency, and an IT specialist for CERN

"I agree that people will be more comfortable with the lack of privacy and more tolerant of indiscretions thus revealed. Other than that, I am highly skeptical of improvements to general moral character suggested by the scenario." —**Rollie Cole**, director of technology policy, Sagamore Institute for Policy Research, a nonprofit, nonpartisan think tank headquartered in Indianapolis, IN

"I'm going to run counterintuitive here, anticipating perhaps a McLuhan-style media reversal. Consider that TV show where people must answer embarrassing public questions while connected to a polygraph, an imperfect but often uncanny system. Or imagine a sci-fi world where everyone is psychic and can detect duplicity. Would this lead to a more open and honest culture, or a more guarded one? Consider Winston Smith, in *1984*. Under the surveillance of Big Brother, was he embracing the openness and transparency, or did he become excessively guarded of his every move and gesture? I believe the answer is the latter, and the best reason I see for it is a McLuhan-style reversal. Does this lead to more forgiveness of indiscretions? Or greater paranoia over the ordinary things one does day-to-day, that can be construed as indiscretions in the heightened awareness of a surveillance society? Picking your nose? Scratching various body parts? Checking the change cup of a vending machine? I'm inclined to think, as does Foucault in *Discipline and Punish*, the potential observation in the panopticon, the fixing of us all under glass like a butterfly in a case on a pin, creates more discipline and self-censorship, and by extension, less movement of any sort, less action, because movement and action themselves contain risk, and in a discipline society, risk must be minimized, as surely as Major Major Major Major tried to minimize risk in *Catch 22*. Greater integrity and openness? Hardly. Greater fear and suspicion and paranoia? Most definitely." —**Christine Boese**, PhD, researcher and analyst for Avenue A-Razorfish and Microsoft

"I think people's notion of private vs. public is a relative concept; its construct may change over time, but the notion of what is private vs. what is public will always be contested. Thus, a sense of vilification over a sense of reward will always be tied to a relative sense of what is loss and what is gain. Being caught with a skimpy model may not jeopardize a political career because people will have lost faith in digital image documents to be authentic proofs of evidence." —**Shakib Ahsan**, MBA and MA in educational technology, now at Concordia University, Canada

"The adults of 2020 are the youth of 2008. They are becoming insensitive to privacy issues, and they are open to share everything about their lives. To hide some information about past and present life will be irrelevant." —**Rafik Dammak,** software engineer, STMicroelectronics, Tunisia; DiploFoundation participant in the study of Internet

"More likely that digital duplicity will become a high art!" —**Greg Laudeman,** utilization catalyst and facilitator, community technology specialist, Georgia Tech Enterprise Innovation Institute

"Well, enhancement and development of technology doesn't only mean that they are used in a similar way and for the same purpose. Also as the technology advances, there will be a lot of security issues. As Internet for now is very young, a lot of experiments are going on. By 2020, there will be more clarity on what will lead to what consequences and the safety measures. So I don't think there will be less privacy by 2020. Now also, let's talk about WWW. On Facebook, it is the user's choice to allow his friends to see his/her detailed profile or only limited profile. So I think by 2020, there will clearly defined security measures developed and implemented for addressing security and privacy issues." —**Sudip Aryal**, president, Nepal Rural Information Technology Development Society

"My agreement is somewhat tempered by concerns that there is great potential for abuse of all this private information by identity thieves and others. And there is great potential for a public backlash that might well affect future job seekers or candidates for public office." —**Jan Schaffer**, executive director of J-Lab, the Institute for Interactive Journalism; former business editor and a Pulitzer Prize winner for the Philadelphia Inquirer

"We will once again be a small village where everyone knows everyone else…gone will be the days where the cities swallowed up people and hid them from view." —**Garland T. McCoy**, founder, Technology Policy Institute, a think tank focused on the economics of innovation; formerly senior vice president at the Progress and Freedom Foundation

"It is not in human nature to be open, transparent, and forgiving. The problem of the commons (free riding) existed 500 years ago, and it will exist in 2020. People are surprisingly invested in celebrities and living vicariously. This includes watching celebrities 'burn.' I expect this to continue." —**Todd Wagner**, health economist, Health Economics Resource Center, Palo Alto, CA, part of the U.S. Veterans Administration; also involved with the Center for Health Care Evaluation

"Social networks will provide ample opportunity to share as much or as little of one's lifestream as possible. Mechanisms for support, enlightenment, and education will be as accepted as real-world, real-life counterparts. In fact, the distinction between sharing online and sharing in person will be blurred extensively. Business, institutions, organizations will have to follow suit, or they will not be trusted." —**Michael Stephens**, assistant professor, graduate school of library and information science, Dominican University, River Forest, Illinois; an expert on "Library 2.0"

"In 2020 we will have a different mix of cultures than we see today, and cannot take for granted that Western civilization will continue in an independent path, without influence from other cultures that have a different approach to privacy." —**Roberto Gaetano**, ICANN board member; also responsible for software development for International Atomic Energy Agency; an active participant in the ICANN policy-making process

"Integrity doesn't come from telling all to any and everyone who will listen; that is part of someone's character developed by their parents, families, church/synagogue, and other cultural and educational influences." —**Dan Larson**, president and CEO of PKD Foundation, a nonprofit organization working for patient advocacy and education

"As technology increases access into people's lives, people will adapt to protect the information they value. If institutions install spy cameras, individuals will find ways to avoid, jam, or work around them. Jamming and spoofing signals will become a personal defense against invasiveness. People who enjoy sharing their lives will simply find new audiences." —**Buddy Scalera**, vice president for interactive content and market research for CommonHealth Qi, in charge of interactive online strategies, including social and viral marketing

"For a new generation of kids growing up around social networks, MMORPGs [massively multiplayer online role-playing games] and IM, things you can share online tend not to be issued as 'privacy concerns.' Yet I think an indiscrete video on a 2020's YouTube will always be disturbing—and there's no reason why bullying should stop until there." —**Tiago Casagrande**, works with social communications and new technologies for verbeat

"Like providing for retirement, the burden of maintaining privacy will revert more to the individual. As the awareness of what can happen with data on the Net grows, so will a set of personal practices to guard against its pitfalls. The definition of what is privacy may change some, but the expectation of who is responsible for it will alter drastically." —**Jerry McCann**, vice president and director of the social marketing group at Carton Donofrio Parners Inc., marketing and advertising firm in the Washington, DC, area

"Lives will not be completely transparent as digital literacy increases and people get more aware of their online presence. Even though they will share more personal information, they will do so knowing about the effects. Every online identity will therefore, to some degree, be constructed. The ubiquitous access to social networks and being 'always on' will also lead to a change of social norms. For example: Small day-to-day lies like 'I couldn't meet you for coffee because I have been in phone conferences all day' won't work anymore, since your immediate social network will be able to see what you were doing at the time (through your lifestream, status updates, etc.). Instead, slightly more open communication will be socially accepted." —**Peter Bihr**, freelance consultant on Web strategies, communities, blogging, and social media; based in Berlin, Germany

"It's too early to make this kind of prediction. While the Facebook/MySpace phenomenon is real, who knows where the notoriously fickle Internet audience will go next?" —**Joan Connell**, online editor, The Nation magazine, formerly an executive producer for MSNBC.com, senior editor for MSN and a Pulitzer finalist for her reporting

"Yes, people will be more open to sharing personal information. Hopefully, people will be more careful about what they say and do in the Internet. It won't, however, prevent those who are irresponsible at every turn in their lives from putting material on the Internet that is harmful. It's very difficult to take back something that has been virally spread across the Net. Newspapers are able to print retractions where many readers can find them. However, a retraction on the Internet can be totally lost among the billions of kilobytes floating through the ether." —**Mike Samson**, interactive media writer and producer

"The Internet may become the much hoped for 'truth machine' and lie detector. Already, we see users exposing malpractice and poor business service to others. Whilst we are unable to safeguard completely against the criminal element, poor service, lies, and deceit, we will find that people will be able to vote with their pocket if a person/company deserves trust. Intolerance or perceptions of what is/isn't acceptable are already beginning to crumble as a more rigid 'religious-based' lifestyle appears more extreme then one of individual choice and freedom." —**Robert Eller**, Concept Omega, a media marketing and communication company

"We do not need to go forth a decade to anticipate a much more complex (hypercomplex, as Qvortrup calls it) social environment. People will most certainly adopt more flexible identities and more public facets of those identities and that will not produce enhanced transparency; quite the opposite. Still, transparency in that particular sense is not a very desirable goal in itself—it rhymes with conformity, and that runs against the pillars of knowledge appropriation and development." —**Luis Santos**, Universidade do Minho–Braga, Portugal

"In the 20th century, Warhol stated that 'in the future, everyone will be famous for 15 minutes.' YouTube and social networking sites have made this radical statement a reality. 2020, on the other hand, will usher in an era where privacy and anonymity are highly coveted. A 21st-century spin on Warhol is, 'in the future, everyone will be anonymous for 15 minutes.'" —**Ed Dieterle**, Harvard Graduate School of Education; research tied to handheld devices for ubiquitous learning

"The Net is reconfiguring our notions of public and private, as all communication technologies have done in the past. But the Net has even greater impact. We will continue to put our private selves on public display—but I think we will grow increasingly less comfortable with that display being viewed by general audiences, and with data mining efforts to analyze our online behavior. As a result, we will want to exercise more control over who can see our keystrokes, watch our videos, be our Facebook friends. We will want more sophisticated privacy controls to maintain a balance between communication and exhibitionism." —**Dennis Baron**, professor of English and linguistics, University of Illinois, runs the Web of Language site and researches the technologies of communication

"Facebook recently received a major backlash for its sharing of personal shopping information. Ask.com's AskEraser is a huge success. People will become more comfortable sharing with people in their 'circle' and will find ways to restrict access to outsiders, even if it means posting fictitious content." —**Theresa Maddix**, satisfaction research analyst, ForeSee Results

"People simply must come to accept the fact that privacy is a myth." —**Don Ranly**, PhD, professor emeritus, University of Missouri School of Journalism

"The people who care about their reputation the most will still be subject to Swift Boat attacks. It will still be difficult to correct the record in 2020." —**Brian T. Nakamoto**, cofounder of MrJoy Inc. and product-line manager for Everyone.net (a leading provider of outsourced e-mail solutions for individuals and companies around the world)

"Transparency will grow in the coming years, but I think the sheer barrage of personal information is ultimately what has caused any perceived 'forgiveness.' People will use the same tools to manage their identity—i.e., creating online profiles, sites, and posts that express the public version of themselves they wish to share—and will likely rely on that information outweighing any other information available online." —**Jasmine Sante**, Sante Strategies, independent Web strategy consultant in the Washington, DC, area

"Keep in mind that a scenario like this that develops a new sense of privacy is only the positive side. Almost certainly, crime and fraud will be a big aspect of this new tolerance as people learn to manipulate the new freedoms." —**Jennifer Jarratt**, principal, Leading Futurists LLC; works with formalized methodologies to assess and interpret potential futures

"David Brin (*The Transparent Society*) was right. More openness, forgiveness. But I suspect many micropolitical problems will occur in silence—a job candidate turned down with a MySpace image a quiet reason, a relationship soured due to a discovered photo. And religious conservatives are prepared to accept a fallen world, which they can condemn; greater openness is something they've been coping with for a few hundred years." —**Bryan Alexander**, director of research National Institute for Technology and Liberal Education, blogger, expert on computer-mediated pedagogy; based in Ripton, VT

"While I mostly agree with this scenario (much of this is already taking place today), it does imply a prevailing mode of 'good intentions' among future societies...Expecting future societies to be 'more forgiving' is a bit too idealistic. I think that in parallel to such scenario, a one where 'anonymization' technologies would be more ubiquitous should also be considered." —**Fadi Salem**, research associate, Dubai School of Government; research focuses on e-government and development in the Middle East and North Africa

"A forerunner in this regard was Arnold Schwarzenegger when he ran for governor, and people raised past indiscretions. He said, 'Old news. Move on.' There is no doubt that our whole sense of privacy is changing radically." —**Bruce Henry**, Concordia University, Montreal, Canada

"In Brazilian society, there is no paranoia, as seen in North America, about the 'invasion of privacy.' I sometimes think that this easy-going tolerance of what in some other societies would be unthinkable is essentially the result of centuries of middle- and upper-class families having domestic servants who were totally privy to the ongoing trials and tribulations of the family that owned the house. Perhaps the 'compadresco' system of the extended family (godparents, and 'special relations' between persons), which sometimes included domestic servants having well-to-do godparents, and vice-versa, helped reduce any pretense of family secrets or skeletons in the closet. As a result, there is very little identity theft or even seminars on privacy in the world of information technology in Brazil. I believe that this is a healthy view of personal information in a complex society and probably will prevail over time." —**Fredric M. Litto**, consultant for Pearson Education Global e-Learning, president, Brazil Distance Learning Association

"Identity theft will become rampant. Many will totally withdraw from technology as a result (the percentage of the total population withdrawing will increase). New laws will be enacted to protect privacy, but the rate at which those laws are broken will become epidemic. Still, privacy will continue to be valued, although increased transparency will also be accepted." —**Dixon Hutchinson**, software engineer

"The timing of 2020 is probably too early for the scenario described…As for people acting more responsibly due to their lives being under the spotlight, people will continue to make mistakes and kids will be kids, I suspect. But if people are more understanding of others' indiscretions, then you may also get more people to acknowledge their mistakes, and that would be a positive thing." —**Heath Gibson**, manager of research and market analysis, Big Pond, a competitive intelligence company and provider of broadband customer Web sites in Australia

"People will have a heightened awareness of the openness that digital identity enables, but there will also be more people interested in controlling information about themselves, and more regulation to enable such control, so that it will really be a choice and not the default to live your life as an open book." —**Paul Hyland**, executive producer, edweek.org; formerly a member of the board of directors for Computer Professionals for Social Responsibility and director at Media Matters for America

ANONYMOUS COMMENTS

The following section includes a selection of contributions to the discussion of the future of the Internet and privacy, identity, and forgiveness from survey participants who chose to remain anonymous. Among them:

"The Internet at present allows us to create a persona that the world can see; this is not often how we are in real life; I don't think there will be a drastic change."

"The Internet could well become the next greatest world power as political movements go global online."

"The fear of forcefully putting the dark side of people into the light is not the way to build personal integrity. There will be a point of diminished returns in terms of benefits from public exposure. There will be backlash and people will more and more treasure the little privacy they have."

"It will have to be otherwise; future generations will not get jobs."

"People will be more careful about how they craft their online reputations. Just as in the real world, people will judge you on your words and actions. The Internet is your permanent record."

"People will manage their 'online' image as an integrated part of their overall 'offline' image. The distinctions between online/offline will blur. People's behavior will be like that of someone living in a small town—hidden indiscretions may still happen, but people will continue to be conscious of their actions knowing the public nature of what they do. There will be more diversity in interests because people can find communities that mirror their values and interests and will not have to subscribe to 'majority think.' Most people will lead two lives—those on the grid and their private selves off the grid."

"Social-networking-type info sharing is not a universal trend. Endless 'sharing' actually gets quite boring. People are sophisticated interactors, the social-networking aspect is merely a tool, one of many, and will not be the be-all and end-all."

"My view is much more dystopian. Cheap video and audio monitoring is a means of control by powerful organizations like governments and large corporations. Hopefully, technological advances will allow some preservation of personal privacy."

"Much like any other time, society will have its celebrities who thrive on notoriety. Most people posting their party pictures on Facebook now will come to view them as bad tattoos, and there will be similar technologies developed for erasing them. Most people will always want to control how much information they expose, when, and to whom. Information about public figures, such as politicians, will be even more judicious about the information posted. Transparency would help in things like corporate governance and the legislative process, but the people involved in the deal-making may try to block transparency so as not to be held accountable for the compromises necessary."

"The average person leads a very dull life. There should be a point of fatigue that prevents this from catching on."

"While I do agree that people will more freely share personal information through social networking, I also believe that we will see heightened rates of identity theft and privacy breaches that may compromise an individual's professional reputation."

"Keeping secrets will become very difficult in 2020."

"People will take less responsibility for their actions."

"Transparency is great, as long as it doesn't involve me—as in the old NIMBY [not in my back yard] paradigm. For example, current presidential contenders believe in full disclosure, but not of their own plans, positions, or histories."

"Life without privacy is hell. The best people (or 'best' people, if you prefer) will be known for *not* being reachable on the Internet, for *not* having their personal information available. This is true today and will be more true in the future. The best/best people will use alternative channels, like F2F [face-to-face] and RW [real world]."

"Moral standards and moral integrity have been diluted because of the status quo. Creator transparency leads to a saturation of otherwise unacceptable behavior, which in turn dulls our senses to unacceptable behavior."

"People's attitude vis-à-vis threats or breach to their privacy will continue to be context and person dependent. The information overload in both exposure and/or rectification might lead to a reduction of the free disclosure of personal information."

"Privacy will never go away. 2020 will see a system that protects privacy for those who want it."

"I'd agree if the 'transparency' involved those in power! It's not 'transparency' but unilateral stripping of abilities (note I didn't say 'rights').

"Transparency will reduce crime."

"There will be more transparency, but I'm not convinced there will be more forgiveness."

"Nothing will change human nature. We will always want to laugh at the follies and misfortunes of others (as long as they happen on an individual scale), and players in politics and business will always seek any edge to place a seed of doubt about the integrity and intelligence of their opponents."

"Chaos will rule, but people will take mudslinging less seriously, as it will happen to all."

"Privacy concerns need to be default strong to ensure that cases where there is abuse can be dealt with, and to ensure governments and corporations are held responsible for security and integrity."

"Anonymity will not be relinquished as suggested by this scenario. People like it too much. They like not being accountable."

"Facebook and other social networks will change the relationships we have with our ever-widening circle of family, friends, acquaintances, colleagues, etc."

"There will be an about-face now, as more and more people feel violated as the sharing increases."

"Humans are paranoid animals."

"Given the current power of cover-your-ass and the Patriot Act, I believe it will take more than 13 years for people to become that much less judgmental about many things; what this means is there is a risk of more people being vilified or even jailed even as they allow more information to be gathered about themselves. Perhaps by 2020 someone will write a 'megatrends' blockbuster that gets people to start reexamining their assumptions."

"All of the information aggregators connected to social networking software and databases like Google's empire, Facebook, etc., will lead to other computer users', and, more importantly, *corporate* knowledge of our time spent online and what we describe of our time spent offline. This will not, however, enable 'more responsibility' for personal actions. More judgments will likely ensue, and changes in what is OK. Changing notions of what constitutes an indiscretion seem the more likely scenario for future shifts in 'morality.'"

"The more we share, the easier it is to share, the more human we all become."

"This will be accompanied by an increase of lawsuits for defamation, etc."

"This is the philosophy driving many of the moves to Standard Business Processes in Government—the more transparent, the more secure."

"Web content will be filtered more often, weeding out all the bad to make room for the good."

"Our info has become a commodity. Sure, everything is about commoditization, accounting, and cash. It seems IMPOSSIBLE to imagine things won't be embarrassing, which would imply no morals but likely might be a social blindness to some kinds of past events for some people, while others will be held more accountable."

"There will be a learning curve for people sharing their information, but eventually people will figure it out."

"I am more persuaded by [David] Brin's transparent society analysis than any other competing analyses."

"There will be both a great push of some people aiming for media attention, giving away all privacy in hope of fame, and a great withdrawal of many online users into their private online space they share with a few friends."

"People will have many different online 'personalities' and so will be less concerned about the transparency."

"People will learn to better encrypt or learn not to upload as much personal information to the Internet."

"Great privacy problems are just beginning and might not have been solved by 2020."

"Technology does not change human nature. The close-minded and unforgiving have been forced back from particular positions over the last several centuries, but have always latched onto something else that threatens the end of civilization as we know it and continued to oppose it in the same old ways. Once it was Copernican astronomy; then civil liberties and the end of slavery; then evolutionary biology; then prejudice against the poor and the foreigner in general. Now we have gay rights and stem cell research, and a background of evolution denial."

"People will never become more forgiving. The damages inflicted by online embarrassment may become more short-lived, but there will still be damage."

"The policy-aware Web should be ready by 2020 to solve this issue!"

"Information is no longer secure. There have been so many public details lost in the U.K. over the past few months. Are people worried about this? Personally, I keep my personal Web interactions to a minimum, as I don't want my personal details misappropriated. There's a big problem with identity theft, and cyber identity theft could be a major issue over the next few years."

"People are mainly unaware of how much of their privacy they have lost; once it becomes clear, I believe there will be a backlash. Improved privacy software will fill the current market void eventually."

"This is close to 50–50. I doubt that anything will ever make people more responsible for their own actions. I expect people to become simply more jaded so as to not care what is 'outed.'

I also expect that most people are modest only because it is fashionable and fashion is changing, people are becoming too free with their private information."

"There is a generation gap. The younger generation will continue to make their lives transparent, but as they mature, there is a greater chance that the transparency will reduce. It only takes one event to change the tide on transparency."

"I'd like to believe in the integrity-and-forgiveness scenario, but the U.S. political climate for the last 8 years or so has made me more of a cynic."

"Technology has nothing to offer to morals."

"The risk of an attitudinal pendulum swing exists that would mitigate this sense of openness."

"We have not yet seen the worst of the future repercussions of what is currently being shared on the social networking sites."

"I highly doubt that any piece of technology will change human nature. YouTube and Facebook allow people to create half-truths and fictitious versions of themselves (note the recent media coverage of the MySpace suicide)."

"There are huge differences in norms among online users. Social norms for privacy will not move nearly as fast as the frontier for tolerating lack of privacy."

"The older population—that has been most aware of privacy issues—will decline in comparison to the youngsters just beginning to use the Internet and social networking tools. The fact of youthful indiscretions uncovered from YouTube or other sources will have less of an impact on overall reputation."

"A generation ago, middle-class girls from 'good families' appeared nude in *Woodstock*, the movie (as well as in Woodstock, the event). Now they do so in 'girls gone wild' events. Privacy is changing independent of the number of transistors per cm squared. While the nature of privacy is changing and people's use of technology reflects that change, technology isn't necessarily driving that change."

"Quite the reverse is happening—misinformation dominates."

"There will be a backlash against all organizations that do not maintain an individual's personal information in a secure manner, and this has already begun."

"This could go either way. The scenario provided is a positive spin on a future. The darker version is that people get slammed by the transparency, and go underground via closed networks and sites rather than public forums."

"As identity theft and governmental abuse continue, people will start demanding their privacy back."

"I don't know about the corrections, but we are already half-way there. A lot of people (a majority?) have given up on the idea of anonymity and privacy."

"Sadly, people are people are people. The benefits of anonymity are benefits not only to users with integrity and open-mindedness but also to those with closed minds and bigoted views."

"The public notion of privacy will remain roughly the same, though there will be more opportunities to disclose information in return for services. A high-profile case of mishandled personal data will cause a public backlash against the collection and archiving of such information. Users will become (or will be educated to be) more mindful of disclosure and will demand privacy controls in online applications. People will maintain several distinct personas online, i.e., professional plus several interest-based/cultural/religious identities, and insist on privacy controls to keep them separate to some degree. Greater tolerance and forgiveness is a function of increasing societal liberalism, and the 'outing' sounds like celebrity gossip to me—the public appetite for that will likely remain constant."

"With more opportunity for public expression, a more pronounced dichotomy will develop between the public self and the private self, a kind of split personality, or even multiple personalities/profiles…An individual's MySpace profile is never an accurate representation of the self, but rather a representation of how a user wishes to be perceived."

"What is considered acceptable will change."

"Technology may change, but human nature will not evolve so quickly. Scarce resources and the normal fear/flight protection syndrome combined with the comfort of being with those whom we identify as like us will still prevail."

"I think this would be true except that the terrible economy of 2008–2015 will create a culture of competition, greed, and hoarding in America. People need to 'have enough' and be actualized in their jobs in order to share and collaborate."

"The scenario described is a post-9/11 surveillance wet dream."

"People don't change their judgmental attitudes anywhere near that fast. Much more likely: People gradually see friends and neighbors destroyed when past indiscretions become the Internet meme of the moment and slowly realize they have to be just as careful of online appearance as real life. (Besides 'benefits' of sharing data? The benefits accrue mostly to the network owners.)"

"We're going to have a massive correction on this fetish of exposing ourselves on the Web, as kids grow up and have to explain why mommy shaved her head or forgot to wear panties when she left the house."

"Our current fascination with being transparent will change in some way. And YouTube will be so far out of date it isn't even funny."

"This is the direction we are heading, although I think this leads to a lot of nondocumentation, where people will simply not journal or record certain activities because of the repercussions."

"I don't see technology fundamentally making people more forgiving. They are and will continue to be not so smart about protecting their personal information."

"More data will be shared, and YouTube videos outing people will no longer be a major drama, but I don't think human nature or Western culture encourages personal responsibility or kindness of the kind described here."

"More open means a more informed citizenry and better chances for social democracies."

"In 10 years, people will start being affected by consequences of the free-for-all information and understand the value of information. In the future, people will shy away rather than be more transparent."

"I don't think people will have a choice in the future regarding personal information—in order to receive a service, etc., one will have to opt in regarding personal data, without much of a choice. Sandra Braman is absolutely correct when she speaks and writes of the 'Informational State,' the state [government authorities] will indeed make use of this data. This does not bode well for the future, as the public will become increasingly more transparent while government will become increasingly less so."

"I agree there will be even less privacy. However, I disagree that this will lead to tolerance and acceptance."

"Civilization has already become desensitized to sex, drugs, and rock 'n' roll through prime-time television. Morals and ethics are diminished; criminals have more rights than law-abiding citizens due to the greediness of the legal profession. Privacy is almost nonexistent now, and government control is increasing in the perceived face of terrorist attacks and identity theft incidences. People give up privacy in order to protect themselves financially."

"Legal provisions and reputation systems will be used to guarantee more privacy, also on past-publicised events."

"High-profile horror stories will unfortunately probably be necessary to remind people that they need to be careful."

"I doubt this will contribute to a civil society."

"Equating transparency with honesty and integrity? People will be more guarded about what they value most if they feel they cannot be secure in their anonymity. The children and teens will become more cautious as they grow up and move away from use of sites that cater to their culture."

"Forgiveness might actually decrease, as the original offense is always out there and can always be resurrected."

"Oddly enough, it isn't that I think people are becoming forgiving, they are becoming numb. What was once appalling is now just surprising. In 2020 people will actually be *more* protective than they are now. What will likely drive this is some action of egregious misuse of personal information, e.g., terrorism."

"Privacy is and will be under attack. Social networking is voluntary; however, other forms of information gathering and aggregation are invasive and will only escalate in the future."

"As the consequences of sharing personal information become clearer, people will share less. Also, with the increased amount of information, there will be decreased opportunity for fact checking deliberate lies and distortions inserted into pervasive media."

"Future generations will be comfortable, but I don't think Boomers and Generation X will be as comfortable."

"Given the fact that various religions have preached this message for generations without significant success, it is difficult to believe that technology will be successful. Privacy and the public's attitudes toward it may change, but integrity and forgiveness are not likely to change as a result of that."

"The mainstream of open information will create even more determined strongholds of privacy as backlash."

"Transparency will degrade individual integrity. There will always be opportunists who will see this openness as a means to personal gain. There still must be a dividing line between private and public. I do not see people becoming more responsible for their own actions. If anything, the opposite may be true, looking always to shift blame. This transparency could facilitate this more easily."

"A line will be drawn by each individual to delineate what's private and what's not. I just don't know how small the resulting circle of privacy will be. It will be much smaller than it is today, and much smaller than what I'm comfortable with."

"Human nature stays the same no matter what technology enables."

"There is a backlash coming—the pendulum is much more likely to swing in the other direction (i.e., people insist on more privacy/anonymity) than not."

"The more we know about each other's humanity, the more integrity, forgiveness, and compassion."

"We will see people pull back on what they make available on the Internet."

"Privacy and security will continue to be an issue. The scenario defined here is very utopian."

"There will also be a market for 'predictive speculation,' evaluating all the personal information available about someone and predicting publically what they are likely to do in the future."

"Decreased transparency will follow from the increased governmental and industrial attempts at controlling the use of both technology and the information content. An unfortunate response will follow: less information actually available to users."

"Yes, but it is a coevolution with cultural acceptance of different behaviors. As a society, we are more diverse and, therefore, better able to handle the coworker seen drinking on YouTube. I still hope proper discretion will be exhibited by all individuals."

"Corporations will fine-tune data mining in social networks in order to emphasize consumerism. Governments are currently salivating over the potential to control their citizens even more through the use of such transparent social networks and will soon conduct new types of online surveillance."

"Humans need their little secrets. It is encoded in our DNA. We cannot and will not feel comfortable naked, even if everyone else is."

"Some people may become more immune to it, but it will still be shocking for others, just like reality TV."

"Despite all the positive attributes of the Internet, cell phones, and other related technologies, people will still be concerned about privacy."

"Distasteful as this future sounds, it seems likely that it is confirmed—Orwellian and full of reputation-corrective measures. Sounds nice for lawyers."

"There will be an enlargement of the definition of the individual to accommodate various roles that a person can project."

"Personal responsibility and forgiveness are two human traits that will never become very pervasive in the general population."

"Privacy shall never be jeopardized. The lack of privacy is the door to many abuses and mostly favours dictatorship of all kinds."

"There will be less stigma for past indiscretions, yes, but I don't think data will be easily correctable. I think there'll be less stigma because the number of people who have been screwed by a similar circumstance will have increased, not because it'll be easy to correct (and correcting is basically going back and rewriting history, right? Who's to say people won't airbrush their past?)."

"Although people will be even more open to sharing their personal information, etc., this will not lead to more forgiveness; and I am afraid that it will be a lot longer before 'carefully investigated reputation corrections and clarifications' are seen on major media sites."

"Palates are becoming jaded, and the assumption is made that the more you see something, the more it is 'OK,' so by 2020, the dark secrets that used to (perhaps rightfully) be things you should be ashamed of will be 'OK' because people will be desensitized."

"There will be a lack of privacy but a bigger importance of online reputation both on the personal and working sides."

"Much of this will hold true—especially the idea of forgiveness for past indiscretions—but let's not mistake the fad of today's MySpace/Facebook tsunami for a societal change. There *will* be a backlash against such wide-open intimacy. However (as Pew research has shown), the upcoming generation will still be more willing to share what was once considered personal information than previous generations. But not to the extent shown here."

"Most folks have an online persona, their representation of themselves to the online world. If they are smart, the profiles will contain no intimate details that are true. Online payments will be made through anonymous accounts with prepaid limits. The very idea of meeting in real life is preposterous, and an invitation to do so will be considered bad manners."

"This is further than 10 years away. Expect to still be in a backlash against transparency in 2020. Ultimately, people may get worn down enough not to care, and there is a large section of the population who already don't care about the invasive nature of technology, but if this 'transparency' comes as part of a government mandate (at least within the U.S.), then there will be significant outrage."

"History has shown that access to information about personal behavior of large numbers of people leads too easily to oppressive uses."

"People can still hide behind their online images, which may be good facades or may reflect reality."

"This seems a highly imaginary scenario for 2020. 'Reputation corrections'? Are you kidding? Not in this world of titillation. Privacy remains extremely important as instances of either invasion of that privacy or foolish surrendering of it clearly demonstrate the unpleasant (to put it mildly) consequences."

"The idea that a reduction in shame will produce an increase in personal responsibility seems incoherent. Major media outlets in 2020? What will those be?"

"We still need some privacy ;-)."

"More identity management platforms will arise to combat identity theft, manage reputation by polling, as well as claiming links and IP."

"This will be led by the millennial generation and fought by all others."

"Things will come to a point where all information or data on the Internet is disposable."

"People will be more cautious, not less, in the future."

"Facebook, etc., have already forced people to be more consistent between online and offline identity, since people increasingly interact with the same people on both."

"Those in power will use information to maintain power. As with concerns from historic fascist states, the power of information will be used to reinforce the position of fascists. Society will become more divided as the things that distinguish us one from another become ever more under a microscope. In the past, when indiscretions might be forgiven because no one knew the true story, here the 'true story' will be on YouTube, and the forces of division and alienation will continue to grow. This will breed a counterculture that is far better at managing personal information and is able to subvert the system."

"Unethical people will continue to do unethical things, regardless of (or even because of) who they think may be watching. This is primarily because they know that it is easier to beg for forgiveness (ideally with crocodile tears and earnest apologies on widespread media) than ask for permission. People will simply become more tolerant of increasingly outrageous behaviour."

"This is a pendulum, and it will swing back to desiring more privacy as the millennials find their indiscretions coming back to haunt them later in life."

"Transparency doesn't necessarily overcome prudery, biopolitics, or normative expectations."

"Increased cases of online harassment and loss of privacy are the result of the increased connectivity."

"Global criminal organizations will be perceived as a major threat to security, far beyond privacy. As long as drug-associated crime increases, people will not share private information."

"Forever there has been the need for having seen/unseen parts of your life, because that helps you actively manage what it means to be 'you.' If there is no boundary, then a person will believe they have lost control over their free will. My other concern is that it creates a rich/poor chasm, because the wealthy will pay for ways to screen/wall out the prying eyes. That's not good."

"I don't think that 'careful investigation' will be any more responsible or 'careful' in 2020 than it is in 2007."

"It will be 2030 or later before we see this happen."

"People will still love embarrassing others. I do think, however, that what it takes to embarrass someone will get ever more extreme."

"Privacy is gone forever; in a dozen years, people will come to accept this fact."

"Forgiveness is not the nature of the beast called media. It is unlikely that much will change. People will be held more accountable for past actions since everything is recorded, changing opinions will be more and more difficult. However, people will become more skilled at interpreting and understanding privacy rules. There will be a few high-profile cases which will make providers of personal information think twice about the future consequences of what they put online; people will probably become more protective of certain aspects of their identity while continuing to push the boundaries for other aspects that have proven to be less problematic in the face of eternal registration of everything people do and say online."

"As people interact with each other more and information is exchanged more freely, we will be less embarrassed about revealing truths."

"The notion of privacy will be more informed, but people will not want to divulge info and more than they do today."

"A 'backlash' will have occurred by 2020. People will be actively seeking ways to *limit* access to their information and looking for opportunities to mask their identity."

"Propensity to share does not equate lowering of privacy concerns. Given the proliferation of media, corrections will be too time-consuming."

"I would disagree that individuals will become more forgiving because, as transparency heightens, there is less individual control, which may make people feel more trapped and isolated. These feelings could create less forgiveness and empathy."

"Much of the current social networking phenomenon will die of boredom soon. Most people's personal information, opinions, and emotions just aren't all that interesting to anyone but themselves. Right now people are enamored with the technology and process, but I don't think people genuinely connect with strangers over the Internet. People, as a mob, are not forgiving, and there isn't enough time between now and 2020 for them to evolve into a new species. Same with integrity—not enough time to evolve more of it. They may, however, learn to be more careful. I hope people will come to understand we all run the risk of losing our personal liberty until we come to expect and demand personal privacy not only from other people but from our government."

"This will have to happen; otherwise, the Internet could face a 'nuclear winter' where people are forced offline in order to protect their privacy from each other and from authorities and companies that use the information for their own agendas."

"Our perceptions of what is acceptable are widening in the digital age."

"Personal disclosure is a fad with a psychic reward that will fade. There will continue to be a wide range of personality types with only some choosing a lifestyle of personal branding and messaging."

"The future of what 'privacy' means will evolve into something very different than what people understand it to mean today. Everyone will share, and it will be clear that everyone makes mistakes and overcomes obstacles. We will be much more forgiving, and what will our 'heroes' look like then?"

"People will not be more forgiving of pasts of others. Society is becoming more selfish."

"The value of connectivity will become more important than privacy. We will have to all work to make sure this open space is not corrupted by identity theft and corporate piracy."

"The opportunity cost of privacy will become increasingly prohibitive."

"Honesty is a good thing. Find your comfort level."

"There's going to be a backlash against sharing so much of one's personal life. And I hate to be cynical, but I doubt the media in 2020 will have 'carefully investigated' anything."

"While people are sharing online, much of it is not reality. The online world has opened up a Pandora's box of situations where untruths and doctored photos, videos, and written false statements and comments prevail. Many people are reluctant to put their true feelings/thoughts online. There is no reason to believe many of the things you read online."

"Nothing can change human nature."

"Human curiosity and human sense of self-dignity include the quest for privacy and the desire to be left alone. Since governments are not becoming more transparent, the fight over sharing information with the state will continue, unless someone instigates some human catastrophe that will scare people into submission to the overwhelming power of the state."

"Reputations are meant to represent, not entertain."

"Kids simply believe they are invincible and, as such, will subject themselves to the most embarrassing things, including being videotaped. Also, there will continue to be problems with security breaches that result in fluctuations in the notion of privacy and just how secure a person's information is online."

"Whatever happened to privacy? It seems like many people don't even weigh it anymore."

"The abuse of personal information will most likely outweigh the benefit of sharing such information with parties unknown. There has been no example that indiscriminate sharing leads to any benefit. There will constantly be pressure to mine personal information due to commercial and political incentives, but even those who favor such have not shown any ability to, in fact, take advantage of such information. It is likely that an incident that harms many people's interest will happen and increase the awareness. As for the tolerance of past indiscretions exposed on the Internet, it'd be in general no different from the same in other media or social context."

"You have only to look at the growth of scam e-mail, the growth of identity theft, or the increased prevalence of child predators to see how the loss of privacy has allowed evil to thrive. While there will certainly be increased openness, there will rise a whole industry built on partitioning and protecting the data we want to keep public from that we wish to share."

"Yes, people will be more open with their opinions and emotions because it is more comfortable to speak to a machine than it is to an actual human being when it comes to touchy subjects."

"There is a backwash to major privacy invasions and revelations. Citizens of conservative worlds want more anonymity as governments become more restrictive."

"Personal information revealed online will become content that is owned by the person generating the content, just as other content is. It will no longer be available for harvesting without paying fees to the content owner."

"And we will be absolutely inundated with the minor peccadillos of very minor celebrities."

"Information provided in public has no expectation of privacy. Avatars will still be used as nonprofessional personas, so that people can manage what information becomes public about themselves. Transparency in professional life will, however, bring greater accountability and integrity."

"Indiscretions vs. comfort in exchanging privacy for convenience are more distinct. I also see more restrictions (mostly due to commerce, not shame) on 'amateur' sites."

"I don't see a backlash against the tell-all exhibitionism and reality-show era we're in now."

"Being 'outed'—although not perceived by the anonymous viewers as so serious—will be the source of a great deal of pain for victims."

"The generation growing up with a very open attitude toward sharing personal information and emotions online will still not have enough power to change the overall perception of privacy by the year 2020."

"There will be a significant backlash against the 'look at me' generation. If anything, the relative openness that we have seen in recent past has been affected. People create the persona that they can't live in real life and 'manage' their online presence in a way that is very fake. That will only increase to the point where most people will have little trust on how people present themselves."

"Society and culture continually swing between poles of privacy and transparency. Assuming that there will be an abiding trend toward transparency is a bit too linear."

"Verification of what is presented as the truth will become a common activity. The question then becomes, who is to say what is true or not?"

"Individuals will be so inundated by content that they will become numb. The bar will be raised for outrage, but it will still exist. Expect that within microcommunities, standards and expectations will vary."

ENDNOTES

1. From *Testimony of Chief Privacy Officer Hugo Teufel III Before the Senate Committee on Homeland Security and Governmental Affairs: Protecting Personal Information: Is the Federal Government Doing Enough?* by U.S. Homeland Security, June 18, 2008, available at http://www.dhs.gov/xnews/testimony/testimony_1215461682919.shtm.
2. From *Privacy: Agencies Should Ensure that Designated Senior Officials Have Oversight of Key Functions*, a report from the U.S. Government Accountability Office, May 2008, available at http://www.gao.gov/new.items/d08603.pdf.
3. See the DataPortability Web site, available at http://www.dataportability.org/.
4. See *A Race to the Bottom: Privacy Ranking of Internet Service Companies*, by Privacy International, September 2007, available at http://www.privacyinternational.org/article.shtml?cmd%5B347%5D=x-347-553961.
5. See the Pudding Media Web site, available at http://puddingmedia.com/.
6. See "Information Searches That Solve Problems," Pew Internet Project, December 2007, available at http://www.pewinternet.org/pdfs/Pew_UI_LibrariesReport.pdf.
7. See "Digital Footprints," Pew Internet Project, December 2007, available at http://www.pewinternet.org/pdfs/PIP_Digital_Footprints.pdf.
8. See *Search Privacy Practices*: A Work in Progress, by the Center for Democracy and Technology, August 2007, available at http://www.cdt.org/privacy/20070808searchprivacy.pdf.
9. See *Privacy and Human Rights 2006 Executive Summary*, by Electronic Privacy Information Center and Privacy International, December 2007, available at http://www.privacyinternational.org/article.shtml?cmd%5B347%5D=x-347-559060.
10. See *The Future of Reputation: Gossip, Rumor, and Privacy on the Internet*, by Daniel J. Solove, December 2007, p. 11, available at http://books.google.com/books?id=42nJLWB0ztYC&pg=PA11&dq=Solove+%22Future+of+Reputation%22+%22scarlet+letter+in+digital+form.
11. From "Policies Designed for Self-Interested Citizens May Undermine 'The Moral Sentiments': Evidence From Economic Experiments," by Samuel Bowles, published in *Science*, June 20, 2008, abstract available at http://www.sciencemag.org/cgi/content/abstract/320/5883/160.
12. See the Semantic Web site, available at http://www.w3.org/2001/sw/.

Scenario 5

The Evolution of Augmented Reality and Virtual Reality

Prediction: *Many lives are touched by the use of augmented reality or spent interacting in artificial spaces. In 2020 virtual worlds, mirror worlds, and augmented reality are popular network formats, thanks to the rapid evolution of natural, intuitive technology interfaces and personalized information overlays. To be fully connected, advanced organizations and individuals must have a presence in the "metaverse" and/or the "geoWeb." Most well-equipped Internet users will spend some part of their waking hours—at work and at play—at least partially linked to augmentations of the real world or alternate worlds. This lifestyle involves seamless transitions between artificial reality, virtual reality, and the status formerly known as "real life."*

Expert Respondents' Reactions	
Mostly Agree	55%
Mostly Disagree	30%
Did Not Respond	15%

All Respondents' Reactions	
Mostly Agree	56%
Mostly Disagree	31%
Did Not Respond	13%

Note. Since results are based on a nonrandom sample, a margin of error cannot be computed. The "prediction" was composed to elicit responses and is not a formal forecast.

Respondents were presented with a brief set of information outlining the status quo of the issue 2007 that prefaced this scenario. It read:

> While most current Internet interaction is found in the user-generated content and social networks of Web 2.0, the 3-D Web-computing ecosystem is developing quickly. Augmented reality enables the enhancement of real-world information through the use and confluence of the Internet, RFID, GPS, smart-tag networks, and portable/wearable information technology. 3-D environments, which are just beginning to be more efficient and accessible, offer ideal design spaces for social and economic experimentation, rapid prototyping, and customized and decentralized production. Every item in the physical world is being mapped, tagged, and databased, as humans build mirror worlds (data-enhanced virtual models of the "real" physical world, also known as digital Earth systems or the geoWeb), and innovate in new, virtual worlds (Second Life, Cyworld, World of Warcraft). MIT's Fall 2007 Emerging Technologies conference had a headline session titled "Second Earth: Second Life, Google Earth, and the Future of the Metaverse," with the explanation: "Social virtual worlds such as Second Life and mapping tools such as Google Earth are beginning to overlap, perhaps foreshadowing the advent of an immersive, 3-D "metaverse." A 2007 Gartner study estimated 80% of all active Internet users will have virtual selves by the end of 2011.

OVERVIEW OF RESPONDENTS' REACTIONS: A modest majority of respondents agreed with the idea that time spent leveraging augmented and virtual reality for various uses will continue to grow; some noted that by 2020 AR and VR will have reached

the point that reality itself will be blurred. Many projected that this will enhance the world, providing new opportunities for conferencing, teaching, and 3-D modeling, and some added that breakthroughs to come may bring significant change, including fusion with other developments, such as genetic engineering. Some respondents feared negative ramifications, including possible new extensions of the digital divide, an increase in violence and obesity, and the potential for addiction or overload. Because of this, some respondents noted that people may begin to "opt out" of using AR and VR tools. Many of those who disagreed with the scenario said VR will not reach the scenario's level of acceptance or sophistication by 2020 or indicated that its primary users will "still be geeks and gamers."

More than half of the respondents mostly agreed with this scenario while just under a third disagreed.

"Our beloved mobile handsets (no longer 'phones') will make seamless traveling within electronic circles of our own creation eminently possible," wrote **Susan Crawford**, OneWebDay founder and ICANN board member. "We won't see the difference between RL [real life] and other life—our presence will be felt whether we're there or not. It already is." **Nicholas Carr**, author of *The Big Switch: Rewiring the World, From Edison to Google*, noted, "By 2020, the virtual world will have blended with the physical world; to speak of them as separate spheres will seem anachronistic." **Rollie Cole**, director of technology policy for the Sagamore Institute for Policy Research, responded, "This scenario comes closer to my predicted result than any of them so far." And an anonymous survey participant predicted, "Virtual identities inside worlds like Second Life will become as ubiquitous as e-mail addresses. Telecommuting to virtual worlds for nontech jobs will become more popular as housing and commuting costs continue to rise."

Jamais Cascio, a coauthor of the *Metaverse Roadmap Overview*, a report on the potential futures of VR, AR, and the geoWeb that was released in 2007, commented, "The striking aspect of this scenario is that, for these everyday inhabitants of the metaverse, this is real life. We in the present don't think of ourselves as living in 'cyberspace,' even though people of a decade previous would have termed it such. Of the various forms of the metaverse, however, the majority of activity will take place in blended- or augmented-reality spaces, not in distinct virtual/alternative world spaces."

Adrian Schofield, manager of the applied research unit at Johannesburg Centre for Software Engineering in South Africa, responded, "Much will depend on the ability of the hardware and power sources to keep pace with the software that enables the metaverse/geoWeb. It also remains to be seen if the proliferation of wireless has a negative impact on the human body."

Clay Shirky, author of the book *Here Comes Everybody* and a professor in the Interactive Technologies Program at New York University, sees success for augmented reality, not for virtual worlds. "Augmented reality is in many ways the opposite of virtual worlds," he wrote. "Fusion of data and physical space will succeed, VR alternatives to it will not."

Jerry Michalski, founder and president of Sociate, a technology consulting firm, commented, "I see worlds like Second Life as of very limited interest. However, gaming environments from Webkinz to World of Warcraft are extremely popular and teach more valuable lessons than the early generations of single-player video games. Also, the tagging and instrumentation of the world as an augmented reality will soon find some useful applications, making it increasingly common."

Hal Varian, chief economist for Google, predicted, "The transition will be driven by gaming, but I hope to see scientific and educational spillovers."

Bryan Trogdon, president of First Semantic, a company working to leverage the Semantic Web, wrote, "Wall-sized televisions supporting blazing-fast data transfers, voice recognition, and a fully realized Semantic Web will blur the lines between real and virtual. This 'teleliving' will fundamentally change the way we shop, work, learn, and live."

John Murphy, director of USA.gov technologies, noted, "The existing virtual worlds are similar to playgrounds; future virtual technologies will have been adapted to productive uses, which will drive even more acceptance and use."

Garland McCoy, founder of the Technology Policy Institute, a think tank focused on the economics of innovation, responded, "Humans have always looked at themselves in the mirror and daydreamed…now the reflection in the mirror can look back at us…and daydream."

Barry Chudakov, principal with the Chudakov Company, commented, "David Gelernter saw this coming a decade ago, and much of what he wrote in *Mirror Worlds* will be commonplace by 2020. We are augmenting our ability to see and imagine our world; we are literally walking into the mirror and exploring the reflection. This has huge implications for what we see there and how we see ourselves when we're in these mirror worlds. In most of human history, we have not had simulations to describe and invent ourselves other than texts and two-dimensional representations. These mirror worlds are multidimensional experiences with profound implications for education, medicine, and social interaction. 'Real life' as we know it is over. Soon, when anyone mentions reality, the first question we will ask is, 'Which reality are you referring to?' We will choose our realities, and in each reality there will be truths germane to that reality, and so we will choose our truth as well."

An anonymous respondent noted, "It will be a business and social tool and also a way of testing hypotheses. And it will be a way of separating social and business/fun and serious/angel and devil-me identities."

Jason Stoddard, managing partner at Centric/Agency of Change, predicted, "Augmented reality will become nearly the de facto interface standard by 2020, with 2-D and 3-D overlays over real-world objects providing rich information, context, entertainment, and (yes) promotions and offers. At the same time, a metaverse (especially when presented in an augmented-reality-overlay environment) provides compelling ways to facilitate teamwork and collaboration while reducing overall travel budgets."

Those who disagreed often shared the point of view expressed by **Joanna Sharpe**, senior marketing manager for Microsoft, who wrote, "I don't think most well-equipped Internet users will spend some part of their waking hours, at work and at play, linked to augmented, virtual-reality worlds. A smaller subset of the well-equipped Internet users will spend time as outlined in this question, but it's going to be a smaller percentage of Internet users, 5–10%, tops, not most."

AR, VR, and the GeoWeb Will Enhance Our Lives; Blurring Will Eliminate Distinctions

Many of the respondents who mostly agreed with this scenario said it will offer positives that will benefit people in some way. **Fred Hapgood**, technology author and consultant, noted that the lack of regulation thus far in virtual worlds is an attractive feature. "If you want to throw a rock concert online, you don't have to post bonds, buy insurance, rent portable toilets, and so on," he explained. "There are no closing costs associated with buying virtual real estate. As time goes on and the thicket of regulation in the physical world gets denser, this feature will become more and more important."

Cliff Figallo, social innovator and original member of the first online community, The WELL, now of AdaptLocal.org, wrote, "Virtual worlds will help local communities plan their adaptation to the impacts of climate change." **Jill O'Neill**, communication director for the National Federation of Abstracting and Information Services and author of the Infotoday blog, commented, "This will happen on the basis of economics and any forthcoming fuel shortages. It is easier (and far less costly in terms of time and money) to have people interact in a virtual world rather than have them traipse around the world."

A number of survey participants said in their responses that virtual worlds will revolutionize training and education—all forms of knowledge sharing. **Debbie Murray**, associate director of the University of Kentucky's health education extension office, noted, "Many of our problems can be solved inexpensively by being able to simulate real-world conditions and manipulate those conditions to arrive at projected outcomes." **Jane Sarasohn-Kahn**, founder of THINK-Health, responded, "The metaverse and augmented reality will have a transformational impact on health and health care." And **Peter Kim**, a senior analyst for Forrester Research specializing in e-strategy and management, wrote, "Educational applications of virtual reality will prove to be highly valuable. Individuals will be able to learn in new ways and improve their physical beings through virtual experimentation."

Tze-Meng Tan of Multimedia Development Corporation in Malaysia, a director at OpenSOS, responded, "The virtual world removes all barriers of human limitation; you can be anyone you want to be instead of being bound by physical and material limitations. That allows people to be who they naturally are, freed of any perception they may have of themselves based on their 'real life'—it is the power of removing the barriers of your own perception of yourself."

Beth Hespe, vice president for Garfield Group Public Relations, predicted, "The notion of a mirror or virtual world will be replaced by another version where both are merged. They will not be separate. It will hard to define where your real self and virtual self end as GPS/LBS [global positioning system/location-based services] functionality are merged into devices of all kinds."

Some respondents wrote that people online will blend real-life and virtual applications. **Gbenga Sesan**, an Internet-for-development consultant for Paradigm Initiative in Nigeria, commented, "The difference between 'real' and 'virtual' is becoming less obvious/important. By 2020, anyone without a search result through Google may be assumed dead (or to be using a pseudo name because even dead people will have information at least on Wikipedia). Real life in 2020 will not be very different from what was known in 2007 as 'virtual life'!"

An anonymous survey participant predicted, "Being in a 'metaverse' will not be a separate activity for most people. Just as in today's world there is less concept of online/offline (because some people are always 'online'), there will be no sense of 'going' to a metaverse—in their daily activities, people will interact with information-augmented physical places...People's presence will be shared ubiquitously—where they are, what they are doing."

"Is the future of the Web 3-D and integrated with the real world? Of course it is," wrote **Alexander Halavais**, professor and social informatics researcher at Quinnipiac University. An anonymous respondent wrote, "Interface design in general is moving toward the metaverse, which means that everyone who interacts with a computer will encounter augmented reality." And another wrote, "Just as e-mail today augments other forms of communications, artificial spaces will augment real spaces." **Jim Witte**, a professor at Clemson University who researches Second Life and the differences between online and offline society, responded, "Mobile devices will act as the means to access and seamlessly bridge artificial and virtual worlds and maintain a sense of blended reality."

Giulio Prisco, chief executive officer of Metafuturing Second Life, formerly an analyst for the European Space Agency and IT specialist for CERN, responded, "This scenario is correct, but people (especially young people) will just see this as part of their 'real' life." **Rob Boostrom**, a PhD student at Southern Illinois University–Carbondale and creator of the Society for Word of Mouth, a group for educators to share ideas about sharing ideas, wrote, "Many of the distinctions that are popularly used now to suggest a great divide between 3-D Web technology and 'reality' will no longer be used. It will be recognized as another aspect of interacting in reality much like the telephone is today."

DJ Strouse, an international relations and computer science student at the University of Southern California, predicted, "People will stop calling metaverses 'virtual reality' and begin referring to them the same way they refer to Miami, Tokyo, or Capetown."

Rafik Dammak, a software engineer for STMicroelectronics in Tunisia, responded, "Existence will defined by a person's presence in all-virtual or almost-virtual spaces. The absence from artificial reality will be a sort of death. Virtual spaces will be the continuity of real life."

An anonymous respondent commented, "The distinction between 'real' and 'online' life is the result of both novelty and technophobia, and it will disappear. Just like other technologies (telephone, TV, etc.), the Internet will be accepted and integrated as part of real everyday life—rather than being seen as separate from or competitive with it."

Don Heath, an Internet pioneer, former board chair for the Internet Society and member of the U.S. State Department Advisory Committee on International Communication and Information Policy, wrote, "There is little doubt that the ability of human senses to be realized with Internet interaction will occur. How widespread the use will be is subject to question, and that depends on the purposes for their use."

Some People Say They Are Already Augmenting Reality and Living in VR

Many respondents noted that the transition to individuals' cultivation of more life experiences online has already begun. "*Augmented reality* and *artificial spaces* are apt terms, and they're already blended into our noisy environment nearly everywhere; it's bound to get more cluttered," wrote **David Allen**, PhD, Temple University.

Those who mostly agreed with the scenario said it especially rings true for younger people. "The youngest generation of Americans, the children under 12, are the fastest adopters of virtual-reality games," wrote **Janet Cohen**, a blogger, futurist, and trend analyst. "Parents buy them stuffed animals that include a subscription to multiplayer games. This generation will be the first to be completely comfortable in virtual worlds."

Josh Quittner, longtime technology writer and executive editor of *Fortune* magazine, added, "As computing power increases and our ability to render lifelike (and dreamlike) graphics matures, more believable forms of virtual worlds will take hold. While current iterations of virtual worlds (Second Life, etc.) still have enormous room to grow, a whole generation of children is growing up on Club Penguin and Webkinz. They will continue to socialize in more sophisticated virtual worlds as those worlds evolve."

Catherine Fitzpatrick of the Open Society Institute, a lecturer on humanitarian issues, noted that people are already commonly using

MapQuest and Google Earth, and some are highly engaged in Second Life and World of Warcraft, so AR and VR are already being explored. "However," she wrote, "the cost of RFID-ing everything and securing it from destruction by hackers and vandals will be too great. At some point, the quest to make earth/real life be the basis for the metaverse will be forsaken as too impractical. Instead, the metaverse, while mirroring aspects of real life, will be stand-alone as it is now in Second Life, with some portals or turnstiles between real life/Second Life but not any kind of actual meshing as you indicate. More and more transactions and interactions on the Internet will take the form of streaming virtual worlds or Flash miniworlds."

An anonymous respondent said she works in an environment that uses social networking to create a virtual space in which relationships can be developed and supported. "But," she added, "we are seeing very little uptake on the virtual-worlds concept— even by the youngest of our service and worker populations. Gamers enjoy the virtual-world experience—the rest of the population does not. For many of us who work with innovative communication technologies, it offers no benefit to building social relationships or work relationships. I expect that the gaming crowd will grow and that we will all use different technologies than we do now (more ubiquitous RFID and GPS), but do not think that by 2020 the use of these technologies will create a significant change in our lives that will cause us to spend more hours of our day in the artificial-reality worlds."

Another anonymous respondent who agreed with the scenario predicted that advances will be made in interfaces and reported already feeling immersed in AR and VR. "You feel like you know someone when you see their avatar. It's closer to face-to-face than e-mail or text-based chat. Plus, my 'real-life' friends and I communicate over Facebook more than I thought we would. We play Scrabulous and throw 'snowballs' at each other, and it's fun."

Maz Hardey, a social analyst and blogger completing a doctorate funded by the Economic Social Research Council in the United Kingdom, wrote that the divisions now seen—with men spending more time than women in Second Life and women spending more time than men on social networks such as MySpace and Bebo—may change. "By 2020, the scenario could be that there are more sophisticated technologies that make such 'virtual' realities compelling to both men and women Internet users. Moreover, these 'life worlds' are likely to be accessed not just through a computer but other devices that cut down on the 'interface' and 'user' divergences. If this is the case, then a presence in a 'metaverse' may in turn respond to the 'real' digital presence that an individual already shares across SNSs [social networking sites]. However, it is unlikely that these will take the place of 'real' connections. What is likely—as we are seeing now—is that the intersection of 'real' and 'virtual' will be outdated."

Military applications currently in use were mentioned by several respondents. "We are in the last generation of human fighter pilots," wrote **Dick Davies**, a partner at Project Management and Control, Inc. and a past president of the Association of Information Technology Professionals. "Already, drones in Iraq are piloted in San Diego. What will improve is the ability of the artificial spaces to control physical reality, to expand our reach more effectively in many aspects of the physical universe."

Many Note That Communication Is Just Being Refined

There are varying definitions of virtual reality, and even augmented reality can be seen as different things to different people. Much of the variety in responses was due to varying definitions of the terminology. Some people considered cave paintings, books, and television to be forms of virtual reality, and they saw most Web 2.0 relationships as already representing VR. Some people defined VR as more of an out-of-body immersion than one gets when using these "old technologies" or new ones such as Facebook (with profile photos serving as avatars) or Second Life (with its cartoonish renderings of avatars). A number of respondents noted that people in technologically advantaged areas of the world are already exploiting AR and VR, and more will likely participate as the tools are made easier to use.

"Ever since we could communicate beyond the reach of face-to-face, 'virtual' worlds and relations have existed," **Hamish MacEwan**, a consultant at Open ICT in New Zealand, pointed out. "A map is not the territory and a letter is not the person. We have always had multiple facades, for most, most common, work, home, and play. The extension into more immersive 'unreal' worlds is going to happen."

An anonymous respondent commented, "Isn't e-mail an artificial communication reality? Isn't reading a book a virtual reality? Humans have always craved escape and alternate forms of communication. This is just a natural extension of millennia-old trends."

Scott Brenner, a technologist, Web developer, and consultant for Fortune 100 companies, responded, "We've been using augmented reality for decades. Radio broadcasts, TV shows, long-distance phone conversations, video conferencing, etc., are forms of augmented reality. The key point is interaction, which will become easier and more natural. Some will use the technology for play, but the big impact will be those who use it for business or as a means of enhancing their personal lives."

Paul Jones, director of ibiblio.org at the University of North Carolina–Chapel Hill, wrote, "A few years back, I gave a talk about body invasion by augmentation technologies. I asked if the audience thought that was at all weird or scary. No one thought so. It was a 50th class reunion, and all the members were augmented in some way. Hearing, heart, kidney, hips. It made perfect sense to them to be further augmented."

An anonymous respondent commented, "Virtual worlds have come and gone since the MOOs [object-oriented multiuser domains], and there is no reason to believe that this will change."

"Using the Internet to find out how to get from here to there was near-miraculous when it first started happening a decade ago," wrote **Howard Rheingold**, author of *Virtual Reality* and *Virtual Communities*. "Now it's part of daily life for hundreds of millions of people. And just as location-aware, mobile navigation systems are used by relatively affluent enthusiasts today, it won't be many years before cheap toys know where they are. Mashups, simulations, virtual worlds, geotagging, and applications that don't seem possible today will just be part of the environment, like dial tone."

Many people reacted by pointing out the fact that "virtual reality" has been used by people to varied degrees during all of human existence through different formats and approaches. **John Abell**, new-media project director for the Committee of Concerned Journalists, noted, "This is role playing by another name. There has always been a (limited) demand for this, but I don't think it will ever significantly increase."

Breakthroughs Will Change How We Live

Some respondents were optimistic that technological development and the improvement of user interfaces will allow many to enjoy opportunities offered in AR and VR settings. "The browser that we know will be replaced by a 3-D platform, and Internet will become a 3-D environment where people will 'live' more than surf," wrote **Fernando Barrio**, senior lecturer and program leader for the Master of Art in e-business regulation at London Metropolitan University.

Todd Spraggins, strategic architect for Nortel Networks and president and chairman of the board of directors of the Communications Platforms Trade Association, noted, "Such seamlessness between these worlds will be brought about by advances in personal area networks and sensor technology." And **Jay Neely**, founder of News Armada, a Boston-based Internet news company, responded, "Designing portable/ubiquitous hardware that enables augmentation of the physical world with data stored in virtual worlds will be both the biggest challenge and opportunity in realizing this prediction."

Steve Goldstein, an ICANN board member who is retired from the U.S. National Science Foundation, where his job in the 1990s was to diffuse the Internet internationally, predicted, "My intuition tells me that the evolution will be strongly influenced by fusion with other developments such as genetic engineering, creation of artificial life forms (through a merger of genetic engineering and microelectronics, for example), global warming. (Will it force humankind indoors more and lead to more isolated and/or speculative existences, and how might that affect augmented reality evolution?)"

Kathryn Greenhill, an emerging technologies specialist at Murdoch University, responded, "The worlds will not mirror geographic lands, and avatars will not necessarily mirror people. The potentials of online virtual worlds are greater than exact replicas of what exists, and the real potential lies in their twisting and warping the mirror to make things better, stronger, faster, more beautiful."

Fadi Salem, a researcher of e-government at the Dubai School of Government, foresees the need for new laws and standards. "Long before 2020, many businesses will make presence in the 'metaverse' mandatory for employees. Many governments will have a regulation system in place for such presence by then."

Vancouver-based technology reporter **C. R. Roberts** anticipates social adjustments will have to be made in response to the 2020 scenario. "In a reaction to the virtual world," he wrote, "entrepreneurs will establish 'virt-free' zones where reality is not augmented. In various heavily connected areas, there will be sanctuaries (hotels, restaurants, bars, summer camps, vehicles) which people may visit to separate themselves from adhesion or other realities."

Stan Felder, CEO of Felder Communications, commented, "Robots, interviral reality, sexuality will change but still be the same. The stage will simply be bigger and society and religion will create a new set of taboos." **Peter Bihr**, a freelance consultant on Web strategies based in Berlin, Germany, added, "It will be interesting to see if—and how—standards will evolve that allow for the movement between the different kinds of virtual realities." **Luis Santos** of the Universidade do Minho–Braga, Portugal, wrote, "An increased degree of man-machine interweaving seems acceptable."

Some respondents saw major developments to come in the realm encompassed by this scenario and chose to look out beyond 2020. "I can envisage whole segments of society virtually cocooned in their virtual existence," predicted **Robert Eller** of Concept Omega, a media marketing and communication company. "Fully body-suited, fluids and nourishment being fed or removed, and more or less hardwired into the interface. Whilst this may not be a reality in 2020, I do believe that this will be a possible reality extending to downloading one's conscious self to one's cyborg counterpart. This will, in effect, mean immortality. The present steps into the 'second' life are only a beginning, and whilst this may not be mainstream, there will be a large niche group getting their interactive fix this way."

Havi Hoffman, senior editor for product development for Yahoo, wrote, "There are niche communities where this could emerge first: aging baby boomers in affluent nursing home/robotic retirement environments interacting with dispersed friends and family via virtual-reality environments that are much easier on fragile carbon systems; people [who are] pioneering settlement in hostile environments interacting socially in a virtual world created to help maintain communications while isolated in a space suit, or survival pod of some sort, living in deep ocean or polar regions or in a space station or lunar outpost; infected people could also use virtual environments while in quarantine. I can visualize...dystopias emerging; mirror worlds being used, as in *Total Recall*, by the powerful to control the behavior of the many. But I can also picture free zones, enclaves of affluence and innovation like Silicon Valley, and its counterparts around the world—still thriving, precarious as ever, and still subject to cycles of expansion and contraction. I can see metaverse/multiplayer gaming become the prevailing metaphor for workplace problem solving. It would be nice if nation-states would duke it out in the metaverse instead of in the meatspace. Avatars, after all, are easy to replace."

An anonymous respondent predicted, "Evolutionary steps will be multiplied for anyone who shows up to try them out. The diversity of possibilities will take over education, self-development, and for many, might be better substitutes for life itself. Exploration of the darker sides may morph some into evil deeds in regular life, but increasingly, those actions will be censored to the virtual world, making the external worlds far nicer. Some may prefer to spend their lives in virtual enclaves. If so, so what?"

RL Is Not Likely to Be Overtaken by VR Anytime Soon

Many respondents used the word *overrated* to describe synthetic online worlds like Second Life or described massively multiplayer online role-playing games as addictions or distractions. "For some reason I've never been able to comprehend, certain pundits can seriously propose that the wave of the future is chatting using electronic hand puppets," wrote **Seth Finkelstein**, author of the Infothought blog, writer, and programmer. "Flight Simulator is not an aircraft, and typing at a screen is not an augmentation of the real world."

"The 'second self' hobby has been widely overrated," responded **Geert Lovink**, a professor and expert on culture, sociology, and the Internet who is based in Amsterdam. "It is pushed by a specific group of artists, academics, and entrepreneurs who believe in cyberculture as some parallel universe. Most people are not interested in avatars. They have trouble enough managing their first life. What the metaverse faction refuses to see is that they operate in a niche. It is only a specific social group that is interested in this online activity. Having said that, technology, of course, is on the side of the metaverse gurus and their followers. There is more bandwidth, more storage and computer power than ever before—and it has to be utilized for something. The overcapacity will not be used by blogs or Web 2.0 applications. 3-D is the perfect industry solution and is pushed accordingly, mainly by bored manager types who do not have a first life."

Some respondents were not so critical of VR worlds, but they just did not think they will be a dominant force in 2020.

Social media researcher **danah boyd**, of Harvard University's Berkman Center for Internet and Society, commented, "Predictions in this vein tend to emerge every 5–10 years. Remember VRML [virtual reality modeling language]? Remember the days of MUDs [multi-user domains]? *Snow Crash* is great science fiction, but dreadful social prediction (although lovely technology motivation). Many things will prevent us from focusing on immersive environments or 3-D engagement. At the simplest level, people don't want to be immersive—they want to be mobile and to maintain connections with their friends, family, and loved ones when absent. Mobile supports these connections; immersive systems take them over. (And, then, there's my way-early research on how 3-D systems will always be sex-biased because depth-cue prioritization is dependent on the levels of sex hormones in your system…in other words, there's a reason why women get sick going into immersive environments, and there's no good way to solve it.)"

Bruce Turner, director of planning services for a U.S. regional transportation commission, pretended to have his avatar file his response to the scenario: "Bruce Turner's alternate self Nevadaweasel here: Bruce and I are in each other's presence no more than 5 minutes a day, usually to respond to other proxy selves. In 2020 this will probably be the case: (1) Some will reject it altogether, first as a progressive, then a regressive movement, (2) those who do participate in augmented reality will do so routinely, and only a very limited group will spend as much time as the current gaming geeks. The technology that seems so cool to us today will, as it becomes routine, be very much accepted by the majority but play a decreasing role (time-wise) in their lives. E.g., Nevadaweasel may shop for Bruce and be his public Net avatar/persona but not become an obsession. A reality world will simply be to the current WOW [World of Warcraft] world like Windows is to DOS: a convenience that improves accessibility to existing functions. Anyone for a stroll through the Amazon warehouse with Nevadaweasel?"

Michael Botein, founding director of the Media Center at New York University Law School, wrote, "Second Life and related phenomena seem little more than unilateral egoistic forms of stress reduction—electronic substance abuse in a way. Unlike traditional forms of 'acting out' with other people, these ultimately lead towards isolation. Although some are brokers of information among people, they do not seem to promote long-term affiliations. I doubt very much that we'll see a political or cultural revolution arising out of 'MyPage.'"

Milton Mueller, a professor at Syracuse University and researcher of technology policy, responded, "Do you consider (meatspace) football games to be a virtual reality? If so, then this prediction is true, as greater leisure time and wealth lead more people to spend time in recreational/creative activities. Augmented realities such as World of Warcraft, Second Life, however, are now and will always remain the preserve of a small band of aficionados willing to do an enormous amount of work to play. Most people will save their energy for the real world and be more passive about virtual worlds."

An anonymous respondent quipped, "I won't be hanging around any other world unless someone can add more time to this one."

Jerry McCann, vice president and director of the social marketing group at Carton Donofrio Partners Inc., located in the Washington, DC, area, had a similar response, adding, "Virtual space takes an enormous amount of time and energy. To the extent that virtual spaces can offer effective alternatives for real-life enterprise and social life, they may continue but are not likely to dominate. TV was supposed to kill the movies and didn't. The Internet was supposed to kill TV and didn't. I doubt virtual reality will die but equally doubt it will dominate."

Many respondents who have doubts about significant advances in the numbers of people using AR and synthetic online worlds day-to-day framed the 2020 user interface in today's terms. "This vision falls apart with 'intuitive technology interfaces,'" wrote an anonymous participant. "We still haven't got 3-D navigation sorted for the unwashed masses at home in front of their PCs—and controlling this from our cell phones? I doubt it."

Other respondents also noted that VR worlds will not be in popular use by 2020, although they will be of use to some people. "The technology may not be far off, but the demographic assumptions that supported the previous question on wireless directly contradict those here," commented **Tom Vest**, IP network architect and consultant for RIPE NCC Science Group and consultant for OECD and the Internet Society. "Techno-economic elites of 2020 may be exploring the fringes of this new reality by 2020, but few others."

"This 2020 scenario is appealing to the geeks and the gamers among us," wrote **Susan Mernit**, an independent consultant who was formerly an executive with Yahoo and America Online, "but I don't see the seamless transitions that this posits happening this quickly—it's elitist and too far out of the mainstream for many Americans, especially those with less free time. Having said that, I do think there are sectors of society that will use the metaverse to play and to train in disproportionate numbers—and that we will see a rise in virtual worlds as entertainment spaces outside of gaming (think sex, travel, historic simulations)."

Scott Smith, consultant, writer, futurist, and principal at Changeist LLC, based in North Carolina, predicted, "As we've seen with use of the Web and blogging, participation in general metaverses may decline in duration and variety after a short-term peak in usage as users seek to rebalance toward the 'real' and authentic and see fewer benefits in being active in metaverses. This is not to say that function- or interest-specific metaverses may not continue to flourish, based around certain applications or activities, but a mass market spending significant time in virtual worlds on a daily basis is less likely."

Anthony Townsend, research director for the Technology Horizons Program of The Institute for the Future, wrote, "Separate, 'virtual' worlds will be much less important than augmented realities. The real world is a fascinating place—overlaying information and cues from digital spaces will make it even more compelling—for socializing, traveling, playing games, and working. It will still be real life [but it will be enhanced] in the sense that people who wear eyeglasses still see real life, just a refocused version of it."

Walt Dickie, CTO for C&R Research, one of the principal developers of CheckMetrix in the early 1990s, responded, "It won't look like Second Life (already melting down) but the convergence of the physical world, data-rich representations of it; social networks and gaming/entertainment will create an augmented reality that will become the environment in which an entire ecosystem of connected hardware allows humans to let their 'real' lives and 'metalives' interpenetrate."

Karen Schneider, a researcher and thought leader in the library and technology community based at the College Center for Library Automation in Tallahassee, FL, wrote, "This might be the new TV. I'm waiting for the breakthrough reality show where I can be on some island from my living room. Well, no, I'm not really, but I'm sure it's imminent."

There Will Be Economic, Generational Divides; Some Will "Opt Out," Become Addicted, or Be Unproductive

As is the case with most looks at the future of a technology, some people are predicting that these developments will cause a divide between the "haves" and the "have-nots," and others are saying there will be people who have access to this technology who choose to opt out. "Real life remains real life," wrote **David Maher**, senior vice president for law and policy for the Public Interest Registry, the Internet top-level domain registry. "Other 'realities' will more likely interfere with rather than augment real life."

Brian Dunbar, an Internet manager for NASA, wrote about the digital divide: "The physical infrastructure required to make these features available to large numbers of people will restrict their widespread use to affluent sectors of developed nations." An anonymous respondent commented, "The current increase in income differences might well mean we have even more of a division between the folks participating in the metaverse and those eking out a living in the world's poorer areas." Another wrote, "Social responsibility will call any requirement for virtual-reality access to be a barrier to full participation." And another commented, "People who work in white-collar/office/education fields will be able to seamlessly jump between work and online 'life.' People who have service or manual labor jobs will have a limited amount of time to participate. This limited time will result in limited skills and abilities, further deepening the divide."

Or, perhaps, some suggested, while alternate realities can help people escape negative conditions, addiction to virtuality might be a future root cause of unemployment and/or withdrawal from productive society. **Leonard Witt**, author of the PJNet.org Weblog and an associate professor in communication at Kennesaw State University, GA, predicted, "These virtual environments will be used to help lift people out of mental poverty, even when their real world is immersed in physical poverty. The big next question: Will virtual worlds become the opiate of the masses?"

Many respondents saw a generational divide, with younger people readily moving into the world of the scenario, while older people generally do not participate. “Today’s preadolescents are likely the oldest to experience such a fully immersed virtual reality,” responded **Jade Miller**, a researcher of global flows of information and culture and PhD student at the University of Southern California. “Older Internet users may have virtual selves but will likely use them only sparingly or to spy on their children.” An anonymous survey participant commented, “There will be an enormous generation gap.” Another noted, “How people use these technologies will be age dependent. Younger people will like and use them to be more creative and interactive. Older people will simplify their environments to be more efficient; they will still use these technologies, but in a more focused way.”

A number of respondents predicted that people will decide to “opt out” of the virtual and augmented opportunities available. “I believe, bottom line, that people have only one life to live, no matter how many avatars you create, and that people will weary of the virtual and yearn for the real world,” wrote **Jan Schaffer**, executive director of the Institute for Interactive Journalism and a Pulitzer Prize winner.

Charles Ess, a professor of philosophy and religion and researcher on online culture and ethics at Drury University, responded, “While it is certainly true that these expressions of CMC will become more important, it is equally true, as the current turn *away* from Second Life suggests, that people are also getting tired of ‘the virtual.’ I might have an augmented self in some virtual world by 2011—but my suspicion is that that ‘self’ will be a largely dressed-up version of a very mundane self that needs to check on bank balances, make appointments for a haircut or automobile inspection, etc. The genuinely pedestrian tasks of daily life will not clearly be enhanced or made better by building avatars around them. I also have a strong suspicion that as these technologies increasingly dominate our lives, there will also be a strong—perhaps overly romantic—reaction against them. People will be willing to pay real money to talk with a real person rather than a voicemail system. And until we get more-or-less infinite bandwidth systems that include every dimension of 'being there' in fine detail—including smells, touch, etc., I suspect more and more people will find that they enjoy getting out of 'The Matrix' that already seems to increasingly dominate our lives in the developed world. (We are in love with the technologies of our enslavement, Neil Postman said in 1984. But perhaps even machine-reinforced love can only go so far?)"

Dian Schaffhauser, founder of Sourcingmag.com and a writer and editor for *Computerworld*, CampusTechnology.com, and other publications wrote, "Virtualization is great for machines and those who have limited ability to interact with First Life. But it's just an interesting hobby or game for the rest of us. I was rather hoping that the green movement would persuade the great majority of us that what we truly need to value is the real Life—the one we can touch and smell and walk on."

Jim Lucas, Web manager for CACI, an intelligence and security solutions company, commented, "A rebound effect will occur that drives people to treasure actual human contact more." **William Winton**, product manager for digital media for 1105 Government Information Group, noted, "The 'slow-life' movement has grown in direct response to the disassociated, amorphous, and out-of-touch societies that are emerging in the developed world. Encouraging family, friends, and neighborhood, the movement seeks to restore the tangible social bonds that the Internet cannot replicate. People will discover that the 'real world,' for all its faults, is much more interesting than any 'virtual world' could ever be." An anonymous participant wrote, "As virtual worlds become overrun with 'real-world' problems, people will abandon their use."

And an anonymous respondent predicted the following 2020 scenario: "A significant elite swears by augmented reality and cannot remember any detail, direction without their AR spectacles. The rest are much more comfortable with limiting the digital information to e-ink surfaces but are also unable to offer the technology, process it—and ostracism ensues. Democratisation of the technology is the main pacification path considered by the political strata, largely AR enthusiasts. Spending time in NAR (*not* augmented reality) is a common ascetic experience."

Some Say the Scenario Is Likely to Happen, and We Must Be Wary of Other Dangerous Implications

Respondents who mostly agreed sometimes concentrated their elaborations on the fears they have for such a future. "Although this appears to be almost sci-fi–like, it will only take some major cataclysmic event to reverse this trend, e.g., young people's identities being manipulated by others to persuade them to do immoral things or even commit suicide en masse," responded **Robin Gunston**, consulting futurist for Mariri Consulting, a strategic and business planning company.

Ed Lyell, a pioneer in issues related to the Internet and education, expressed concerns about violent VR triggering negative behaviors in the real world. "Some young people are unable to separate violent acts in an artificial world from violent acts in the real world," he wrote. "We need to ensure that more people in the world are educated in the ability to discern multiple layers or types of reality. One of my mentors was S. I. Hayakawa, a leader in General Semantics. Being able to separate object and referent, to see multiple roles, layers, viewpoints without seeing any of them as absolute will become a more necessary skill."

Joe McCarthy, principal instigator at MyStrands, and formerly a principal scientist at Nokia Research Center in Palo Alto, CA, commented, "It's not clear to me whether/how immersive online worlds will augment or enhance the offline world, and I fear that the time and attention consumed in such worlds will come at the expense of actions that might make the offline world a better place."

An anonymous respondent predicted that in 2020 "the 'metaverse' is a dangerous distraction to facing the problems of the real world, and, to the extent that it becomes corporatized, it becomes just another marketing wasteland." Another anonymous respondent commented, "This dynamic has already started. It is a threat to tangible innovation. Innovating in virtual worlds will start to deprive tangible innovations in the real world. Our brainpower will be devoted to a virtual world and not the real world."

Lisa Carr, director of strategy for Targetbase Interactive, wrote, "As people enter the metaverse and virtual realities, they will lose touch with actual realities. The level of obesity will rise concomitantly."

Clement Chau, manager for the Developmental Technologies Research Group at Tufts University, predicted that adoption of virtual identities will be simple, but it will raise problems. "Adopting a virtual identity will be as seamless as the adoption of a professional identity in the 20th and early 21st centuries," he wrote. "However, we will have problems and concerns keeping our multiple virtual and real-life identities consistent. We will begin to see both positive and negative implications of such potential inconsistencies seeping into different aspects of our lives."

Timothy McManus, a vice president with Nuance Communications, a software technology company known for speech-recognition work, noted that there are privacy implications tied to the development of most aspects presented in this scenario. "This scenario… reinforces the case for more controls on privacy and more limits on access to personal information, because people will have one or more personalities or lifestyles in a virtual world that is fundamentally different from the physical world," he commented.

The dystopian film *The Matrix* was mentioned by a number of survey respondents. "This scenario paints a 'Matrix' model which is eerily true already for some people," wrote **Michael Castengera**, a senior lecturer at the University of Georgia and president of Media Strategies and Tactics Inc. "Research shows that many people care as passionately about their virtual life and friends as their real-world life and friends. People are actually getting married in Second Life. Two questions come to mind. One is—is this retreat into a virtual world actually an escape for a limited number of people who don't have the social skills to make it in the real world? Second is—will the global-warming, environmentally degraded real world lose its attraction (less fresh air, no singing birds, no sweet smelling flowers), thus making a virtual world more attractive or at least more acceptable?"

AR and VR Easing From Status as Accessories, "Toys," and Entertainment Into the Mainstream

Respondents' elaborations in the *Future of the Internet III* survey are dependent on their individual definitions of the terminology used in each scenario and their understanding of the issues being addressed. This allows for more layered responses that serve as a realistic indicator of current attitudes and expectations about emerging network technologies and their influence. The AR-VR scenario drew the broadest variance in respondent reactions, probably because people's definitions vary. For example, to some, "virtual reality" may range from experiencing daydreams, reading a book, watching television, IMing, Twittering, or being "on" Facebook to assuming that only a totally immersive 3-D world with perfect haptics, optics, and so forth, is VR. In the realm of possibility in defining "augmented reality," the yellow first-down line overlay applied on the football field on U.S. television broadcasts may fulfill some people's concept, while others have a much higher expectation.

This section of this chapter outlines the current status of and some expectations for the elements of this scenario. Government and commercial interests are continuing to invest heavily in VR and AR. The most significant, cutting-edge work being done in this realm is still being contracted by the U.S. Defense Advanced Research Projects Agency (DARPA)—the organization that birthed the projects that led to the origins of the Internet, which celebrated its 50th anniversary in April 2008. Other governments have similar R&D organizations working toward new breakthroughs in VR and AR, and most are already implementing it in many forms in military and other applications.

Some people who shop and buy products online today consider it a virtual-reality or augmented-reality experience; in the United States alone, consumers spent $29.2 billion online from November 1 to December 31, 2007, up 19% from the same span in 2006, according to online research company comScore, Inc. The fastest growing online retail category is video game hardware, software, and accessories, for which sales were up 129% in 2007.[1] A significant percentage of computer-based gaming still takes place offline, but the development of the technologies used to produce these games and the human skill sets being developed in the playing of the games are important elements in the advance of online AR and VR. NPD Group reported in early 2008 that the video game industry recorded $17.9 billion in overall sales in 2007, 43% more than in 2006 ($12.5 billion, which was also a record), bringing in new consumers, not just young men but also families and people over 40; much of this is attributed to the introduction of Nintendo's Wii system, with a highly intuitive user interface (UI) that interprets users' body movements.[2]

Many of the respondents to this survey noted that the UI is key to successful adoption of AR and VR tools. (Because the technologies that are most readily adopted are those that are easiest to use, interface design is addressed in its own scenario—scenario 6—in the next chapter of the *Future of the Internet III* survey.)

Online games come in many formats. The most popular MMORPGs include World of Warcraft, EverQuest, and Lineage II. A study published in the May 2008 issue of the *Harvard Business Review* by Byron Reeves, Thomas Malone, and Tony O'Driscoll indicated that these games engage people in the practice of useful pursuits, including rapid response, working in ad hoc teams, distributed decision making, and leadership through collaboration instead of the traditional top-down authority structures. The study suggests that participation in these VR worlds might provide insights to the future of leadership.[3]

The positive evolution of VR aspects in game systems, the gradual convergence of television and the Internet, and further refinements of high definition (HD) and 3-D visual presentation and user interfaces continue. The hardware and software are important, but humans' implementation of them is the driving force for change. From its earliest days, the Internet has been a tool people use to augment reality and immerse themselves with others in forms of virtual reality, overcoming time, space, and geography through the use of e-mail, bulletin boards, MOOs, and other person-to-person communications.

David P. Reed, an Internet pioneer who worked on development of TCP/IP, noted at the turn of the century that the Internet is natively group-forming, driving generative, collaborative processes. Reed's law notes that the utility of large networks can scale exponentially with the size of the network.[4]

While MMORPGs are retaining a position of importance, these synthetic gaming worlds and the synthetic social virtual-reality worlds such as Second Life (with 12 million registered users, but only about a million or less who access the site at least monthly) are not nearly as popular as Facebook (with 123.9 million unique visitors and 50.6 billion page views worldwide in May 2008 alone) and MySpace (with 114.6 million unique visitors and 45.4 billion page views in May 2008). One of the primary reasons the avatar-populated worlds have not caught on as well as social networks is the fact that they are hard to use and require attention that people do not feel they have the time to give them. New, more natural interfaces could revolutionize these platforms and inspire new ones.

Facebook, an organized collection of searchable databases that leverage the social graph, is a textbook example of one of Reed's "group-forming networks." Launched by Harvard student Mark Zuckerberg in 2004, Facebook took only a few years to become one of the top-trafficked online sites in the world. It is available in at least 15 languages and has more than 80 million active users worldwide. They upload more than 14 million photos daily, and they "meet" each other, declare their identity, "friend" one another, form affinity and work groups, plan parties, and play games such as Scrabulous together there. While it is classified as a "social network," many people leverage it as a political and economic tool. It became popular because it offers people of like interests the ability to meet and share information together easily in collaborative processes.

Analysts at firms such as Gartner are assessing how and why people are attracted to spend time on Internet-enabled human networks. Those who spend the most time networking online are finding it fulfills key needs and even assists them in reaching a point of self-actualization, a concept conceived by neuropsychology pioneer Kurt

Goldstein to describe reaching the pinnacle of our desire to be effective, contributing, creative, mature, capable, moral, and complete human beings. Gartner analyst Adam Sarner has predicted that by 2015 more money will be spent on marketing and selling to multiple, anonymous, online personas than marketing and selling offline.[5]

Researchers are in the process of figuring out how to achieve the ease-of-use and group-forming utility of Facebook in avatar-populated spaces like Second Life. IBM is one of the companies at the forefront of this development, with its Emotiv Systems work and the blending of 3-D technologies into Lotus Connections. It owns about 50 virtual facilities in Second Life, and thousands of its employees have worked within it. Another company—Unisfair—is developing ways to include social networking tools in virtual conferencing software for businesses.

Corporations, NGOs, and governments have been experimenting with the use of virtual spaces such as Second Life for marketing, conferencing, and training in spite of the fact that the user interfaces for most of these VR worlds are still cumbersome and not intuitive enough to feel "real." People generally negotiate their avatars' interactions inside these worlds by using keyboard commands, and conversation must be typed out.

Starwood Hotels built a replica of its new Aloft hotel concept in Second Life. Duke University and Virtual Heroes Inc. are working together to create a 3-D virtual environment for health care training that is being funded by the U.S. Army. The University of Maryland and the I-95 Corridor Coalition built a virtual-world simulation for assessing responses to highway emergencies through the use of the OLIVE Platform from Forterra Systems.

Synthetic worlds continue to draw close scrutiny by marketing analysts, as businesses assess how much to invest in them and how. A January 2008 report by Forrester Research, *Getting Work Done in Virtual Worlds*, encourages businesses to explore VR possibilities for training, remote collaboration, and opportunities for 3-D modeling.[6] Gartner released a report in 2007 that estimated that "80 percent of active Internet users will have a 'second life' in the virtual world by the end of 2011."[7] Analysts at Gartner noted in 2008 that there has been a hiatus in business use of virtual worlds due to a lack of customer interest, but Cisco Systems president John Chambers held a June 2008 press conference in Second Life in which he predicted that virtual interfaces "will forever change business models and entertainment…We may disagree on the time frame. It's not a question of if, but when."[8]

Other virtual environments being constructed today include "virtual globe" programs such as Microsoft's Virtual Earth, Google Earth, and NASA's World Wind. These virtual forms were labeled "mirror worlds" in the 1992 book *Mirror Worlds, or the Day Software Puts the Universe in a Shoebox* by Yale University computer scientist David Gelernter. They allow people to examine models of real human environments, feel as if they are soaring through photorealistic models, share data gathered by global sensors and other systems, and even count how many frequent-flyer miles they might earn by flying from one destination to another. More user-created content is being applied in digital mapping software programs in these mirror worlds that are intended to help people understand the real world, allowing them to share, for example, their travel photos and blog posts tied to specific locations.

The *Metaverse Roadmap*, a 2007 foresight project report produced by the Acceleration Studies Foundation, had this headline: "What happens when video games meet Web 2.0? When virtual worlds meet geospatial maps of the planet? When simulations get real and life and business go virtual? When you use a virtual Earth to navigate the physical Earth, and your avatar becomes your online agent? What happens is the metaverse." The report predicted, "On the social side the most obvious persistent trend will be identity experimentation, self-revelation, and role-play and the creative variation of social norms around gender, ethnicity, social class, etiquette, and group values and goals…and the ensuing social change [is expected] to bring both positive and disruptive effects."[9]

Some experts speculate that when computing, storage, and other required technologies become advanced enough, a Second Life–type environment will somehow be integrated with a Google Earth–type system to form a Second Earth, a world of 3-D representations of everything on the planet.

Such a place would be a new "reality" on a massive scale. Augmentation of reality on a smaller but highly useful scale is continually evolving. Many of the most popular new digital devices shown at the Consumer Electronics Show 2008 in Las Vegas fit this role—helping people live their lives more efficiently and allowing them to share knowledge. One such device was a handheld tool demonstrated by Intel Corp. president and CEO Paul Otellini during his keynote speech. It is capable of serving as a camera and language translator (performing real-time translation between two people talking to each other in different languages), as well as speech recognition and wireless Internet connection. He also demonstrated a new Intel "system-on-a-chip" that is expected to allow people to easily access the Internet through a new generation of media players and TVs. The chip is expected in 2008, but Otellini said the wireless-broadband infrastructure has to be more highly developed for its potential to be fully realized.[10]

As more free wireless broadband access is made available and the mobile devices used for online access are refined, more people are expected to begin weaving higher percentages of augmented and virtual reality into their day-to-day lives. Economics drive change. According to TNS Media Intelligence, the top 100 advertisers in the United States shifted about $1 billion in spending from TV and newspapers to the Web in 2007.[11] Industry-research firm IDC reported in June 2008 that it expects global spending in online ads will reach $65.2 billion for the year, and it estimated spending in 2011 will reach $106.6 billion, about 14% of the total advertising market. It is expected that advertisers will continue to subsidize more free Wi-Fi access in exchange for customers' attention, and some governments will adopt free wireless access for localities as an economic stimulus worth investment.[12]

Among the proof of heightening AR evolution already in evidence: Automobile manufacturer Chrysler has said it will offer wireless

Internet connectability beginning with its 2009 models; GPS technology has become small and cheap and even more ubiquitous—GPS and cell-tower triangulation are being used in some handheld devices to allow augmentation of reality, providing people with information through tools such as Google Maps Mobile that allow them to figure out where they are, plot a course, and even track the locations of friends, relatives, pets, and cars. People are enabled to be more location aware.

Smartphones and cell phones are being used by consumers as GPS-enabled AR lifestyle devices thanks to a boom in new applications that offer geographical location synchronization. More than 70 programs can be downloaded from Where.com, including a flight-time updater, gas station locator, and airport finder. Google's Android is expected to offer Wikitude, a Wikipedia-style description of points of interest near your geographic location. Synchrospot can automatically load your shopping list onscreen when you walk into your supermarket. CallACab allows you to summon a taxi with the push of a button. Loopt and Pelago allow you to use your social network to locate friends and family or find stores and restaurants recommended by them.

Web tools are being used to augment reality and make a difference for global good. The UN, Cisco, and Google joined forces in late 2007 to launch a poverty-tracking project at mdgmonitor.org, the Millennium Development Goals Monitor. The site aggregates information and packages it as "global-awareness" overlays available on Google Earth. Progress toward the UN's goals to reduce poverty and hunger and address issues tied to gender, illiteracy, disease, environment, and lack of access to clean water is being recorded and put on display. Visitors to the site can use Google Earth to see how countries are tackling problems. This is just one of many such initiatives of Google Earth Outreach. The Jane Goodall Institute and Earthwatch, a research organization established to protect the environment, are among the participating groups. The U.S. Holocaust Museum helped create a Google Earth layer illustrating the atrocities committed in the Sudanese government's attempt to wipe out non-Arab tribes in Darfur; it includes graphic photographs and testimony from survivors.

Mobile computing is not expected to be limited to devices. "Wearable computing" is drawing the attention of international researchers working to provide augmented-reality experiences. Among the devices shown at a fall 2007 conference were threadlike sensors that can be woven into shirts to provide wearers with information about their posture; people with back trouble can be reminded to assume the best possible position. Graduate students from the Massachusetts Institute of Technology's Media Lab wore black badges that included sensors to measure motion and speech patterns in order to detect the levels of emotion people exhibit in a conversation; this has many potential applications, for instance, doctors might use such a device to help diagnose patients with depression.[13] A mobile augmented-reality device already on the market is the Ambient Umbrella, an "intelligent" umbrella that tracks the forecast for any of 150 U.S. locations. When rain is expected in the next few hours, it illuminates its handle with a feature called "glowing intelligence." Patterns change to indicate rain, drizzle, or thunderstorms.

Human infrastructures continue to become more networked in what is sometimes called the "Internet of things" or "ubiquitous computing." Highways, shipping pallets, products, appliances, equipment, lighting systems, bridges, dams, and other items perceived to be of value are being equipped with data-generating sensors, and people are finding ways to data mine the information and post it in a useful, accessible form. This expanded Internet is one of the reasons for the move from Internet Protocol 4 (IPv4) to IPv6. The tracking and management of these object networks will continue to create more opportunities for useful information datasets that can be applied in AR and VR applications. The Semantic Web—a set of design principles and enabling technologies now being implemented throughout the World Wide Web—is also refining the accessibility of data and defining data relationships. Experts say as people begin to leverage virtual information to augment real-world needs, the "blurring" of the virtual and real will accelerate. (See more about IPv6 and the Semantic Web in the chapter on Internet architecture, scenario 7.)

Mobile augmented reality is expected to allow people to leverage mirror-world data while in the "real" world. For instance, while visiting a historic site, you use your GPS-enabled camera phone to connect to databases with maps that provide graphic overlays explaining the past of the place that lies before you. Or as you are hunting for real estate and you walk past a "for sale" sign, you use your digital device to visualize 3-D floor plans, the price and property taxes, and the assessed values of neighboring structures.

A group of panelists at the Cebit show in Germany in March 2008 looked ahead to an augmented world where the virtual and real blur. Nils Müller, CEO of TrendOne, said the Web 4.0 era—an "always-on" world in which people can "self-upgrade" through technology extensions—is already beginning to show what it has in store. He said the 3-D printers now used for design and prototyping and 3-D holographic projection systems like the one recently unveiled by a U.K. company, Musion, will be available for home use, and interfaces might be accomplished through thought control as based on research currently being done to help severely disabled people function normally.[14]

Along with the projected positives of emerging AR and VR advances come what some perceive as the negatives; for instance, the head of U.S. Air Force intelligence condemned Google Earth as a danger to national security because it can be used just as easily by people with an opposing agenda as it can be by those pursuing peace-motivated goals. Many other agencies of various governments are opposed to the public display of the world's satellite imagery.[15]

Among the other reports of potential negatives of AR and VR in 2007: The German Society for Scientific Person-Centered Psychotherapy advocated for a ban on violent computer games, reporting a link between fantasy games and violence and adding that the addictive quality of the games also takes time away from important pursuits like exercise, reading, and good nutrition.[16] A study published in the *Journal of Adolescent Health* reported that "exposure to virtual violence increases the risk that children and adults will behave aggressively."[17] Computer games were blamed for England's drop from 3rd place in

2001 to 19th in 2006 in children's reading skills in the International Reading Literacy Study.[18] Reports in the United States loosely linked children's Internet use to obesity and a lack of bone density. The journal *Pediatrics* published a study that reported when children play video games within a few hours of bedtime, they do not get enough of the slow-wave sleep that helps them form factual memories.[19] An epidemic of suicides in the region around Brigend, Wales, in the United Kingdom was found to possibly be linked by the 17 victims' participation in the social networking site Bebo. Many of them had posted messages on the public memorial pages of those who preceded them in committing suicide. The site's administrators shut down some of the pages. Suicide-prevention experts warned that the use of social networks in "glorifying young people who have taken their own lives" can encourage others to do the same in a viral response.[20] The 2007 *Metaverse Roadmap* report warned that mirror worlds can become instruments of state control, and many aspects of AR and VR raise significant questions about the future of civil liberties.

Further extension of AR and VR will bring corresponding challenges in the areas of trust, security, reputation, privacy, identity, and transparency.

ADDITIONAL RESPONSES

Many other participants in the survey shared valuable input on the augmented and virtual reality question. Among them:

> "The Internet is a medium that enhances communication of all types of information. Virtual reality is simply one form of information with an inherent creative component that knows no bounds. Art is ever evolving and mutating, so we must expect the same of virtual developments, too." —**Tim Grafton**, market research director for UMR Research Ltd., a market research company based in New Zealand

> "I agree that time spent in virtual reality may well be commonplace for many by 2020, providing broadband capabilities

increase as well." —**Jennifer Jarratt**, principal, Leading Futurists LLC; works with formalized methodologies to assess and interpret potential futures

"Augmented reality will be very complex, layered, and uneven. AR should show up in politics, in struggles over what information is attached to what spaces and who controls it." —**Bryan Alexander**, director of research National Institute for Technology and Liberal Education, blogger, expert on computer-mediated pedagogy; based in Ripton, VT

"I wonder how many who don't participate will be left behind socially and in the workplace." —**Ruth Martin**, National Chengchi University, Taiwan

"I think this will happen, but 2020 may be a little too soon. The Gartner study of 80% of all active Internet users will have virtual selves by the end of 2011 is way too aggressive. That would mean that over 1.4BB global Internet users today (http://www.Internetworldstats.com/stats.htm) and growing, 80% would have to have virtual selves. I don't buy it in 3 years. I think there will be a small part of the population, starting with today's Second Lifers, who drop in and drop out so to speak. They become the Gibsonesque wireheads." —**Chris Miller**, senior vice president, digital operations and new business for Element 79, an advertising agency

"It is perfectly reasonable to imagine greater integration of work and play within present and representational environments." —**Oscar Gandy**, author, activist, retired emeritus professor of communication, University of Pennsylvania

"The proposed 2020 scenario is already a fact of life for many people I know both online and as acquaintances here at university. It strikes me, therefore, as quite plausible that for as long as the metaverse can provide for a human need for quick and convenient contact, more people of all ages shall flock to it. I do not, however, believe that the metaverse will ever become more than an augmentation of offline life for more than a very small minority of users." —**Francis J. L. Osborn**, futurist and activist, philosophy department, University of Wales Lampeter (formerly St. David's University College)

"Most of the use will remain in the realm of entertainment. There will be continued concerns over balancing life 'in the real world' against any of the virtual ones." —**Sam Punnett**, president, FAD Research Inc.; has worked in the field of interactive digital media since the 1980s, for the last 9 years on strategy, marketing, and e-business development

"In the future, people will move between the 'virtual' and 'physical' worlds in seamless ways—but I don't think we really have any clue as to what this will really look like or mean. Furthermore, I don't think that this transition will occur quite as fast as some believe. It will be more than a decade before we begin to have the capacity to deal with our current technology and our current data overload. If I were to look for future scenarios of how 'virtual reality' and 'real life' interact, I think I would look to the science fiction and fantastic genres of literature and arts, especially to those visions of the future that see our bodies being physically connected to 'virtual realities' through true 'neural nets.' Which of these visions of the future best predicts the future, however, is something I don't think we can even guess at." —**Benjamin M. Ben-Baruch**, senior market intelligence consultant and applied sociologist for Aquent, working at General Motors Corporation

"Absolutely—although I think what will happen is that our existing connections within the Internet will simply become more enriched with these sorts of technologies, so we will not be choosing to move into a different space for these experiences, we will simply become accustomed to richer interfaces." —**Richard Osborne**, Web manager for the School of Education & Lifelong Learning, University of Exeter; research focus is e-learning

"With a little luck, most people with these tendencies will have received sufficient counseling that they choose the real world over computer-generated ones." —**Anthony M. Rutkowski**, cofounder of the Internet Society and a founding trustee; longtime leader in International Telecommunication Union; vice president for regulatory affairs, VeriSign

"Until we can feed our children by spending time in an 'alternate universe,' it will be little more than a time waster for most of us. Where it will play a huge role is in bringing the physically handicapped into the mainstream, allowing a bed-ridden lawyer to present closing arguments, an impaired (e.g., Stephen Hawking) college professor to teach class at three universities simultaneously, and/or allow an amputee soldier to pass his skill/knowledge on to his outfit in the field from a bed at Walter Reed Hospital." —**James Gorman**, principal, Working Technology Partners, a company offering technology solutions to businesses

"While agreeing these technologies will become important, people vastly overestimate the value of 3-D, 'virtual-world' presentation to these interfaces. The virtual world will deliberately look virtual and not try to model our real 3-D world as much as people think. 3-D is hard." —**Brad Templeton**, chairman of the Electronic Frontier Foundation, also of the Foresight Institute, is an Internet pioneer who has been active in the computer networking community since 1979

"The metaverse will be important for many, but recalling the first question and how mobile devices will become an important connection tool for most people, those using the metaverse will be a significant minority of the total. Sustained life (work/education/socializing) inside the metaverse will happen but, like telecommuting, not as quickly as we might expect. Real life's appeals will be stronger." —**Adam Peake**, executive research fellow and telecommunications policy analyst, the Center for Global Communications (GLOCOM); studies the intersection of public policy and the Internet

"Virtual reality is more entertaining for many people than their real lives." —**Camilo Elias**, university student

"Virtual worlds blur the line of fact and fantasy and pose a danger to our society." —**Dana Rostern**, student, Hofstra University

"It seems unclear to lump together augmented reality and mirror worlds. They can have very different contexts, uses, points of interaction, and expectations. The geoWeb will be a powerful force and may increase the sense of physical place. The metaverse will have to deeply increase sensory experience, wayfinding, and orientation—so it may ride on the back of situated augmentation, but not necessarily be coupled with it." —**Duane Degler**, user-centered designer and strategist for Design for

Context, writer and editor for IPGems, focused on knowledge management, semantic integration, and performance improvement

"The opportunity to test one's own environment with a bigger universe is too exciting to turn down." —**Nancy W. Bauer**, chief executive officer and editor-in-chief, WomenMatter, Inc.

"For vivid fictional depiction of realistic future worlds of immersive augmented reality, see: *EARTH* (1990) and *Rainbow's End* (2006)." —**David Brin**, futurist and author whose 1998 nonfiction book *The Transparent Society: Will Technology Force Us to Choose Between Privacy and Freedom?* identified key issues of concern

"A *Snow Crash*–esque world is coming, but it's farther off than 2020. We'll see advancements towards this and some early adopters, but many people will be slow to accept this new augmented reality." —**Jay Buys**, vice president for digital development, Fleishman Hillard, an international marketing and communications company

"Over the next 12 years, artificial spaces will find their home. We do not yet know what form these will take and how they will be used, but they will become a part of the 'normal Internet' in some way. They will continue to be used for entertainment purposes, and also find increased use in the academic communities. Where else will they find a foothold?" —**Ross Rader**, director of retail services, Tucows Inc.; works with ICANN in the Registrars Constituency, part of the Generic Names Supporting Organization (GNSO)

"I'm surprised there is no mention of gaming in this scenario. If people spend time in artificial spaces/virtual worlds, it will likely most often be in game environments." —**Steve Jones**, professor of communication and associate dean of liberal arts and sciences, University of Illinois–Chicago, cofounder of Association of Internet Researchers

"We may make some use of virtual reality or holographic projection in our user interfaces." —**Fred Baker**, Cisco Systems Fellow, former Internet Society (ISOC) chairman of the board; Internet Engineering Task Force (IETF) leader; an architect of the Internet

"Probably this trend will continue. But is it really completely new, or only new in the degree of the augmented reality?" —**Norbert Klein**, member of ICANN's GNSO Council and Internet Society leader who works with Open Institute Cambodia, a company whose primary focus is on information

"This one, I probably would say 'neutral' if that were an option. Of course the technologies will advance, but it's also true that the applications that prevail are often those with the simplest interfaces and functionality (e.g., Twitter). So, the crucial thing will be to what extent these tools will truly add value to our work and happiness." —**Richard Hall**, professor of information science and technology and codirector of the Laboratory for Information Technology Evaluation, Missouri University of Science and Technology

"Individual elements of a society will eventually become part of a global society." —**Hakikur Rahman**, chairman, SchoolNet Foundation; coordinator of Sustainable Development Networking Programme in Bangladesh; active in Internet Society Board, South Asia Foundation

"I think that this vision is overhyped, but I wait to be proven wrong." —**Micheál Ó Foghlú**, research director, Telecommunications Software & Systems Group, Waterford Institute of Technology, Ireland advisory committee; member of W3C; blogger

"The Internet is a communication media, not an alternate universe." —**Christian Huitema**, distinguished engineer, Microsoft Corporation, pioneering Internet engineer (on the Internet Architecture Board 1991–1996; Internet Society leader 1995–2001)

"When people are able to fulfill 'real-life' conceptions and also fulfill activities within virtual worlds, this will bring ease of access to its optimum height, whereby many social activities will be maintained through a virtual existence." —**Sam Ozay**, e-learning and e-communication specialist and solutions architect at Postmodern (Asia/Pacific); formerly general manager at European Language Centre

"Mostly agree, although full adoption will take longer than implied in the statement above." —**Brough Turner**, chief technology officer and cofounder of NMS Communications; oversees evolution of technology and product architectures

"I shudder to think of a future in which one cannot distinguish between virtual and real." —**Hinda Feige Greenberg**, PhD, director of the information center for the Robert Wood Johnson Foundation, dedicated to improving health care for Americans

"This will be just too much fun to ignore. And one can deny a lot in Second Life." —**Dan Lynch**, founder of CyberCash, Inc. and Interop Company and an Internet pioneer; board member of Santa Fe Institute; director of computing for SRI International in the late 1970s

"Augmented reality will be an important recreational activity in niche markets, but fully integrating augmented reality into 'real life' will still require too much commitment for too little reward to become entirely mainstream." —**Michael Edson**, director for Web and new media strategy, Smithsonian Institution

"This future seems plausible, but not by 2020." —**Aaron Schmidt**, Walking Paper Consulting, a blogger who writes about libraries, technology, and usability

"Successful work/play in the metaverse will require a different set of social interaction skills than those that generate success in 'real life.'" —**Mark Youman**, principal, ICF International, a Washington, DC, consulting-services company that works with government and commercial clients

"Human beings have always lived some part of their lives in their imagination. These are tools that extend our imagination, our daydreams, and help us live today and create tomorrow. Like all technical tools, some people will have more access than others." —**Ramona Nelson**, PhD, codirector of the health care informatics program, Slippery Rock University

"Artificial worlds are just an escape from the real one. The time spent on this could be put to a better and higher use (like reading, working, volunteering, helping others, etc.)." —**Judith Siess**, president of Information Bridges International, Inc. and publisher and editor of the One-Person Library newsletter, author and blogger

"Absolutely. However, by then it will be clear that this situation would have produced substantial drawbacks. In particular, exposure by the young generations to virtual life can produce a lower appreciation of the difference between fiction and reality, as fiction would be too close to reality. This is an aspect that can be governed, as exposure to television has been governed. We have seen that in cultural contexts where families and educational institutions have been less effective, some problems have been created, while when this fiction-vs.-reality situation has been governed by some sort of 'adult supervision' it has worked for the progress of mankind." —**Roberto Gaetano**, ICANN board member; also responsible for software development for International Atomic Energy Agency; an active participant in the ICANN policy-making process

"If our imaginations can get lost in something as basic as a good book, it seems very plausible that Internet will be our new fantasyland." —**Bill Warren**, vice president of government relations, Walt Disney World; founding editor of the Orlando Business Journal

"In the future, virtual environments will allow people to participate in rich experiences, including workplace situations, remotely with an avatar." —**Buddy Scalera**, vice president for interactive content and market research for CommonHealth Qi, in charge of interactive online strategies, including social and viral marketing

"While I think that the scenario holds true without a doubt, the level of participation in virtual worlds is where I mostly fall a little. The only thing that militates toward some of the day in a virtual world is the enormous popularity now of Webkins and the Penguin Club (a Disney Property) and the foresight of Disney—all of which mean that the generation that will have grown up by then will be used to virtual worlds and augmented reality. The verdict on Second Life as more than a fad is almost out—but not quite yet—and it seems to have some adult staying power. But Webkins and Penguin Club are HUGE! So this scenario, given who the adults will be, rings somewhat true." —**Paul Greenberg**, president, The 56 Group LLC; BPT Partners LLC; MyCRMCareer.com, social media companies

"The urge to live a better live via augmented reality will be too tempting." —**Don Kasprzak**, chief executive officer of Panaround.com, a Web-solutions design company; former system engineer at Apple Computer

"In the fields of architecture and design, immersive 3-D has been explored for some time, now influenced by 3-D CAD [computer-assisted design] to modeling realities and the advantages of mapping on Google Earth. Design firms operate virtually, networked by the Internet with leading-edge design universities offering educational curriculums fully online from multiple physical ports. The metaverse is real for architecture and design. It is key to transforming understanding about change of physical realities of the built environment." —**Janice Stevenor Dale**, president, JSDA Inc., The Design University

"Second Life has had its day and is now fading—in the current manifestation, such artificial realities are little more than a fad. Google Maps, however, represents a significant new medium by which to bridge the virtual and actual worlds, creating new forms of expression, communication, and learning." —**Joan Connell**, online editor, *The Nation* magazine, formerly an executive producer for MSNBC.com, senior editor for MSN and a Pulitzer finalist for her reporting

"These modeling tools will be used to create new ways of learning and simulating possible options to assist in decision making. I don't see it creating this parallel world like Second Life where people spend an overwhelming amount of time. There is too much to gain from 'real' life—including an income and food." —**Cameron Norman**, assistant professor in the department of public health sciences, the University of Toronto; actively engaged in use of the Internet to help tackle tough health issues, director of evaluation with the Centre for Global eHealth Innovation

"The Internet has made it possible for millions of people to express their creativity like never before...people who in the past would have never sat down and drawn a picture or shot a video. This has spawned many new forms of art expression, virtual reality, and interactivity. The real genius will be the person who uses this technology to help us realize and participate in the real world around us." —**Mike Samson**, interactive media writer and producer

"Virtual reality will be driven greatly by economic and environmental concerns. Students will be able to attend classes in the virtual world. Corporations will rely heavily on virtual environments, reducing building costs by allowing employees to telecommute. Families will even be able to take virtual vacations at a fraction of the cost of the 'real world' equivalent." —**Jamie Richard Wilson**, journalist and freelance Web developer

"For as long as the Internet has been around (and even before), a 3-D metaverse has been predicted—and it keeps not happening. Immersive virtual environments are appealing for a certain percentage of people (reasonably constant over time), but mostly people do and will use increased ICT capabilities in ways that are more and more integrated with the rest of their lives, not disconnected." —**Jacob Kramer-Duffield**, blogger and student, University of North Carolina School of Information and Library Science

"I agree with this scenario, although virtual worlds must evolve into an essential aspect of our daily lives by offering an essential 'service.' Otherwise, they will become boring and cumbersome." —**Susan Thomas**, S2 Enterprises LLC

"Second Life is the most overhyped alleged phenom of the century, so far. There's plenty of reality to go around; we don't need artificial versions of it." —**Jeff Jarvis**, top blogger at Buzzmachine.com; professor at City University of New York Graduate School of Journalism; contributor and consultant to the Guardian; advisor to start-ups

"While levels of engagement will vary, 3-D technology will take off the more personalized it gets. Being able to have a virtual coffee with a friend in a different city, where it looks like you're sitting at a café, can see a photorealistic representation of your friend, can select your favorite songs on a virtual jukebox, and talk in real time via headsets, will be a major driving force. In other words, the more senses a 3-D world can stimulate beyond just vision, the more people will gravitate toward it." —**John Jordan**, associate professor of communications, University of Wisconsin–Milwaukee

"Are we headed toward *The Matrix*? If so, then we should remember that while the first movie was inspiring, the next two were disappointing flops. It's far too early to engage in the kind of futurology this particular question prompts." —**Dennis Baron**, professor of English and linguistics, University of Illinois, runs the Web of Language site and researches the technologies of communication

"Augmented reality will be pervasive in 2020 as both a social norm and a business strategy. Virtual reality tools will be used for entertainment, as they are today, as well as for remote training, education, consulting, and management. Geography will continue to mean less to businesses, and new forms of entertainment will make it less likely that you travel to your vacation by 2020." —**Gerard LaFond**, founder and chief strategy officer, red TANGENT, and cofounder of Persuasive Games; works with leading brand companies such as Sony, Ubisoft, and Kraft

"Virtual worlds are the next evolution in the social networking, and once it becomes easier to integrate content from an external environment into the virtual, the development and integration will become as ubiquitous as the Web." —**Karmon Runquist**, ITS manager, Western Illinois University; formerly a consultant at Alliance Library System

"Virtual worlds will remain the domain of the escapists and of professions, such as architecture and modeling, where there are real-world applications. One caveat: This idea is very *Star Trek* holodeck. If the holodeck can be mass produced, it will alter our existence." —**Theresa Maddix**, satisfaction research analyst, ForeSee Results

"While 'many' will be involved, the majority of Internet users and especially the majority of real-world people will not be active in metaverses. Already, I often hear/read of backlashes for the antisocial nature of places like Second Life. Antisocial meaning we are interacting via our computer, not face-to-face." —**Mike Driehorst**, messaging strategist; leads social media for Hanson Inc., an interactive communications and video production company

"Shudder! Yes, I think this will happen. The only upside is I will be able to shop and try on clothes without leaving my home." —**Mariana Almeida**, product manager of Web products for health care, Kaiser Permanente

"It will still be possible to withdraw from 'real life.' In fact, more people will totally disappear from their 'real-life' social circles than happens today." —**John Jobst**, IT specialist, U.S. Army Corps of Engineers

"Virtual worlds still won't beat the sensations possible in real life. However, we're probably moving closer to having avatars that can be virtual personal assistants instead of being idle when they're not in use." —**Brian T. Nakamoto**, cofounder of MrJoy Inc. and product-line manager for Everyone.net (a leading provider of outsourced e-mail solutions for individuals and companies around the world)

"Much depends on improvements in interface design. Virtual worlds will only become ubiquitous once they can communicate more information with less hassle than simpler mediums like Web pages or teleconferences. 'More information' might mean things like body language, or allowing users to use their real hands to control virtual hands, for the purpose of building or interacting. Meanwhile, videoconferencing itself will only become ubiquitous once interface technology allows teleconferencers to make eye contact with one another." —**Ivor Tossell**, technology columnist/journalist for the *Toronto Globe and Mail*, known as "the blogging journalist" and a social observer

"I see this as an extension of the gaming phenomenon, so if the quality of [the] 'real world' matches and/or exceeds games, then most online users will be virtual inhabitants." —**Lynn Blumenstein**, senior editor, *Library Hotline, Reed Business Information*

"While a certain percentage of people will be inordinately consumed by artificial and virtual reality, they will only be a small minority. Most people will still be involved in making a living and living ordinary lives." —**Steven Hausman**, president, HausmanTech Consulting

"This seems to be a logical projection of current trends." —**David F. Salisbury**, associate director for science and research communications, Vanderbilt University; formerly science and technology reporter for *Christian Science Monitor*

"People will use technology only if they see a benefit to it. While having an avatar or similar presence in a virtual world(s)

will be common, technology will mainly be a navigation and entertainment tool for life. It seems that divides between RL and VL will remain. Although we will see less of an offline/online split in work and study." —**Ted M. Coopman**, lecturer, San Jose State University

"I agree with this based on the assumption that 'play' within these virtual worlds becomes more intuitive and the software less cumbersome. We are already seeing a huge increase in virtual worlds via mobile devices in China. I expect the same to happen in the U.S. as the technology advances, and costs become cheaper. Adolescents and young adults will be the first to adopt and adapt to virtual worlds." —**Rachel Kachur**, behavioral researcher, U.S. Centers for Disease Control & Prevention

"Absolutely agree and look forward to this freedom. Here is the opportunity to exchange products of the creative mind outside today's corporate relationships. Workers involved in producing creative constructs will be able to operate totally within these confines. Physical product, phones, iPods, etc. will be marketed and sold here. Entertainment may well live here entirely." —**Eric Kreider**, director of Web services, the University of Akron, OH

"But there will be a 'back-to-reality' countertrend." —**Paul McFate**, senior online communications specialist, LDS Philanthropies

"The phrase 'well-equipped Internet users' is operative here. There will be a digital divide, making for Internet hierarchies based on either income or social relationship with corporate sponsorship. On another note, I do not see a lessening in the need for so-called 'real-life' experiences; these will be even more in demand as the natural environment becomes so degraded that it will become somewhat expensive to enjoy true isolation in a natural environment (as opposed to a virtual or built one). Again, those with extra resources will vie for the status and privilege of having the wherewithal to go authentic, while others may need to remain at home in air-conditioned bubbles while scarce resources are rationed. Taken to extremes, this scenario might also mean that even Internet connectivity will be rationed along with the power to drive it. So, seamlessness in the sense described above does depend on the availability of other resources needed to keep the planet alive at the same time." —**Alex Don**, linguist and educator

"The progress of Second Life and its rivals has been surprisingly slow, considering the hype, and the online experience pretty uninspiring—though better 3-D may improve it." —**Roderick White**, editor, *Admap* magazine, World Advertising Research Center

"This participation in the metaverse will be an illusion, merely a persona, a projection of one's ego. But then, how much of what we call 'real life' is just that?" —**Peter W. Van Ness**, president, Van Ness Group, a Web-development company; founded Personal Computer Solutions in 1983; cofounded StockPlan, Inc. and MyStockOptions.com

"Commercial enterprises of any magnitude will have a virtual presence. Individuals will have several alter egos." —**Robert Grant**, chief executive officer, VoyaCare, Inc.

"Growth and extension of metaverses will continue, as well as advances in hardware that will allow ever-increasing use of connected and stand-alone AR, some of which can and will even be connected to metaverses...Military equipment has been using AR for many years with heads-up displays, targeting reticles, and overlaid image enhancements to...extend the senses and available knowledge about whatever the user is doing in real time with the activity. Manufacturing is just now beginning to see the value and benefits of AR...Instructions and design data can be given to workers as they handle the items being built or serviced rather than referring back and forth to separate documentation; wearable computers have allowed data to be presented to the worker as an overlaid display of data added to the real view of what they are doing as they do it...Most civilian AR is still stand-alone, but those too will continue to develop and will be fully integrated with Internet services, and their use will increase exponentially. Metaverses (separate from games) continue to gain in popularity and relevance, as real-world (RW) business and education are conducted within their bounds, not just entertainment...A highly developed metaverse that mirrors pieces of RW can provide abilities to examine replicas of RW things in details and in ways not possible in RW and provide insights.

Virtual meetings occurring within metaverses are already one of the most popular and effective uses of a metaverse environment…people log into a metaverse and attend the meeting with their avatar telepresence, with which they can listen to the speaker, view a presentation, and interact...Metaverses are in their infancy still, but the tremendous potential almost guarantees that it will continue to grow." —**Shawn Kelly Apochromantic**, configurations manager, designer, technologist, futurist, General Atomics, and volunteer builder in Second Life

"While the direction is correct, the time frame seems way premature." —**J. W. Huston**, president of Huston Consultancy and futurist

"I don't think it will be a majority. I do, however, think that the lines will continue to blur and that this distinction will be less clear in 2020 than it is to us now: I know what you mean when you differentiate between my use of Second Life and, say, Flickr. But maybe by 2020, those two experiences will be not so drastically different." —**Matt Gallivan**, senior research analyst, National Public Radio

"NASA is trying to figure out how to use a Second Life–type tool to allow future Mars explorers to stay in touch with the mother planet. They think they can lick the latency problem from Mars to Earth. That speaks to the possibilities that exist in these tools." —**Douglas Schulz**, managing editor for online publishing, America's Health Insurance Plans; formerly a Web team director for a biotech industry organization and manager of Internet and Web services at the Council of Better Business Bureaus

"Most people will prefer 'real' interactions to 'virtual' ones." —**Mike Langum**, Web developer, U.S. Office of Personnel Management

"Augmented reality will become easier and easier and become seamless parts of our lives." —**Lawrence Swiader**, chief information officer, U.S. Holocaust Memorial Museum

"By 2020, most people will scarcely remember what it was like to live only in the 'real world.'" —**Elizabeth Cleary**, student, National College of Art & Design, Dublin, Ireland

"While there very likely will be a small community of early adopters of 'augmented reality,' it will be many years (30,

40?) before its use becomes prevalent. Cost of use is certainly one of the 'braking' effects holding back rapid adoption; but it probably will be the result of 'unrecognized benefits' accruing from 'augmented reality' that will keep the number of its users in low figures for a very long time. I witnessed this same phenomenon in the case of various technologies: fax, high-definition TV, supersonic flight for passengers, among others. It's the old tug-of-war between technological 'push' versus market 'pull.' Nothing new under the sun." —**Fredric M. Litto**, consultant for Pearson Education Global e-Learning, president, Brazil Distance Learning Association

"My only comment to this is that the metaverse is a real place, just not located in the 4-Ds that we are used to. Continued enhancement of the 'feelies' (*Brave New World*, Aldous Huxley) will make the metaverse more and more a part of the universe." —**Richard Hammond**, knowledge management team leader, United States Environmental Protection Agency; knowledge management expert currently examining Semantic Web and RFID

"This is already happening to some extent. In Second Life, Science Island and NOAA's island are good examples of virtual world delivery of their content to a generation that has little or no interest in static Web content. Ten years ago, it was 'you'd better get a Web page,' now (certainly in 2020) it's 'you'd better get a virtual world presence.' I love the thought of a virtual planet!" —**Patti Nelson**, a webmaster who works on U.S. government sites

"The accidental loss of important privacy through creating virtual selves will drastically reduce the move towards a metaverse. The hope is for effective and practical means of practicing safe virtual play." —**Dixon Hutchinson**, software engineer

"Techno-dreaming." —**David Hakken**, Indiana University School of Informatics and a professor of anthropology who studies social change and the use of automated information and communication technologies

"Many elites will interact with each other and the 'metaverse' this way, but I don't believe that this technology will be so widespread that it affects most people." —**Paul Hyland**, executive producer, edweek.org; formerly a member of the board of directors for

Computer Professionals for Social Responsibility and director at Media Matters for America

ANONYMOUS COMMENTS

The following section includes a selection of contributions to the discussion of the future of the Internet and augmented reality and virtual reality from survey participants who chose to remain anonymous. Among them:

> "Man and machine will become one, as prophesized by Ray Kurzweil (*The Age of Spiritual Machines*) and Bill Joy. Artificial reality and virtual reality will replace real life, which ultimately will lead to the demise of human free will. Whoever owns the Internet will ultimately control the virtual beings—i.e., the world. Kurzweil's predictions of downloading the human mind into a machine seemed ridiculous in 2000; however, improvements in intelligent robots and technology in general should make us rethink what we are doing and consequences for humanity."

> "Second Life will be history. Metaverse indicates another world. Transmedia will come up, offering opportunities to transverse back and forth between environments (not worlds)."

> "Yes, and Second Life (v.2008) will seem quaint!"

> "We will mostly be wearing our networks, interacting with them without even thinking about it."

> "I still do not have a complete version of my real self."

> "You have touched on one of those possible inflection points that may rapidly change society. This may be one of them that evolves in ways we cannot imagine."

> "Sadly, this will likely be very true and will continue to break down social structures and relationships that previously existed in the 'real' world."

> "Immersive online pornography—as a commercial market sector—reaches a size and scope rivaling all other media forms combined."

"It is already happening now. People have forgotten how to interact in the real world without constantly looking at their phones and PDAs."

"Absolutely; look at how popular Wii is for children. Virtual reality will become the next way to socialise, work, shop, learn, and communicate."

"Virtual spaces provide opportunities to realize goals or dreams that may be difficult in 'real life.'"

"We will work in worlds that allow us to test theoretical limits, and such tests will aid in the search for cures for cancer and poverty and violence, etc."

"I wonder how well people will be able to distinguish between the real and the virtual, and the effect it will have on flesh and blood interactions. We are the compilation of our memories (which are coconstructed in interaction with others), and we interact within the context of an assumed shared past. It'll be interesting to see the effects of unilateral experience of 'shared' moments on the interactions that follow."

"I hope this doesn't happen. What happens to true human contact?"

"It will take more than 12 years for everyone to be comfortable with an avatar and for proper guidelines to exist within a virtual environment."

"This will be the socially illegal drug of the future. For that reason, these forms of drift from reality will more frequently occur."

"Virtual worlds will play a much more significant role in people's lives as bandwidth and computing issues are solved."

"Virtual worlds have been overhyped and underdelivered, and will continue to be for some years to come."

"The holodeck is almost here."

"Virtual realities will allow users to explore different cultures, countries, organizations, and technologies in a relatively safe and inexpensive manner. These virtual realities will be a primary source of education where users can vicariously experience different social situations and develop advanced knowledge and needed skills for eventually negotiating real situations."

"Rather than our real world being enveloped by the virtual worlds, aspects of the virtual world will become a greater part of our real world. The virtual technologies will be used to create TV shows and movies, used for workforce collaboration, and will turn into a huge boon for people with disabilities."

"This will probably happen but be driven more by online shopping sites than social sites. Also, I think 2020 is a little early for it to develop to the degree described."

"This statement ignores the fact that virtual worlds are inhabited by real people represented by their avatars. So when I play World of Warcraft with a guild of players, I chat about children, work, and a host of other things related to the real world. So 'seamless transition' is really 'seamless living.'"

"The geoWeb will grow and become more accurate due to the ubiquity of mobile computing. However, the geoWeb is an improved way to interact with geographic data, not a metaphor for the real world. Online gameworlds will continue to grow as well, but I don't see a place for a truly ubiquitous metaverse. People will have to make decisions about how much time, effort, and money they put into an entirely fake experience that only complicates the shared experience. Most will choose to keep it real instead."

"This is just video gaming in disguise and will take away far too much from productive time to be anything other than a leisure activity."

"All this and more was nicely predicted in nonfiction form in the 1991 book *Cyberspace: First Steps* from MIT Press. A rich literature grew around that book, itself inspired by the work of William Gibson and Vernor Vinge."

"As far as 'advanced organizations and individuals' are concerned, that is probably no more than 20% of mankind at that time, if it still exists."

"It won't be a single metaverse or geoWeb, but multiple ones."

"A tiny fraction of the moneyed Western world populates 'virtual space.' Yeah, I read about Chinese game players, too. It's a tiny fraction."

"Not by 2020, maybe by 2030."

"The fact that people increasingly live their life in an artificial space can alienate people from the desires, emotions, and experiences of human interaction and lead to an escape from interacting with and changing the real world."

"Sounds great in cyberpunk books of the '80s, but it's not going to happen."

"Probably will happen within 3 years."

"We are getting closer and closer to virtual reality with every software update."

"Our immersion in further augmented realities is reflected by the state of nonaugmented realities; a green, wild, natural augmentation might be preferred over the concrete-and-steel urban landscape."

"Virtual worlds allow anonymity to be maintained while duplicating real-life situations. This has practical value for both work and entertainment."

"Augmented reality will continue to be a pastime of a few groups. Its impact will most likely be realised by market forces. Imagine a world where augmented HUD displays show marketing and pricing information for real-world objects—this is probably where the majority of people will encounter it. In the supermarket or in the street."

"Predicted way back by Michael Heim."

"I hope most people will not be consumed by this virtual world. At most, perhaps 20–30% of the population with good Internet access will have a serious presence in AR, etc. From an organisational perspective, they may use RFID, GPS, etc., but it does not need to be in an AR—fundamentally what business driver will there be for commerce in AR? Some niche markets are already springing up, the ability to do some shopping worldwide, or to order takeaway in AR, etc., is viable, and of course virtual participation in meetings may be of some interest."

"Many people ask 'what's the point?' For sure, there will be overlays of tagging and mapping the real world, but most time people spend online will not be in Second Life–type applications."

"In order to stay up to speed, people will be forced to participate in this venue as a way to communicate—the next generation of social computing."

"Augmented reality will grow, yes, but I think there is a limit to mirror worlds and virtual worlds. Linden Lab's ambitious plans to grow more popular than the World Wide Web won't be achieved by 2020, if ever."

"Increases in fossil-fuel prices will make travel more and more difficult. Virtual spaces allow us to still feel that we're together without the expense and pollution of long-distance travel."

"The drive toward networking is fundamentally human, and we will operate on multiple levels at multiple times."

"'Augmented reality' will, like mobile phones or e-mail, merely be added to existing forms of social interaction rather than replacing them."

"These technologies, despite all the hype, are more than 10 years out for the bulk of the population."

"We have always incorporated tools into our lives. This is just a tool. There is only a real life, which may have more and more aspects that take place in an artificial 3-D environment. That is just as 'real' as a phone call (which is also virtual). The whole virtual thing is a crock-of-hype BS as being somehow different than other technologies that have come before it."

"As the MTV generation makes more virtual worlds, the text-messaging generation will use them, and eventually people will live secondly in VR."

"[I agree with this scenario] if, and only if, the interface into virtual environments is as easy as interfacing in physical environments, with natural language interfaces and intuitive information-access displays."

"Most people don't have time for this stuff unless it provides real additional value, i.e., gaming."

"Google Earth and the geotagging concept hold the most interest. I actually believe we are in the middle of more of a 'get a life' backlash, as forces of social control and hegemony attempt to rein in the geeks, whom some perceive have gotten a little too mainstream and are not keeping to their designated place. Not that I have any intention of becoming any less eccentric and 'geeky' in my obsessions on technology and learning. If these things are wrong, I don't want to be right!"

"There are flat-worlders in today's time as there were in Rome. There might be more in common between them than the actual shape of the world in the last 2,000 years. Virtual realists who transcend both of the worlds in their make-believe are similar creatures; they were always a few and equally capable to perform altruistic levels of stoicism, despite the physical barriers around them."

"This will only affect a small proportion of the world—the rest have real work to do (and lives)."

"Not everyone uses a computer in daily life and not everyone will want to interact outside of real life. An immersive 3-D metaverse will be available, but I don't think that a large proportion of people will be involved in it."

"Certainly, I expect the current teleconferences to evolve into this sort of thing, and that is where most people will experience it. Of course, there will always be a gamer community, but in the same proportion as now."

"Most people in the developed world will be linked in regardless of their intent in that they will be augmenting 'enhanced' models of the real world. Most people will also 'view' multiple of these enhanced models, e.g., seeing the traffic in the city, as a very simple, basic example...Other 'realities' will come along, in medicine for one."

"More addiction to online worlds. More augmented reality. No standards for online worlds."

"It could mean many different things—more Second Life than 'Neuromancer.'"

"I see great adoption of this in the education field, and continuing in the gaming field, but not widespread use throughout society."

"Hopefully by 2020, our 3-D worlds will be more tied together so that our avatars can go in and out of different spaces."

"Maybe in 100 years, but not 12."

"World of Warcraft is different but is essentially still a game. Gaming in 3-D worlds will probably be bigger than ever in 2020, but why would you want to go do your banking in some 3-D world, when it's easier to just use a regular Web site?"

"We're wet-ware for a reason. The human connection is never going to disappear. It will remain the primary force in people's lives. Virtual worlds are great for specific tasks, like architecture, engineering, simulations, and gaming, but those interactions are inextricably connected to real-life needs."

"Absolutely will be true, and already a source of addiction that will certainly increase."

"You could make a pretty good argument that the words you use for 2020 apply today. You could also go back 20 years ago and find descriptions that sound just like your 2007 status quo."

"2020 seems a bit fast for this for me, since this has been the prediction since about 1994. Seamless interactive 3-D menus and displays, yes. Entire metaverses and worlds (like in Neal Stephenson's *Snow Crash*) by 2020? Not quite. Fairly soon after that, maybe. And I highly doubt that the status known as 'real life' will be that far sublimated for most people.

"This or something like this is going to happen. Whether it is true by 2020 or surpassed by something else in 2020 is hard to say."

"Other drivers to this trend are increased pollution, too-high fuel prices, etc. People are programmed to be social and safe."

"It will become very blurred."

"This may not be a good thing. Life in the artificial world opens you up to reactions that may be partially hidden in the real world. Avatars have a way of being a lot more direct—and some people may have trouble dealing with that honesty (or whatever)."

"The 10-year-old kids who are growing up with computers, Internet, and electronic games will have been graduated from college in 2020. They will have devised all sorts of applications for work and play by that time and probably will have much of a seamless life. My hope is that virtual reality is used to promote the human race, not destroy it."

"Gaming will continue to be a big attraction and will undoubtedly get more sophisticated. The scenario you describe, an alter, cyber universe, would definitely blur reality as we know it and have detrimental effects on education and work productivity. It may become a problem for some people, but I don't think it will become the standard."

"I'm already seeing early adopters reducing their time on the Internet/wired world because the true human interaction is lacking, and it's too easy to get sucked back into the virtual world."

"This 2020 scenario scares me. I think we have enough complexity in our physical world—I'd hate to see a virtual world become a path for denial of challenges in the physical world."

"Slightly overplayed. Although VR will be increasingly important, the idea of 'seamless transitions' to real life makes me think that some people ought to get out a bit more."

"Bad times like the coming global recession will force people to look for new escapes—talking movies boomed during the Depression and WWII, virtual worlds will boom during our third-millennium recession."

"We no longer have the time and/or no longer know how to connect in the real world. Due to that malady, we will seek other ways to satisfy the human need to connect."

"The increasingly virtual universe will motivate an interest in returning to basics—as exampled by the increase in brand marketing technologies (from e-mail to advergaming), and this has also motivated a renewed emphasis on physical direct mail simply because, at the end of the day (real or virtual), the objective will be to have an impact."

"Absolutely accurate."

"AR technology is still primitive and is not ready for consumer use…True AR is much less useful to the average person than the ability to use MapQuest on their telephone. Find a consumer application for this technology, and you'll have a 'metaverse' of sorts; until you do, this stuff is science fiction."

"For me, this is a horror scenario. Augmented reality will also augment crime and favour the isolation of individuals."

"As most technologies have evolved, they have been driven by the pornography industry, so I would expect augmentation to begin there before moving mainstream."

"I don't think the infrastructure/costs of using the infrastructure will have advanced sufficiently for this to be an option for most individuals."

"We're a looooooong way from virtual worlds. Maybe in 2040."

"3-D worlds will be a huge part of the 2020 Internet, and the interfaces will be improved a lot, according to the bigger computational power of PCs and Macs."

"Augmented networks will exist for certain functions (perhaps conferencing), but in general they are too cumbersome for the multitasking typical user. Improved computing power will make augmented realities better than ever but will not be the major paradigm for networking."

"This is already happening."

"I think this scenario may occur in the future, but 2020 is not quite far enough out."

"Real life will be secondary."

"I definitely agree with this one. I only hope that it doesn't become a repressive, dystopian nightmare."

"The real question is whether being so connected has value for the majority of people. It has potentially significant entertainment value, but limited economic value. We are already seeing a fundamental assumption about the Internet debunked by business economics—namely, that the Internet would reduce the need for, and value of, face-to-face business. Economic research has shown the exact opposite to be true."

"The couch potato will win. Holodecks will become real."

"People's real-world geography will still be much more important in their daily lives than their online presence."

"Games have had this 'immersion' for many years already."

"These artificial spaces are ways to learn new ideas…However, those who are off the 'metaverse grid' will stay off—about 20–30%—as they simply don't need it. And more people will find they don't need it to gain their own spiritual 'boost.'"

"Virtual worlds will remain a niche interest for certain people."

"As the technology improves to the point that it nearly disappears, this will indeed happen. What it will be like, how people will share and split their real and virtual lives, is way beyond me."

"Distance education becomes an experience where everyone is virtually in the room together and can virtually experience each other. As the fuel crisis reaches increased levels, face-to-face meetings can take place in virtual realities…We will live in a world where traveling great distances as we do today will become more the exception than the norm."

"There has always been a lag between technology and human will. Just because it is physically/technologically possible doesn't mean that folks will readily take to it. I believe (and hope) this level of technological integration into reality will invite some fairly in-depth thought and discussion on the nature of reality and the universe and, ultimately, the nature of man."

"Technological infeasibility and cost mean that probably only social elites will have access to these kinds of technologies."

"Second Life hasn't gone away, as was predicted. People will move towards the 'geoWeb' as they did to the once-new World Wide Web."

"Real life will become ever MORE satisfying as people see how unsatisfying cyberlife is."

"We are in the middle of a tremendous revolution."

"For experts, ICTs will become more and more integrated with everyday life, and the boundaries will blur; for the basic users, ICTs will continue to be tools more or less removed from the rest of their lives."

"Augmented reality will be an essential tool for communication, but also to replace lots of items we use that don't need to be physical."

"Most likely, life will be this way, and it is really a misfortune for the human race—being exploited for profit. That is the bottom line of these artificial spaces."

"This relates to a previous elaboration on driving forces competing to create multiple identity 'sets.' The dilemma in education to produce critical thinkers and skilled laborers that transformed the physical institution in previous centuries now takes a cue from border-free spaces of 2020 to enable learners as voyeurs to experience, study curricula vicariously. Mathematics, e.g., retakes the questions of philosophy."

"The overall level of involvement will not be changed substantially."

"The real world will continue to be the space for 'living and working,' and the alternative virtual worlds will be another escape valve."

"Living is more than information flow."

"Hopefully, access will no longer be an issue with mobile technology integrated with 3-D Web."

"This scenario is most likely to become true after 2030 because the technology is still not affordable for everyone."

"I am nervous about what this means for 'real life.'"

"This scares me, but I agree that it may be the case in 2020. People want to find an outlet, and the Internet provides a very easy one."

"These demographics are already doing most of what is projected by 2020."

"Only if the user interface becomes much easier to use."

"This is commonly predicted by many, and that alone is a good indication that it would not happen that way, based on the scorecards of future predictions from the past. Some elements of this will be there, but most likely it'd look markedly different than this vision."

"The author(s) of this scenario assume a level of competence and insight on the part of most, that I think goes far beyond reality. When you look at the lack of competence in intellectual activity as demonstrated in public education, postsecondary education, and the complaints of employers as they attempt to extricate productivity from a dysfunctional workforce, I cannot even conceive of the projections of the Gartner study, let alone the scenario you suggest."

"It will be mostly universities and companies."

"We have been talking about virtual reality since the '90s, and Second Life is the most effective presentation. It's not very good, overall. Instead, social networking will encourage more face-to-face meetings and more travel."

"I disagree. In part, this a hope that we don't end up in escape environments regardless of whether this is *Brave New World*, *Star Trek*, or Second Life. Yes, I do recognize there is a great likelihood that there is a growing income potential for those who can cross-populate traffic from Second Life or other social network sites back to their own Web sites."

"I am only interested in such a scenario if it helps me more smoothly conduct my real life here on this (not Google) earth."

"Artificial environments remain artificial; folks can tell the difference."

"To meet consumer needs, companies will make a big push to enter this arena."

"More people are interacting in artificial spaces on a daily basis. Wii is another good example. With the elderly playing games in these artificial spaces even more, people are interacting in them and are involved. It will inevitably become a part of our daily lives."

"Online interaction is mostly favored by men with poor social skills. Everyone else will always prefer in-person interaction. Humans as a species are adapted to small-group, in-person interaction and talking."

"Having a secondary 3-D body is not the way to deal with the limitations of a physical universe. Having a secondary 3-D body will hardly enhance someone's spiritual life; if anything, it might confuse a whole lot of people on the priorities in life."

"Yes, large organisations must have a presence, etc., and the Koreans are leading the way in showing the great extent to which a population can take up using avatars and interacting in online worlds. BUT, the real world is still a beautiful place and people need exercise."

"These technologies will be very useful to people dealing with complex datasets involving multiple experts and simulations—things like urban and environmental planning, medicine, engineering, and design. These technologies are useless to anyone negotiating human issues; who can tell when your avatar is bullshitting? Direct human interaction is ultimately the way we judge honesty and develop relationships. Avatars might work after that. Outside of simulations and interactions, most people today are overwhelmed by data stimulus—e-mail, cell phones, texting—which are low bandwidth compared to the extra stimulus provided by artificial spaces."

"In 2020 as people continue to move away from traditional physical interactions with each other in favor of remote electronic interactions (via e-mail, blogs, virtual worlds, etc.), psychological studies will come along suggesting that humans have a deep-seated instinctive biological need for human touch and a direct interaction with other humans who are physically present. Since biological and technological evolution occur at vastly disparate time scales, these concerns will slow the headlong rush toward virtual worlds."

Endnotes

1. See *2007 Online Holiday Shopping Season Surpasses $29 Billion in Sales*, a comScore press release, January 7, 2008, available at http://www.comscore.com/press/release.asp?press=1990.
2. See "NPD December: $18 Billion Year for Game Industry," by Chris Kohler, published in *Wired*, January 17, 2008, available at http://blog.wired.com/games/2008/01/npd-december-18.html.
3. See "Leadership's Online Labs," by Bryon Reeves, Thomas W. Malone, and Tony O'Driscoll, published in *Harvard Business Review*, May 2008, available at http://harvardbusinessonline.hbsp.harvard.edu/hbsp/hbr/articles/article.jsp?value=BR0805&ml_subscriber=true&ml_action=get-article&ml_issueid=BR0805&articleID=R0805C&pageNumber=1.
4. See "Group-Forming Networks Resource Page," in *Reed's Locus*, by World Summit on the Information Society, David P. Reed, available at http://www.reed.com/dpr/.
5. See *Gartner Says 'Generation Virtual' Will Have a Profound Influence on Culture, Society, and Business*, press release by Gartner, November 13, 2007(a), available at http://www.gartner.com/it/page.jsp?id=545108.
6. See *Getting Work Done in Virtual Worlds*, by Erica Driver and Paul Jackson, report published by Forrester, January 7, 2008, available at http://www.forrester.com/Research/Document/Excerpt/0,7211,43450,00.html. Authors interviewed 14 vendor and user companies; report notes that virtual worlds "can reduce costs and improve the work experience."
7. From *Symposium/ITxpo 2007: Emerging Trends*, by Gartner, published April 24, 2007(b), available at http://www.gartner.com/it/page.jsp?id=503861. The research firm noted "collaborative and community-related aspects of these environments will dominate in the future...the majority of active Internet users and major enterprises will find value in participating in this area in the coming years." It also noted there will be a "struggle to develop appropriate and relevant business models."

8. See "Cisco Chief Lauds Coming Virtual-World Technology," by Matt Hamblen, published in *Computerworld*, June 24, 2008, available at http://www.computerworld.com/action/article.do?command=viewArticleBasic&articleId=9103079&source=NLT_AM&nlid=1.
9. See *Metaverse Roadmap Overview: Pathways to the 3-D Web*, by John Smart, Jamais Cascio, and Jerry Paffendorf, published by the Acceleration Studies Foundation, June 2007, available at http://www.metaverseroadmap.org/overview/.
10. See "Intel CEO Sees Net Boosting Consumer Electronics" by Scott Ard, published by *CNET News*, January 7, 2008, available at http://news.cnet.com/Intel-CEO-sees-Net-boosting-consumer-electronics/2100-7353_3-6225031.html?tag=nw.1.
11. See "Top 100's Ad-Spend Growth Grinds to Halt," by Bradley Johnson, published by *Advertising Age*, June 23, 2008, available at http://adage.com/article?article_id=127793.
12. See *Worldwide Spending on Internet Advertising Will Soar Past $106 Billion in 2011*, a press release published by IDC, June 25, 2008(b), available at http://www.idc.com/getdoc.jsp?containerId=prUS21304208.
13. See the Web site for the 11th International Symposium on Wearable Computers, Massachusetts Institute of Technology, held in Boston, October 11–13, 2007, available at http://www.iswc.net/iswc07/.
14. See "Is the '4.0' Era Upon Us?" by Chris Kanaracus, published by IDG News Service in *InfoWorld*, March 4, 2008, available at http://www.infoworld.com/article/08/03/04/Is-the-4.0-era-upon-us_1.html.
15. See "U.S. General Laments Google Earth Capability," by Kristin Roberts, published by *Reuters*, June 22, 2007, available at http://www.reuters.com/article/technologyNews/idUSN2140694720070622.
16. See the German Society for Scientific Person-Centred Psychotherapy Web site, available at http://www.gwg-ev.org/cms/cms.php.
17. From an analysis by University of Michigan researchers Rowell Huesmann and Brad Bushman, December 2007, published in the *Journal of Adolescent Health*, available at http://jahonline.org/.
18. See "Video Games Blamed for Reading Problems," published by *Metro*, November 28, 2007, available at http://www.metro.co.uk/news/article.html?in_article_id=77466&in_page_id=34.
19. See "The Impact of Singular Excessive Computer Game and Television Exposure on Sleep Patterns" by Markus Dworak, Thomas Schierl,

Thomas Bruns, and Heiko Struder, published in *Pediatrics*, November 1, 2007, available at http://pediatrics.aappublications.org/contents-by-date.0.dtl.

20. See "How Grim Was My Valley," by Rod Nordland, published by *Newsweek*, February 29, 2008, available at http://www.newsweek.com/id/117749/page/1.

Scenario 6

The Evolution of the Internet User Interface

Prediction: *In 2020 the most commonly used communications appliances prominently feature built-in voice recognition. People have adjusted to hearing individuals dictating information in public to their computing devices. In addition, "haptic" technologies based on touch feedback have been fully developed, so, for instance, a small, handheld Internet appliance allows you to display and use a full-size virtual keyboard on any flat surface for those moments when you would prefer not to talk aloud to your networked computer. It is common to see people "air typing" as they interface with the projection of a networked keyboard visible only to them.*

Expert Respondents' Reactions	
Mostly Agree	64%
Mostly Disagree	21%
Did Not Respond	15%

All Respondents' Reactions	
Mostly Agree	67%
Mostly Disagree	19%
Did Not Respond	14%

Note. Since results are based on a nonrandom sample, a margin of error cannot be computed. The "prediction" was composed to elicit responses and is not a formal forecast.

OVERVIEW OF RESPONDENTS' REACTIONS: A clear majority favored the idea that 2020 user interfaces will offer advanced talk, touch, and typing options, and some respondents added a fourth "t" —think. Those who chose to elaborate in extended responses disagreed on which of the four will make the most progress by 2020, with a fairly even yes-no split on the success of voice recognition or significant wireless keyboard advances and mostly positive support of the advance of interfaces involving touch and gestures. A number of respondents projected the possibility of a thought-based interface—neural networks, mind-controlled human-computer interaction. Many expressed concerns over overt public displays of ICT use and emphasized the desire for people to keep private communications private.

Nearly two-thirds of survey participants mostly agreed with this scenario tied to advances in network interfaces, with just one-fifth mostly disagreeing.

"It is these technologies that will enable the mobile device to become powerful enough for use in serious applications," responded **Brad Templeton**, chairman of the Electronic Frontier Foundation and an Internet pioneer who has been active online since 1979. **Cliff Figallo**, social innovator and original member of the first online community, The WELL, agreed. "More time on the move, less time sitting at orthodox computer interfaces," he wrote. "The need to communicate and think through handhelds will stimulate growth in use of such features."

"In addition to this," predicted **David Brin**, futurist and author of *The Transparent Society*, "there will be 'subvocal' inputs that detect 'almost speech' that you will, but do not actually, voice. Small sensors on teeth will also let you tap commands. Your eyeballs will track desires, sensed by your eyeglasses. And so on."

Hamish MacEwan, a consultant with Open ICT in New Zealand, wrote, "By 2020, those who desire it will speak or think to technology." And **Dan Lynch**, founder of CyberCash and Interop Company and an Internet pioneer, wrote, "Air guitar gone wild. A good idea. Long die the mouse! It was good for a long while."

Respondents noted that intuitive, human-centric interfaces allow technology to eliminate some social, economic, and physical divides. "Ease of access + usability will entice more people to interact with technology—in other words, it will not only be limited to computer-literate people," commented **Sam Ozay**, an e-learning and e-communication specialist and solutions architect at Postmodern (Asia/Pacific). **Jan Schaffer**, executive director of J-Lab, the Institute for Interactive Journalism, wrote, "I see great benefits for education in the form of alternative learning and assessments of learning for dyslexics and LD [learning disability] children."

Security is always an issue, as noted by **Alejandro Pisanty**, director of computer services at the Universidad Nacional Autónoma de México, an active leader in the Internet Society and ICANN. "They will all be hacked big-time!" he predicted. "Think of using a cell phone for video recording a person who types on her lap while riding a subway."

Many are concerned about social ramifications of new interfaces. "By 2020, I would hope that there is some other way to get information without a public display of any kind through ubiquitous technology," suggested **Teresa Hartman**, associate professor and head of education at the University of Nebraska Medical Center. "Interactions with personal communication interfaces should be less intrusive to others than taking out a notepad today and writing a note. Communication users have allowed the public display of their interactions to continue and even increase due to what they perceive as a 'wow' factor—'look at me, I have a cell phone and know how to use it.' I see the prediction of us air typing to be in the same category.

In the future, using technology (hopefully) won't be a status item, and can be conducted discreetly and with panache. Somehow, interactions with communication/information have to be put back in the individual's world, instead of bleeding over into everyone's world, and not causing any more interruption or notice than a quick cough into a handkerchief."

All respondents expect evolution of some kind. "Yes, yes, and yes," noted **Leonard Witt**, associate professor at Kennesaw State University in Georgia and author of the weblog PJNet.org. "It's all disruptive technology, which means, as Clayton Christensen says, cheaper, smaller, faster, and easier to use. It can't be stopped."

"Solitude will soon become a thing of the past, as no one is ever disconnected," commented **Lisa Carr**, director of strategy for Targetbase Interactive.

YES, TALK WILL BE EFFECTIVELY DEVELOPED, OR NO, TALK CANNOT BE EFFECTIVELY DEVELOPED, AT LEAST BY THEN

Respondents debated the idea of voice as a user interface with some in support of its development to perfection, some saying the technical issues to develop it correctly have been and will continue to be too difficult to overcome by 2020, and some expressing concern over social acceptability.

"By 2020, the voice 'interface' will be more sophisticated," predicted **Maz Hardey**, a social analyst and blogger completing a doctorate funded by the Economic Social Research Council at the University of York. "When not touch typing, voice commands will allow the user to talk to those in the immediate and physical vicinity, as well as to update and 'chat' across SNS."

"Talking is more fun, voice recognition will make typing obsolete," wrote **John Levine**, founder of Taughannock Networks, a leader of the Internet Research Task Force's Anti-Spam Research Group.

"We are already used to the way dictating to devices would sound, since Bluetooth headsets and cells create a similar hearing experience," commented **Paul Greenberg**, president of the 56 Group LLC. "This is not a difficult one to see, given the rates of technological advance, especially in computers and electronic gear, that we are seeing today."

"Wi-Fi– and WiMAX-enabled badges with voice recognition will act as personal assistants—allowing you to talk with someone by saying their name, to post a voice blog, or access directions from the Internet for the task at hand," predicted **Jim Kohlenberger**, director of Voice on the Net Coalition, a senior fellow at the Benton Foundation.

Those who disputed the likely use of talk as a UI by 2020 generally noted how difficult it has been up to this point to overcome the technical barriers in designing a usable talk interface. "Speech recognition and even natural-language understanding are evolving, but it's been a very gradual process over several decades, and it is likely to take several additional decades before we approach Hal-like performance," commented NMS Communications CTO **Brough Turner**, referring to the AI computer HAL (for "Heuristically programmed ALgorithmic computer") in the film *2001: A Space Odyssey*.

"Voice will continue to be the most oversold, overhyped, but unused interface," noted **Walt Dickie**, executive vice president and CTO for C&R Research.

"Voice recognition has been a holy grail of computing since *Star Trek* in the 1960s," wrote **Charles Ess**, a researcher on online culture and ethics based at Drury University in Springfield, MO, a leader of the Association of Internet Researchers. "Like the artificial intelligence that was supposed to make it happen…it has faltered for a host of reasons, beginning with technical ones. Perhaps there will be some sort of technological breakthrough in the next few years that will make voice recognition workable and affordable—but I'm not optimistic."

An anonymous respondent noted, "Voice interfaces are just so damned difficult for anything but simple tasks; they force one to go through menus or to use a command-line interface. Speech being how it is, I can't imagine even natural-language processing overcoming many of the problems."

"Although voice control will progress to where it can be mainstream, it will not surpass other input mechanisms—mostly touchscreen and accelerometers," commented **Todd Spraggins**, chairman of the board of directors of the Communications Platforms Trade Association and a strategic architect with Nortel Carrier Networks.

Clay Shirky, consultant and professor in the Interactive Telecommunications Program at New York University, author of *Here Comes Everyone*, wrote, "Ben Shneiderman's work on the limits of voice recognition and the weakness of the human brain's ability to coprocess other information alongside the spoken word are, in my view, dispositive critiques."

"I worked on voice-activated technologies and AI in the 1980s, and I am familiar with the overblown predictions that were made then," responded **Micheál Ó Foghlú**, research director for the Telecommunications Software & Systems Group, Waterford Institute of Technology, Ireland, arguing in support of the positive future of the talk interface. "Steady progress has been made, and the need to use innovative interfaces on small mobile devices is a good spur for these developments in the next 12 years."

Some noted that those with special needs are most likely to use speech-recognition UIs first. "With increased attention being given to the need of 'specially-abled' people, 'talk and touch' will become more popular (and profitable) as devices that employ such will help empower more people who never had the chance," noted **Gbenga Sesan**, a consultant for Internet development with Paradigm Initiative in Nigeria, adding that 2020 may be too soon for it to be practicable.

Ted Coopman, a technology lecturer at San Jose State University, responded that it is possible the new interface may change the way we use language. "People will use voice for some communication and texting for others as they see fit," he wrote. "It is possible that dictating will become a standard way to create text in private, especially for those who write a lot. This might change the texture and style of literature and grammar generally. At any rate, the written word will more closely resemble the spoken word."

Concerns over the appropriate use of talk interfaces were expressed by some respondents.

"I mostly agree with the scenario, although a rise in voice-driven interactions might lead to social reactions against the use of these devices in public spaces," commented **Paul Miller**, technology evangelist on the senior management team at Talis, a company delivering human-centric Web applications, based in the United Kingdom. "See, for example, the differing attitudes to speaking on phones in restaurants, etc., today. In some places this is acceptable, in others most definitely not."

"The sound rules out using [voice] in many environments," commented **Christine Boese**, researcher and analyst for AvenueA-Razorfish and Microsoft, "(and I even avoid listening to podcasts on the subway because my hearing is so bad already, and the train noise is too loud). Privacy concerns arise with too much spoken technology, or should, when we see people walking up and down aisles at the grocery store, talking out loud on their mobile phones with the ear bud hanging out of their ears."

"People in airports and grocery stores who talk to themselves using those stupid-looking knobs in their ears are already annoying. Imagine an office where people in cubicles are all talking to themselves—composing proposals, sending e-mails, making notes on their next presentation to the boss. Yipes!" noted **Mike Samson**, an interactive media writer and producer.

An anonymous respondent commented, "What would be really great is if a sound-proof force field could be created around people while they are yakking away on their phones about their latest foot fungus. Someone needs to invent that."

"I expect to see some use of these things, but my use of them so far (Amtrak's 'Julie,' for example) suggests that they only work when conflicting sound can be stopped and when talking to a computer is not disruptive to others," wrote **Fred Baker**, Cisco Systems Fellow and a longtime leader of the Internet Society and IETF. "That imposes quite a limit."

Mary Ann Allison, principal of The Allison Group, predicted, "Several entrepreneurs will get rich creating sound buffers and cones." And **Peter Van Ness**, president of the Van Ness Group, a Web-development company, noted, "Voice recognition still has a long way to go. However, if there is a breakthrough and voice recognition becomes ubiquitous, we will abandon typing as fast as you can say, 'the quick brown fox jumps over the lazy dog.'"

TOUCH IS NATURAL AND INTUITIVE, AND IT WILL SUCCEED

While talk drew heated debate from the respondents who wrote elaborations to their answers on this scenario, positive support for the future of the touch interface was nearly unanimous. "Touch is there already, with the Microsoft Surface computer, the iPhone, the Wii," noted **Christian Huitema**, distinguished engineer with Microsoft and an Internet pioneer and active leader of the Internet Architecture Board (IAB) and Internet Society.

"New touchscreens will provide graphical nontextual interfaces," predicted **Ulrich Spalthoff**, director of advanced technology for Alcatel-Lucent in Paris.

Jerry Michalski, founder and president of Sociate, a technology consulting firm, responded, "Touch is the first major step away from the windows/mouse interface, which is very long in the tooth. We're due for some more advances in the next 13 years."

Brian Nakamoto, cofounder of MrJoy Inc. and a manager for Everyone.net, commented, "Touch-input interfaces will hopefully be more innovative than virtual keyboards."

Jonathan Dube, president of the Online News Association, director of digital media at CBC News, and publisher of Cyberjournalist.net, wrote, "Touch feedback will be the primary mode, with voice recognition an increasingly common tool (but not on airplanes!)."

An anonymous respondent predicted, "Touching machines in ways that we have not imagined will become possible."

MANY SAY TYPING HAS ADVANTAGES AND IT WILL ADVANCE

Many respondents see the survival of keyboards as input devices as highly likely. "QWERTY rules," wrote an anonymous respondent. "Typing is faster than talking."

"Most people form a tactile bond with their keyboards and a comfort with their workplace/desktop environments that will be difficult to replace with haptic appliances and voice recognition," noted **Michael Edson**, director for Web and new-media strategy for the Smithsonian Institution.

Jeff Jarvis, blogger at Buzzmachine.com and a professor at City University of New York, predicted, "We will have control environments that don't require us to read buttons. We will also have some means of typing specific wording quickly and accurately without two-handed (or two-thumbed) keyboards. I await their invention."

"Air keyboards or projected keyboards will be a great advancement, as they will allow small devices to become fully functional computers that finally will allow people to work the way they want and with a maximum of convenience," wrote **John Jordan**, associate professor of communications at the University of Wisconsin–Milwaukee.

Survey participants had mixed reactions to the idea of air typing. Many thought it unlikely. "This one sounds too much like the Kitchen of the Future at some 1930s World Fair; I think we'll have better, more adaptable devices, but I doubt we'll be air typing," commented **Susan Crawford**, founder of OneWebDay (celebrated each September 22), and an ICANN board member and law professor at Yale. "Roll-out, flexible keyboards might be the more likely development," wrote **Seth Finkelstein**, author of the Infothought blog and an EFF Pioneer Award winner. An anonymous respondent commented, "Laser-based keyboards are available today but are often inaccurate and inconvenient. It's hard to imagine this situation will change much by 2020." Another anonymous respondent wrote, "Keyboards will remain. So will street signs and the alphabet."

Many who disagreed with the idea of air typing noted the lack of physical feedback one gets when typing in an empty space. "Tactile interaction requires feedback," noted **Richard Osborne**, a Web manager at the University of Exeter. "That's why our hands are designed the way they are." An anonymous respondent commented, "Are you kidding me? Historically, people haven't been able to touch on even flat keyboards without the feel of a 'click.' Touch-typing requires touch. What you describe is insanely difficult."

But Internet sociologist and author **Howard Rheingold** responded, "The point-and-click user interface is 40 years ago. It's time for more human-machine bandwidth. You are a typist, try 'air typing,' and see if it doesn't feel natural very quickly." And **Havi Hoffman**, a senior editor for Yahoo and blogger, noted, "I can imagine air typing of a kind, and a flat and more fluid electronic paper than we've seen yet."

Some respondents disputed the idea that keyboarding will still be a dominant UI.

"I still envisage a replacement for the keyboard, virtual or otherwise," commented **Adrian Schofield**, manager of the applied research unit at the Johannesburg Centre for Software Engineering. "My vision is of a virtual pen that can interpret any type of script."

"We will be beyond the use of keyboards and will be using primarily voice-command or voice-recognition technologies," wrote **John Murphy**, director of USA.gov technologies.

"Typing will be a thing of the past; it seems reasonable that some forms of subvocalization, not to say 'mind reading,' will eliminate the need for a manual interface—kind of like reading 'almost aloud,'" suggested **Oscar Gandy**, author, activist, and emeritus professor of communication at the University of Pennsylvania.

LOOK MA, NO HANDS OR VOICE; COMMUNICATING BY THOUGHT ALONE

Many respondents predicted that brain science will advance to the point at which there will be at least some human-machine interaction conducted through the reading of brain activity. Most who offered this view did not note that they expected this to be true as soon as 2020.

"We'll see thought as a common interface too—as in telepathy," noted an anonymous respondent. "If machines can be 'wireless,' so too will be humans—either through chips or through the ability to understand that the human physiology has the same capacities (or more) than the machines do."

"Future technologies (although perhaps not by 2020) will involve physically connecting our bodies 'wirelessly' to computer/digital networks through true 'neural nets,'" responded **Benjamin Ben-Baruch**, senior market intelligence consultant and applied sociologist for Aquent. "It will literally become possible to interface with these networks via neural nets that connect our nervous systems to the networks. The common technology interfaces will be 'talk-touch-think.'"

"I totally expect even mind-controlled interaction by thought using a simple range of commands, which, in combination, allow 'joystick'-style interaction," commented **Robert Eller** of Concept Omega, a marketing and communication company. "This is already today virtually possible. Some research even shows that we can grow additional synapses into minute glass vials that will connect to wires allowing fighter pilots to steer a jet. Nano and bio technologies should yield some significant advances here making such interaction, if not mind controlled, at least be part of the body."

"Brain waves, just thought processes to get the information out," suggested **Garland McCoy**, founder of the Technology Policy Institute, a think tank focused on the economics of innovation.

"I suspect we will eventually move beyond voice and touch interfaces for computing in the future," predicted **Gary Kreps**, chair of the department of communication at George Mason University. "Instead, we will direct computing directly through our cognitions, through thought."

Steven Hausman, president of HausmanTech Consulting, responded, "Air typing could easily be supplanted by an entirely internal neural interface if advances in nanotechnology continue apace."

Bruce Turner, director of planning services for a U.S. regional transportation commission, agreed, writing, "As brain-mapping technology improves, we may forego the virtual of the real world to direct our consciousness to type inside our brains for transmittal to the surface."

RESPONDENTS SUGGEST ADDITIONAL INTERFACE INNOVATIONS

Some respondents expressed various additional expectations for UI in 2020.

Many noted that gestures and body language (as exemplified in Nintendo's Wii game system) may be more common than talk, typing, or touch. "Air typing (difficult without tactile feedback) will be less commonplace than seeing people make gestures into thin air," noted **Ivor Tossell**, technology columnist and journalist for the *Toronto Globe and Mail*.

Ed Steinmueller, a professor whose research expertise is the industrial structure of high-technology industries, commented, "Although I doubt that the keyboard metaphor is entirely apt, the extension of the interface to gesture seems very likely."

Scott Brenner, technologist and consultant, predicted, "The haptic technologies will prevail, although we'll be getting away from the keyboard method of input. Instead, data getting and giving will be more intuitive, using icons, structured gestures, and a more semantic information universe."

"Other types of inputs, such as simple gestural inputs, may prove more popular than full-keyboard inputs," suggested **Scott Smith**, a futurist with Changeist LLC, consultant, and writer based in North Carolina. "Additionally, interfaces will be more predictive, taking into account contextual information about a user to determine data we might have to manually enter today."

Tiffany Shlain, founder of the Webby Awards, commented, "I could see a whole physical way of communicating with our technology tools that could be part of our health and exercise. A day answering e-mails could be a full-on physical workout ;)."

Smaller movements made by the eyes and face were also noted as possible interface methods by respondents. **Jay Neely**, social strategist and founder of News Armada, a Boston-based Internet-news community, wrote, "Advancements in eye-tracking technology, combined with the miniaturization of components needed to create devices in the same size and form as eyeglasses, make sight a more likely interface for services that only require information consumption and very limited data entry."

"This could also grow to technology that recognizes facial expressions," commented **Jerry McCann**, vice president and director of the social marketing group at Carton Donofrio Partners Inc. "A frown could indicate 'next page' or 'delete' for example. These kinds of advances make a lot of intuitive sense and are to be looked forward to."

A number of respondents noted that devices will interface with ubiquitous computing built into human architecture. "It will be common to see people interacting with signs," responded **Fred Hapgood**, technology author and consultant.

"We will see the display interface device separated from the input device over the next 12 years," wrote **Ross Rader**, a director with Tucows who is active in the ICANN Registrars constituency. "Display devices will be everywhere, and you will be able to use them with your input device. The input device might be virtual, as in the case of the iPhone, or a holographic keyboard, or they might resemble the keyboards and touch pads that people are using today. Likely, some combination of these will prevail. These devices will be able to securely interact with any display device that the user selects, using common standards that permit the user to interact with data in a variety of resolutions and formats."

Chris Miller, senior vice president for digital operations for Element 79, predicted, "Common objects, desks, countertops, etc., will become haptic sensitive and provide feedback and content and send/receive information based on touch. This will correspond to the *everyweb*, which allows appliances, objects, etc., to be networked. The *Minority Report*'s haptic gestures and feedback will be a reality."

Some respondents suggested a cluster of alternative user interfaces. **Sean Steele**, CEO and senior security consultant for infoLock Technologies, predicted that, by 2020, "while air typing and haptic gestures are widespread and ubiquitous, the arrival of embedded optical displays, thought transcription, eye-movement tracking, and predictive-behavior modeling will fundamentally alter the human-computer interaction model. What we think is performed almost in real time, when and how we imagined it to be."

"Even further, these devices will be capable of translating from one language to another, as well as assisting people who have lost sight, hearing, touch, or smell sensation," commented **Steve Goldstein**, ICANN board member. "People might even be able to share emotions or sensations remotely, aided by implantable microbioelectronics. Back in the 1990s, the Sunday magazine of *The Washington Post* had an article that estimated that the sensation of human orgasm could be transmitted with (then unheard of) several-gigabit-per-second bandwidth. We're there, baby! And, already one can buy sex toys that can be manipulated remotely over the Internet (seems that the adult industry is often at the forefront of so many developments that ordinary people can use…and purchase)."

THE SCENARIO IS WRONG; IT IS NOT GOING TO HAPPEN THIS WAY

Those who disputed the scenario expressed a variety of viewpoints in their elaborations. Layered reasoning came into play in a number of the responses.

"Neither of these are particularly efficient interfaces, at least as described," wrote **Jamais Cascio**, blogger, public speaker, and futurist with Open the Future, The Institute for the Future, and Worldchanging. "The social response to mobile-phone conversations in public—resigned/resentful acquiescence—is a likely model for voice interfaces, slowing or even halting their widespread adoption. As for haptics, these seem more likely, but not as described; 'air typing' and similar nonresponsive interfaces have a poor record of usability. More likely is some kind of touch-based interface, possibly even a finger-on-opposite-palm model."

"Products continue to be driven by short product lives and lowest possible cost, and dim displays and flat membrane switches," answered **Tom Jennings**, the creator of FidoNet, the first message-and-file networking system online, and builder of *Wired* magazine's first online presence. "Extreme power management will continue to work apparent miracles in ubiquity and will have unpredictable side effects. People forget that 'lack of interface' also allows for perceptual partitioning and maintaining of separate cognitive spaces. E.g., I can let the phone sit on the table and not answer it, and it doesn't impinge on my conversation. Talk to people over 16 years of age… Oh, I forgot, you're too busy taking their money."

An anonymous respondent noted, "This will only be available for the happy few. And even so, it is likely that the commercial networks will make the use of these tools as cumbersome as cell phone use for us is now as soon as one leaves one's 'regular' environment."

"Future communications devices are unlikely to remain tethered to QWERTY or any other similar relic," commented **Buddy Scalera**, vice president for interactive content and market research for CommonHealth Qi. "The tools are likely to be icon based, batched, and routine oriented. That is, it takes too long to type certain concepts, so taking a cue from programming language, we'll have communication subroutines that we'll drag and drop in highly streamlined conversations. Physical objects tagged with information will be part of an overall, organic language that's able to be virtualized over long distances."

An anonymous survey participant predicted, "In 2020 how we will interact with data is not conceivable, at least by ordinary Joes."

Human-Computer Interaction Is Making Progress Toward Breakthroughs

The degree to which humans are able to use a tool in an intuitive manner influences the diffusion and use of the tool. Technology innovators invest significant effort in getting the interface right. Today's mouse-keyboard-display interface is known to interface researchers as Windows, Icons, Menus, and Pointing (WIMP). In this section of the report, we address advances in the human-computer interface as it is implemented to access and exchange input and output with devices and the Internet.

Two important trends in computing are driving the innovation today. One is the move toward the mobile Internet. The second: Embedded, networked computing devices are bringing human environments alive with possibilities for human-computer interactions in many formats, as people move through each day using home appliances, automobiles, home-entertainment networks, and other items that fulfill their needs. In the future, researchers expect what is called the "Internet of Things" to include intelligent, networked devices that are woven into the world and integrated in such a way as to be barely noticeable but highly functioning.

The integration of networked information processing with the human environment is sometimes referred to as *ubiquitous computing, pervasive computing, everyware* or *ambient intelligence*. It is expected that the devices will be context aware (recognizing who you are and where you are); they will be personalized; they will anticipate your expectations; and they will change in response to your demands. High bandwidth is expected to lead to the creation of virtual-reality and computational video content that will open new interface opportunities. All of these factors are also expected to lead to a new paradigm for human-to-human interfaces; as group coordination and sharing become easier, this trend is likely to inspire change in the nature of all forms of human organization. The acceptance and functionality of all of the promised developments are contingent upon consistent and intuitive user interfaces. The design must be natural to be accepted.

Will such breakthroughs in interface design and the uptake of pervasive computing evolve in a significant way by 2020? Some people think so; some do not.

William Gibson and Bill Gates, two of the visionaries whose early 1990s remarks about the future of the Internet are included in the research database that inspired the *Future of the Internet* surveys, were quoted in the November 15, 2007, edition of *Rolling Stone* making new predictions about the interfaces that are soon to come[1]:

William Gibson was asked, "What are the major challenges we face?" and he answered:

> "Totally ubiquitous computing. One of the things our grandchildren will find quaintest about us is that we distinguish the digital from the real, the virtual from the real. In the future, that will become literally impossible. The distinction between cyberspace and that which isn't cyberspace isn't going to be unimaginable. When I wrote *Neuromancer* in 1984, cyberspace already existed for some people, but they didn't spend all their time there. So cyberspace was there, and we were here. Now cyberspace is *here* for a lot of us, and *there* has become any state of relative nonconnectivity. *There* is where they don't have Wi-Fi. In a world of superubiquitous computing you're not gonna know when you're on or when you're off. You're always going to be on, in some sort of blended-reality state. You only think about it when something goes wrong and it goes off. And then it's a drag...
>
> People worry about the loss of individual privacy, but that comes with a new kind of unavoidable transparency. Eventually, we're going to know everything that every twenty-first-century politician has ever done. It will be very hard for politicians and governments to keep secrets. The whole thing is porous. We just haven't really figured out quite how porous it is..."

Bill Gates was asked by *Rolling Stone*, "What would you say the dream is today?" and he replied:

> "To make computing as pervasive as electricity. You no longer just see the computer on every desk. It's in every pocket, it's in your TV, it's in your car. But there's still a lot missing from computers. They can't see, they can't listen, they can't read your writing. Before long, when you walk into your living room, the computer will see you're there and put up a show you might be interested in or a game your friends want you to play with them. Even in a ten-year timeframe, things like that will be commonplace. Things like speech-recognition turned out to be harder than people expected, but they're very solvable. We're starting to see the emerging edge of this natural interface where speech and ink and motion and touch all become part of the experience and the computer sort of disappears. We're just at the beginning of the impact it's going to have."

In 2008 display designs were maturing to include very large, high-definition displays and small, thin, light, and even foldable/rollable displays that have low power consumption and allow manipulation.

The display screens are becoming more flexible and dynamic. The ability to reach out and touch and organize items on a display, small or large, as with Apple's iPhone and Microsoft's Surface (a $5,000 to $10,000 table-shaped computer), employs "gesture control" and "multitouch" features. Microsoft announced in 2008 that the next version of its Windows OS, now known as Windows 7 or Vienna, will have many touchscreen and gesture features. Gesture-based interfaces are also being included in Pioneer GPS systems and Kodak digital photo frames. In May 2008 Microsoft introduced a new product that is not yet available—the TouchWall. It uses infrared lasers and a camera to project items in a setup that is much more cost-efficient (reportedly just using a few hundred dollars' worth of parts) than the wall-sized Perceptive Pixels' multitouch screen that was first to hit the market and was listed in the Neiman Marcus holiday catalog in 2007 for $100,000.

Projection breakthroughs will allow small Internet devices such as cell phones to display data in a wall-size format. Microvision unveiled a Pico projector in 2008 that can take a cell-screen–sized image, say a movie you want to play on your iPhone, and enlarge it to any size you want, from 5 inches wide to 52 or 60 inches. Microvision has plans to sell it as a peripheral, and its microelectromechanical system could be built right into cell phones—it would add about $60 to the price. Another product invented as an enhancement for small-sized displays is the Myvu Personal Media Viewer, a wearable display that looks like a black set of glasses with earbuds that allows you to enjoy the equivalent of viewing the content playing on your iPhone or other small-screen media player on a 27-inch television about 2 yards away.

The user interface of Nintendo's Wii gaming system—the "Wiimote" as the remote-control device is known—makes use of an accelerometer and optics to detect input from human body movements, and it has become the most sophisticated consumer-product interface device to date that uses this technology, which can sense inclination, shock, and vibration. Cell phones such as the Nokia N95 and iPhone also use an accelerometer to allow switching their displays from portrait to landscape mode. Wii also sells a balance-board input device people stand on, shift their weight on, and step on and off to send data. Innovations in input using accelerometers and optics are exploding following the Wii's success, as people use the accelerometer and optics tools to create their own device inputs. When you use the search phrase "Wii hacks" in a search engine, you get about 2 million hits. More than 8.5 million Wiis were sold in 2007 in the first few months the game was available.

The LtWv Wrist Vmote Lightglove actually looks more like a bracelet, and it features optical-sensor technology like the Wiimote. The interface detects infrared reflection from movements in your wrist and translates them, allowing you to operate a computer, PDA, phone, video game, or television in a rudimentary way. It won the Innovations 2008 Award at the Consumer Electronics Show in Las Vegas. It is not sophisticated enough for a person to do air typing at an efficient speed.[2]

While keyboards have been useful since the 1800s, and the computer mouse has been a significant tool since the late 1900s, these UIs in their current form are not practical for many of the tasks humans will tackle with their tech tools in the mobile-computing and pervasive-network world expected in years to come.

It is not likely that text-based information sharing will disappear soon, if ever. People can ingest data via written text more quickly than they can via spoken content. On average, a typical chat at an effective speed in English averages about 100 to 200 words a minute; the average American adult reads text in English at 250 to 300 words per minute, and some have the ability to read as many as 400 to 1,000 words per minute or more. The average typing speed is in the 40 to 100 words-per-minute range; advanced typists work at a rate of 120 to 200 words per minute. Based on these figures, the most efficient human-to-computer *input* method is the spoken word, and the most efficient computer-to-human *output* method is text. This estimation does not account for differences in comprehension, retention, and other variables.[3]

The most popular software packages for speech recognition on the market today are Nuance's Dragon NaturallySpeaking (which can convert about 150 spoken words per minute to text—not in perfect form, however, with a few errors here and there, depending on your vocalizing style, and if you do not have a head cold) and Philips' SpeechMagic.

Researchers at the Massachusetts Institute of Technology's Computer Science and Artificial Intelligence Laboratory are working on highly advanced speech-recognition and language-translation software.[4] They have used conversational-language programs to operate handheld computing devices, and they are working on using a cell phone as a remote control. People can speak into the phone and instruct the device to operate other equipment, for instance, turning on a video-recording device. They expect smartphones to offer language-translation assistance.

People are already issuing vocal commands to devices, and they have been for a while. The automobile industry has been experimenting with vocalized interfaces for decades. The latest consumer-oriented system now on the market is Microsoft's Sync, which allows voice activation of in-car technology. At your vocal command, it can search for and play any music you have available, download information from a PDA, and send text messages through your mobile phone. Smartphones are already incorporating some speech-recognition interfaces. Tellme (a Microsoft subsidiary), Vlingo, Jott, and Nuance are some of the companies offering software that allows RIM's BlackBerrys and other mobile devices to allow users to employ voice command to locate and "dial" numbers, find movie listings, locate restaurants, and perform other functions.

Speech-recognition programs use statistical information about word patterns to model how people speak, and most software programs have a setup phase during which they are trained to recognize a user's speech patterns. The goal of speech recognition is to have every word exchanged between human and machine understood; at this point, the best systems are only accurate 90–99% of the time and only if the vocal content is delivered with a stable voice for which the software has been set. Imperfections result from a variety of factors, including ambient noise, unclear diction, words spoken with regional accents or speech impediments, inconsistencies in vocal quality due to illness, and other factors. In order for speech recognition to be useful as a global device, all languages would have to be taken into account in programming. Another hurdle for speech recognition: Human-to-human vocalization often involves inflection cues and body language that refine meaning in a way that a computer cannot discern.

Despite the lack of perfection in speech-recognition tools, many people find them to be useful today, and they are being leveraged successfully in human-computer interaction by court reporters, legal and medical transcriptionists, and others who deal with voice content on a regular basis. Today the cutting-edge research being done on speech recognition as an interface is concentrated on "subvocalization," which involves the capture of nerve impulses as you talk to yourself without speaking, just at the point before you actually deliver the signal to your throat muscles and voice box. Several *Future of the Internet III* survey respondents mentioned this possibility in their elaborations on this scenario.

Handwriting recognition is seen by many researchers to be more likely than speech recognition to be a heavily adopted interface because of its approachability and flexibility. Tablet computing utilizes a keyboard and a stylus or pen as input devices. Microsoft is a heavy backer of tablet computing; its OneNote software allows people to search and organize notes that have been inputted by a keyboard, pen, stylus, or digital recorder. Microsoft research has InkSeine—a more sophisticated tablet PC software in prototype stage. It has a pen interface that takes advantage of gesture input.

Pen-based computing is still in its early phase of diffusion as well, with pros and cons to products now available to consumers. One such tool is the LiveScribe digital pen, which records what you hear and links it to what you write as you take notes. You can transfer your notes to your personal computer and consult the written notes and play back the audio clips recorded to match them. You can search the notes as well.

Today's active matrix e-paper displays (EPDs) can be read in daylight and draw little power compared with LCD displays. In addition to being used in e-book devices like Amazon's Kindle and the Sony Reader, e-paper is now finding its way into the design of watches, credit cards, cell phones, security cards, signage, e-shelf labels in stores, and other text-display devices that can be networked to receive and send data.

E Ink's electrophoretic ink is used to provide the text in many such applications. Researchers say while e-paper has come a long way, it is 5 years away from being embraced as a mainstream technology. One current drawback to e-paper and ink is the slight delay in page changes. In addition, typing is awkward because it takes a moment for each keystroke to show up on the display. There's no color in e-paper applications at this point; experts say it will be developed in the next few years.

The keyboard is still a useful interface, but miniature and foldable designs thus far have not been either functional enough or ergonomically pleasing enough to gain widespread acceptance. Researchers have been working for years to perfect a consumer-pleasing system in which users type on projections of keyboards, eliminating the need to tote a real, full-size input device. Many such systems are in various stages of evolution. The I.TECH Virtual Keyboard ($300), with a full-size, laser-projected keyboard was introduced for consumer use in 2004. Interference caused by ambient light and difficulty with touch-typing sensitivity settings are two of the drawbacks to such a system.

Device designers overcome some interface difficulties presented by tiny keyboards by using software—predictive-text intelligence—that helps reduce typing errors caused by people's "fat fingering" on touchscreens or tiny keys, pressing more than one key or a key adjacent to the one they meant to press. It was available on about 750 million phones in 70 languages in 2008.

Companies are already looking a decade ahead in device design. Thin-film transistor technology is allowing companies to make folding screens for cell phones and e-readers, and Plastic Logic Ltd. is using electronic ink in its flexible 10- to 12-inch displays that are envisioned as replacing print newspapers. Fujitsu showed a design concept at the 2008 Consumer Electronics Show (CES) for a proposed device called the Fabric PC; it looks like a soft trifold portfolio that you can fold out to reveal a display with a keyboard that can be folded out so the entire inner surface is a large display. Nokia's concept phone, Morph, is shaped like a bracelet with nano-sized electronics mostly invisible to the naked eye. Samsung has a patent on a phone that can interpret finger gestures. There are also designs for phones that have foldout and rollout information screens that use e-paper. Microsoft and Mitsubishi are partnering to develop LucidTouch, an interface for handheld devices that has a front touchscreen for thumb input and a back touchscreen for fingers to hold the device and type.

Haptics is a word that refers to touch output or delivery of a tactile sensation from a device to a user. An example is the rumbling of a cell phone that has been set to "vibrate" to alert the user without audible ringing. More sophisticated haptic interfaces have been designed to give "touch" feedback to doctors who are performing intricate microsurgery using robotics.

Haptics coordinate sensors, actuators, magnets, motors, and other devices to simulate pressures and textures. Research into this area is extending understanding of the sense of touch. A field of study known as psychophysics is now centered on the analysis of humans' subjective experiencing of sensations.

Gamers feel haptics-engineered force feedback in joysticks equipped to vibrate as they race cars, fly jets, or engage in combat exercises.

In military and industrial settings, haptics research has led to the development of special glove-based feedback systems that allow users to grasp, sculpt, move, and manipulate virtual objects.

The da Vinci robotic-surgery device manufactured by Intuitive Surgical has a force-feedback haptic system developed by Allison Okamura of Johns Hopkins. It provides surgeons with a sensory substitution as they use virtual tools to direct the robot's real surgical moves. Scientists at the University of North Carolina–Chapel Hill have developed a way for humans to feel feedback from nanomanipulation of microscopic-level materials that could not be felt otherwise by human hands.[5]

A significant percentage of consumer-oriented haptics research is devoted to fooling people's fingertips into feeling buttons or other objects that are not really there. For instance, engineers are exploring ways for people to interact with smart screens for different results as they interface—a swirling motion of a fingertip produces the feel of a click, while a swipe feels like scratchboard.

A number of *Future of the Internet III* survey respondents expressed their expectation of the positive evolution of brain-computer interfaces. The concept may seem like a far-future idea, but it has been demonstrated over the past few years in research applications aimed at serving the needs of paraplegics, and in versions tailored for gaming.

One such interface made headlines at the CES in Las Vegas in January 2008. Emotiv Systems demonstrated a $299 gyroscope-equipped, wireless, neural-input headset that uses a dozen sensors to pick up and transmit signals from a player's facial movements and electrical activity in the brain to influence an avatar in gaming systems or in synthetic worlds such as Second Life. NeuroSky is another company with a similar system.

These interface devices (electrodes are used, but no electrode nets or gel are required) tap into the brain's neurons to detect electrical and chemical signals; the systems are built on the principles of an electroencephalograph. The detection of a particular combination of brainwaves triggers the execution of a particular algorithm.

The Emotiv headset is billed as being able to detect more than 30 different expressions, emotions, and actions, including push, pull, life, drop, and rotate, moving objects in a VR world just by thinking of the action. Its developers say it will allow people working in the 3-D Internet and VR to communicate happiness, meditation, frustration, excitement, and other important communication cues. IBM is partnering with Emotiv to develop business and strategic market applications for the technology.

DARPA is funding several researchers' work to develop a sophisticated synaptic interface, a human-computer interface that creates a direct connection between human brains and computers. One of the first military uses would be to do content studies of still and video images collected by the intelligence community. The computers would be used to screen the images with the humans connected, mentally tagging the aspects of greatest importance. The human brain can read and process images more accurately than computers alone. Civilian uses for such a system include air traffic control and assessment of medical imagery.[6]

Breakthroughs in a number of fields tied to interface design are expected as various sciences advance and are combined in the next few years.

When the brain and a computer work together, context awareness reaches new levels. Tufts University computer science professor Robert Jacob noted that reality-based interaction is advancing and augmenting the old WIMP interface, if not replacing it, suggesting that the four "real-world themes" necessary in interfaces to come are body awareness, environmental awareness, social awareness, and naive physics.[7]

Human input to machines is also expected to progress in regard to powering devices. Researchers at MIT are in the proof-of-concept stage on a new energy-efficient chip for cell phones, PDAs, and implantable medical systems that could use human body heat as an energy source.

It is expected that miniaturization will lead to the development of sophisticated phones and other networked devices that can fit into an unobtrusive earpiece or be worn as implants. A poll by Zogby International and 463 Communications in October 2007 asked nearly 10,000 Americans, "How likely would you be to implant a device into your brain that enabled you to use your mind to access the Internet if it could be done safely?" About 75% said they were not at all likely to do so; 11% said they were very or somewhat likely to do so. Men were more interested in the prospect than women; 17% of men and 7% of women said they were very or somewhat likely to try an Internet implant.[8]

As the human-computer interface evolves, the motivation behind design is to make it as seamless as possible. One possible end, then, is a world in which one cannot distinguish the difference between human and machine.

ADDITIONAL RESPONSES

Many other participants in the survey shared valuable input on the user-interface question. Among them:

> "Freeing people from the tyranny of the keyboard will enhance computer use massively." —**Jeremy Swinfen Green**, Telecom Express, an interactive marketing company

> "These virtual typing spaces will be able to take sign language and will also allow for grander hand (or device) gestures to represent scrolling, panning image vistas, moving forward, as well as executing common tasks. Movies like *Minority Report* and gaming consoles like the Sony Wii have foreshadowed and beta-tested the future of interactivity. The applications of this in medicine, communications, and even social media are endless. Imagine wearing an e-suit that is responsive to touch from another user who is air touching you from some distant location via an Internet service. The old AT&T slogan, 'Reach out and touch someone,' will take on a whole new meaning." —**Drew Diskin**, director of e-strategy, Johns Hopkins Medicine

> "Voice will continue to be an important way of communicating simple commands, particularly for mobile, in-house, and in-car contexts. Air typing is just silly, though. A return of chording keyboards seems far more likely." —**Alexander Halavais**, professor and social informatics researcher, Quinnipiac University; explores the ways in which social computing influences society

> "These interface developments are prerequisites to the next significant leap forward in all realms." —**J. W. Huston**, president of Huston Consultancy and futurist

> "The use of the keyboard may disappear." —**Julian Hopkins**, social scientist and PhD candidate at Monash University, Malaysia

"You have to do something with those extra computer cycles." —**Anthony M. Rutkowski**, cofounder of the Internet Society and a founding trustee; longtime leader in International Telecommunication Union; vice president for regulatory affairs, VeriSign

"Touch interfaces, including multitouch and gestural interfaces, will become an integral part of the computing environment of the future, though air typing will probably not be a large part of it without force-feedback gloves that simulate some kind of tactile feedback. Speech will also become good enough to use for most everyday tasks." —**Jason Stoddard**, managing partner/strategy at Centric/Agency of Change, an interactive strategies company; he is also a popular speaker on social media and virtual worlds

"We will be further advanced but cannot envision it yet. These stated devices will seem hilarious to us like 1950s-era, futuristic scenes do today." —**Robert Grant**, chief executive officer, Voya-Carc Inc.

"Technology is almost there, and cultural acceptance will follow. People wearing the first Bluetooth hand-free phones were looked at as strange space invaders; today, millions of people do it." —**Louis Naugès**, president, Revevol, an enterprise 2.0 company with offices in France, Spain, the U.K., and the U.S.; a founder of Microcost, an IT services and hardware company based in France

"Haptic technologies will only be taken up if integrated in mobile phones. What this engineering culture will have to understand is that it will have to move away from the standalone personal computer, which, even up to now, is still situated within the boring and static office environment." —**Geert Lovink**, professor and expert on culture, sociology, and the Internet; based in Amsterdam; author of *Dark Fiber* and *Uncanny Networks*; responsible for the Institute of Network Cultures

"2020 might be a bit soon for 'air typing' to become ubiquitous (and fully secure), but these trends will move forward." —**Michael Zimmer**, PhD and resident fellow at the Information Society Project at Yale Law School; research includes social and cultural dimensions of new information technologies

"Today, the thing—the object, the software, etc.—is in a window of some kind, i.e., a computer screen. Soon the window will go into the thing. We will wear our interfaces. To some extent, we will become our interfaces." —**Barry K. Chudakov**, principal, the Chudakov Company, a marketing and advertising strategies creative consultant who has worked with many major corporations, including Microsoft and Disney

"Talk interface will always have limited appeal and application. Touch will take off." —**John C. Abell**, new-media project director, Committee of Concerned Journalists

"I agree with this—I'm not sure of the time frame though. I've seen displays of this technology online, I believe the work is being done at MIT." —**Janet D. Cohen**, blogger, futurist, and trend analyst

"I'm split down the middle on this one (but being an optimist, selected 'agree' rather than 'disagree'). Yes, voice recognition will be ubiquitous, but people will still type communications that they want to keep private, or to be polite (because they are in a movie, say, or on the train). I'm not really buying the whole 'air typing' thing or any other kind of virtual keyboard, including the soft screen on the iPhone. That's a horrible way to type. If there is a keyboard replacement, I think it'll be physical, not virtual." —**Josh Quittner**, executive editor, *Fortune* magazine; formerly editor of Business 2.0, Time.com, and technology editor of *Time* magazine; also formerly at *Newsday*; freelancer for *Wired* magazine

"Talk is quickly abandoned as researchers realize it is a far better human-to-human communication tool than it is for human-machine interfaces. Touch then thought become the dominant interface technologies." —**DJ Strouse**, international relations and computer science student, University of Southern California

"Our computing tools will evolve to a point where they are a larger extension of our senses using hearing and touch. This new user-interface direction is already being prototyped and working its way in production. Tablet PCs have digital keyboards today, so air keyboards will be in the next wave." —**Joanna Sharpe**, senior marketing manager, Microsoft

"This air typing thing explains one way that the cell phone can overcome interface difficulties. Also note that these types of devices will create a new type of digital divide. While most may be online, the more wealthy will have the most seamless devices." —**Richard Hall**, professor of information science and technology and codirector of the Laboratory for Information Technology Evaluation, Missouri University of Science and Technology

"I agree with touch, but am skeptical of highly functional speech-related inputs, likely because of personal frustrations at current voice-recognition programs on corporate customer service lines." —**Jade Miller**, PhD student, Annenberg School for Communication at the University of Southern California, with a research focus on global flows of information and culture

"Yes to the haptic/air displays. No to voice taking over from regular typing. I just can't see voice recognition progressing enough." —**Emma Duke-Williams**, lecturer in the School of Computing and researcher, University of Portsmouth, U.K.; education blogger

"Unfortunately, I think in 2020 there will be still people who have never seen a computing device or a phone, could not use one as they are illiterate, and have more urgent worries anyway." —**Giulio Prisco**, chief executive officer of Metafuturing Second Life; formerly department head at the European Satellite Centre, analyst at European Space Agency, and an IT specialist for CERN

"Some people will be air typing, but I quarrel with the word *common*. We might have subvocal speech recognition and other technologies that will supplement character-by-character via keyboard input." —**Rollie Cole**, director of technology policy, Sagamore Institute for Policy Research, a nonprofit, nonpartisan think tank headquartered in Indianapolis, IN

"We will see haptic devices, though perhaps we can move beyond QWERTY at the same time." —**Hal Varian**, chief economist, Google, and on the faculty at the University of California–Berkeley; a world-renowned expert on the economics of information technology

"I doubt that talk-technology interfaces will be mature and will stay in progress if there isn't a big technology shift." —**Rafik Dammak**, software engineer, STMicroelectronics, Tunisia; DiploFoundation participant in the study of Internet

"Typing technologies will improve dramatically. In this sense, Apple's iPhone may be a hint of what will be available in the coming years." —**Sebastian Ricciardi**, associate with Jauregui & Associates, a law firm in Buenos Aires; leader in the Argentina chapter of the Internet Society, formerly of ICANN's At-Large Advisory Committee

"This is possible, but I don't think it will happen by 2020. Research and experiments would have happened by then." —**Sudip Aryal**, president, Nepal Rural Information Technology Development Society

"I hope so—in high school, I took theater classes instead of typing." —**Tiffany Shackelford**, consultant who works with clients such as Phase 2 Technology, Stateline.org, Foneshow, WebbMedia, and Daily Me

"That's only 12 years from now. I would love to air type (and sign me up for the air typing competition! I rock!) but not sure it's that imminent (even though I saw it on *CSI*, so that makes it real, right?). Of course, at some point, we have to ask why we're still using the QWERTY keyboard, or even why we're typing, period. Voice, I think, is a little closer, only saying this from knowing people who use these technologies now. They are far from perfect, but that they can even be used tells us something." —**Karen G. Schneider**, research and development, College Center for Library Automation, Tallahassee, FL; expert and thought leader in the library and technology community

"Not much to elaborate. This seems to be a user need, and the technology will follow to fulfill the expectations." —**Roberto Gaetano**, ICANN board member; also responsible for software development for International Atomic Energy Agency; an active participant in the ICANN policy-making process

"Maybe by 2020, you just will have described the old technology here." —**Stan Felder**, president and chief executive officer, Felder Communications, a marketing and advertising firm in Grand Rapids, MI

"Interfaces will have matured to the level where talk and touch will be very common. However, voice-based interfaces are not going to become all that popular while in public, simply because background noise will interfere. Also, talking loudly into your device will be frowned upon; social norms will have evolved by then to control your devices mutely while in public spaces." —**Peter Bihr**, freelance consultant on Web strategies, communities, blogging, and social media; based in Berlin, Germany

"You have to live in a big city to know noise makes the use of voice recognition a difficult application. Haptic technology may be a good solution. Both technologies should remain as a complement to the existing ones." —**João Miguel Rocha Filho**, director, DataOne, a provider of software for connecting to Linux; based in Brazil

"The keyboard is on its way out. It's crude and causes physical problems." —**Cameron Norman**, assistant professor in the department of public health sciences, the University of Toronto; actively engaged in use of the Internet to help tackle tough health issues, director of evaluation with the Centre for Global eHealth Innovation

"Yes, these aspects of virtual reality may indeed see their day by 2020. But it will be limited to those who can afford the technology. It won't quite be accessible to the common man or woman." —**Alexis Chontos**, webmaster, the Art Institute of Pittsburgh

"This isn't a huge stretch from the status quo of 2008. People already walk the streets chatting into wireless headsets, typing on their mobile phones, and tuning out the world with their iPods." —**Jamie Richard Wilson**, journalist and freelance Web developer

"There is no question that keyboarding will remain a strong input option, but that voice and touch will increasingly become the more common modes." —**Susan Thomas**, S2 Enterprises LLC

"It seems to me that we may want to replace the keyboard in this image with the stylus, a much simpler interface device. If voice-recognition software will be that sophisticated by 2020 that it can work with more than 99.5% accuracy, which is really what we need for people to tolerate it, then handwriting recognition should further signal the end of the keyboard age. But don't hold your breath. Right now, handwriting recognition still requires us to go back to the days of the Palmer method, and voice-recognition software's success may ultimately depend on humans learning how to talk like computer voice synthesizers, not the other way around." —**Dennis Baron**, professor of English and linguistics, University of Illinois, runs the Web of Language site and researches the technologies of communication

"These technologies have been around for a long time. Unless Apple can find a way to incorporate them into one of its products, interest will not pick up." —**Theresa Maddix**, satisfaction research analyst, ForeSee Results

"You start out strong—these devices have already been created and are gaining steam with some recent splashy rollouts. But then you just have to take it that extra step into rank speculation with stuff like air typing. What's up with that?" —**Alexis Turner**, Webmaster, Greenwood Publishing Group, New York

"This may not be widespread by then, but it sure is coming. Instead of sitting at your desk for 'desk jobs,' you might be able to 'conduct' your keyboard! Wrist-repetitive stress injuries will be avoided, but new ones will crop up (tennis elbow from editing your reports?)." —**Mariana Almeida**, product manager of Web products for health care, Kaiser Permanente

"Computer-to-mind interfaces will also be available." —**Chris Myers**, webmaster at the University of Michigan

"Tools evolve to the simplest and easiest way to utilize them so everyone can use them. Talk and touch enable this to occur." —**Tim Grafton**, market research director for UMR Research Ltd., a market research company based in New Zealand

"Nice image. Needs better and easier technologies for the average person to use than may be likely." —**Jennifer Jarratt**, principal, Leading Futurists LLC; works with formalized methodologies to assess and interpret potential futures

"Future interfaces will respond to eye movement, allowing much faster interaction with much less effort." —**Lucas Anderson**, Awty International School

"Talk will be a private thing, given acoustics. Touch should be big. Contact lenses might be the major avenue." —**Bryan Alexander**, director of research National Institute for Technology and Liberal Education, blogger, expert on computer-mediated pedagogy; based in Ripton, VT

"Air typing is common in Second Life already ;->, but seriously, the iPhone and voice-activated phones are pointing this way, and there is no resistance from people or materials. Make it so." —**Paul Jones**, director of ibiblio.org at University of North Carolina–Chapel Hill; the original manager of SunSITE, one of the first Web sites in North America

"Can't wait for it to happen." —**Ruth Martin**, National Chengchi University, Taiwan

"In a sense, I saw something like this at the Consumer Electronics Show this year [2008], but I wonder if it's fair to say 'the most commonly used.' I believe the head of Panasonic described the cost to manufacture its 'Life Wall' to be 'priceless.' It's a long way from too expensive for virtually everyone, to be a routine part of everybody's life. (But a lot can happen in 12 years, can't it?)" —**Jim Wiljanen**, president, Evans/Greenwood LLC, MI

"By 2020? Are you kidding? No way. Just last year I was commuting into San Francisco with a wealthy Chinese gentleman (very hot, expensive car) who was talking into his GPS system trying to get directions (GPS system also looked very hot and expensive)...and *it didn't understand* him because of his heavy accent. I was in stitches!" —**Virginia Bisek**, Web content developer and writer

"Maybe one result of this will be a niche market for those looking for 'a quiet place.'" —**Kerry Anderson**, library consultant for the government of Alberta, Canada

"Although I agree with the premise, I think the timeline is too compressed. The implementation, and permeation, of these interfaces will take much longer. Although the acceptance and adoption will be rapid, the devices will not be 'fully developed' by that time." —**Woody Degan**, chief executive officer and operations director, Memphis Sound Entertainment; Consumer First Consulting, IT consulting

"The air typing sounds feasible and sensible." —**Lynn Blumenstein**, senior editor, *Library Hotline, Reed Business Information*

"This should be the case, if technologies continue to drop in price. This is unlikely in a post-oil economy; however, that point is—by most estimations—more distant than the year 2020." —**Francis J. L. Osborn**, futurist and activist, philosophy department, University of Wales Lampeter (formerly St. David's University College)

"It used to be those who walked around talking to themselves were thought to be insane. Now it is thought they are on their BlueTooth. I can easily see this scenario coming to fruition." —**Randi J. Smith**, student, Art Institute of Pittsburgh

"By 2020, the technology will be available for electronics to feature voice recognition as input over traditional text-type input. However, we will see a strong resistance from this revolution. Text input will still be preferred by many, and the implication of talk and touch technologies will be highly debated." —**Clement Chau**, research manager for the Developmental Technologies Research Group at Tufts University

"This is already available in laboratories and test settings. The convenience of having such virtual tools will be a huge benefit to the mobile workforce, reducing the peripherals they now have to haul, and this will catapult this technology into the mainstream." —**Richard Fowler**, auditor specialist, Northrop Grumman

"Absolutely. Some of the first advantages are beginning to show already, and as we continue to exceed the computing cycles needed to run the everyday applications, those same cycles will be used to improve the interface to the applications. The PC, along with its mouse and keyboard, must die to accomplish this, and the sooner, the better." —**Eric Kreider**, director of Web services, the University of Akron, OH

"We have 'individuals dictating information in public' while they talk on cell phones everywhere. Unfortunately, this is being done when and where it is not appropriate, such as during religious services and in entertainment venues. We already have infrared devices similar in nature to the haptic technologies; they will improve by 2020 to where they are as ubiquitous and efficient as described above." —**Thomas Lenzo**, business and technology consultant, Thomas Lenzo Consulting

"Yes, I see this as in mostly general use—again dependent on people actually being able to access the requisite resources such as air, food, power, and water." —**Alex Don**, linguist and educator

"The success of the iPhone is clearly a key indicator here, and improvements to voice-recognition systems will surely help.

This is an area where there's a real consumer need, and the technology seems likely to be able to meet it." —**Roderick White**, editor, *Admap* magazine, World Advertising Research Center

"I agree with all but the last sentence. Voice recognition will be widely used. Air typing will be possible, but it will not likely be a 'common' sight. There will be more accepted ways of using haptic technologies." —**L. Suzanne Suggs**, assistant professor of communication sciences, University of Lugano; research focuses on use of new media and messaging strategies to improve health status

"There is more than a bit of techno-hype in this scenario. The part I agree with is the relatively broad access to this kind of interface in the developed countries; even here, many people will choose a subset of these capabilities. It is also possible that these interfaces will find take up in specialized contexts—for example, for the mobility impaired or for monitoring Alzheimer's patients." —**Amy Friedlander**, director of programs for the Council on Library and Information Resources, a nonprofit that services research and higher education

"Ease-of-use improvements always prosper." —**Jim Lucas**, Web manager, CACI, a provider of national security, defense, and intelligence-related solutions in the interests of the United States

"My guess is that the next big interface will be none of the above. Not sure what, but that's my gut." —**Douglas Schulz**, managing editor for online publishing, America's Health Insurance Plans; formerly a Web team director for a biotech industry organization

"The technology is evolving towards less directly connected interfaces." —**Kathryn K. Goldfarb**, president, KG Communications, an independent consultancy

"I'm a little skeptical that the underlying technology will reach this level by 2020." —**Mike Langum**, Web developer, U.S. Office of Personnel Management

"Haptic technologies in the Apple iPhone and Microsoft Surface have already attracted huge attention and proven successful. Expect that trend to continue. Voice recognition has been around for years but still has a ways to go before it reaches wholesale integration. We may not see that by 2020, but it'll be close." —**Jay Buys**, vice president for digital development, Fleishman Hillard, an international marketing and communications company

"Less a sea change in technological innovation, more an evolution. The growing movement against 'digital refuse' will only advance the movement away from hard interfaces (e.g., keyboards, mice, etc.) towards more soft, environmentally friendly UIs." —**William Winton**, product manager, digital media, 1105 Government Information Group

"Agree, I have been using speech-to-text since before it was practical, and strongly agree we will use keyboarding for precise editing and typesetting." —**Dick Davies**, partner, Project Management and Control, Inc.; past president of the Association of Information Technology Professionals

"For reasons of privacy, nonintrusiveness, and bandwidth, haptic interfaces will dominate voice interfaces in many, if not most, situations." —**James Jay Horning**, chief scientist, information systems security, SPARTA, Inc.; a former fellow at Xerox's Palo Alto Research Center; fellow of the ACM

"What do you mean 2020? How about 2010?" —**Michael Castengera**, senior lecturer at the University of Georgia's Grady College and president of Media Strategies and Tactics Inc., a media consulting firm

"The immediate benefits to be seen in these two applications, voice recognition and haptics, will certainly guarantee their quick adoption, once they are improved from the perspectives of quality and reliability. But who will guarantee to us that the improvements made up to 2020 will be *sufficient* to convince any more than a few early adopters? I'm not putting my investment money into this area. As my mother used to say, 'I'm not buying unripened bananas!' Sorry!" —**Fredric M. Litto**, consultant for Pearson Education Global e-Learning, president, Brazil Distance Learning Association

"Face and mind recognition is the most frequently used input process, for privacy reasons—speaking or typing are considered noisy, indiscreet techniques. Elderly people prefer typing on ergonomic keyboards—but voice recognition (inexplicably for the younger generation) is associated with major disability and very old age by those more than 40 years old and, as such, hardly used." —**Bertil Hatt,** researcher of Internet and social services, innovation valuation; employed by France Telecom and Orange (information technology and services industry) while completing PhD

"I'm agreeing with this scenario because I can envision it happening, but I don't like it very much. It's annoying now to be forced to listen to other people's phone conversations. I can just see people walking down the street or standing in a crowded commuter train air typing." —**Patti Nelson**, a webmaster who works on U.S. government sites

"The keyboard and typing will be obsolete. Voice recognition directed through haptic technologies will become the hybrid interface technology. Technology able to detect thoughts will become experimental." —**Dixon Hutchinson**, software engineer

"iPhone, Nintendo Wii, Microsoft Surface. We're on the way to this today, so it doesn't seem too much of a stretch to see new interfaces, although the keyboard and mouse in some form or another could well be still kicking around." —**Heath Gibson**, manager of research and market analysis, Big Pond, a competitive intelligence company and provider of broadband customer Web sites in Australia

Anonymous Comments

The following section includes a selection of contributions to the discussion of the evolution of the Internet interface from survey participants who chose to remain anonymous. Among them:

"As if cell phone conversations in public aren't annoying enough; but yes, voice recognition and touch feedback should be this far along by then."

"I'm holding out for a wireless synaptic interface to allow for direct input—typing in the air sounds lame."

"Fourteen months ago, the BBC interviewed a scientist at Cambridge who had already used radio-frequency chips to communicate movement and touch sensations between his hand and an artificial limb. His wife had also had a chip installed, and she was able to feel the touch sensations he'd induced through a movement command from his hand to the artificial limb."

"Hmmm—this is too much projecting what we do today into the future—like the 'future' we envisioned in the '50s."

"Touch will be, but speech recognition beyond a few small commands is computationally intractable."

"Even more will happen in the space driven by adaptive-technologies research for the blind and other challenged users."

"Technologies which allow people to limit the amount of baggage they carry will be favored, assuming the medical community is able to refute concerns about wireless technology and the impact on health."

"Probably technologically possible; however, the privacy issues with 'air typing' are unimaginable."

"People will use more and more voice recognition in all electronics, and that will increase productivity and be an improvement safety-wise. I think (and hope) the air typing will not be embraced."

"The accuracy of these systems will have to increase markedly, and I'm not sure they will by 2020. Voice recognition isn't really necessary (or even desirable) in many applications."

"The technology makes some sense and would seem to have little impact on social relationships. If they take hold, the [Nicholas] Negroponte universal computers [One Laptop Per Child project] seem more likely to have an impact."

"Speech recognition continues to improve. So do touch interfaces. These seem destined for increased use."

"I didn't think people would talk into cell phones in public, lest they appear demented. How wrong I was. Air typing, yes."

"Haptic systems will be developed in the next decade, but they will not be that pervasive…The above scenario requires gloves or other structures around the fingers (because most flat surfaces can't push back). I don't think that's happening anytime soon for the majority of the population. While [wearable-computing researcher] Steve Mann is a geek's geek, he's not an icon for the majority of the world."

"The easier things are to use, the more integrated they will be into our daily lives."

"Try any of those so-called voice-recognition devices in any other language than English, and you'll have a clear (!) answer."

"Rather than air typing on a projected surface, flexible keyboards will still be used... Voice recognition and synthesis will become mainstream and perhaps even brainwave-driven user interfaces may be in use."

"Agree, especially if nano and biotech work the same wonders of scientific and research applications in consumer products."

"Old prediction; never happened in the '90s, won't happen in 20 years. These are intrusive to the social spaces of others, and so until they are seamless, they interfere with communicative acts. While texting, etc., is acceptable for young and certain social groups, it is still seen as unacceptable in formal and even in many informal social settings because of the disruption it causes."

"For this to have a serious impact, we need to get today's school children working with these interfaces. Older generations will be less prone to use these user interfaces, so they may have some presence but will not be for the majority of users."

"Voice recognition would not be surprising at all. I'm not so sure on the keyboard projections."

"The written word will remain the cheapest way to exchange information, and therefore the most popular."

"Touch may well be supplanted entirely by other modes of input—gestures, eye movements, or even speech recognition capable of handling subvocalizations."

"We will see other interface mechanisms, as well as those envisaged here."

"Touch interfaces are making progress and will be much more common by 2020, especially in specialized interfaces."

"The iPhone interface, and its success, is a clear indicator that we will see more development of haptic devices."

"People are kinesthetic, auditory, and visual in how they communicate—talk and touch satisfy all."

"These are emerging technologies here and now. By 2020, things will have moved beyond this."

"These have been promised for a long time, and I still haven't seen much convincing."

"The idea of one device is silly. We'll carry more than that. The idea that a keyboard is still our primary input, or even needed, is questionable. Will it be common to have some silicon in public, yes, but then it already is."

"I do not expect voice to dominate as an input method by 2020. It will be used increasingly in constrained environments (control of functions in cars or in household appliances, for instance) but it will not be used for general-purpose computer/machine interfacing. Other input technologies will continue to develop rapidly and be much more suited to many tasks. This is an area of 'invention,' however, and therefore difficult to predict."

"I agree about the new interfaces. But we will communicate with written quite so much? Will access be through eye movement? But I fear that you are right: People will go around shouting into their devices (leaving other people wondering how much they are paid for spouting such drivel or sharing company or client details!). And will we need to travel so much? So we will be in our little own environment interacting via avatars with others in virtual-world meetings."

"Though I think that such interfaces are desirable and are being developed, it will take much longer to achieve."

"Text-to-speech is universally available. Speech recognition exists, but is rather unreliable. By 2020, it should be quite usable. Touchscreens are everyday items today. Haptic technology exists but is not at all common today, contrary to the statement given. But that and gestural computing should be well established by 2020."

"Those types of interfaces will never be as fast or easy to use or reliable as physical ones."

"Agree on the talk, not the touch."

"Not sure how this would work—will everyone be wearing glasses with display units built in? It's quite a possibility!"

"People are annoyed enough by public cell phone usage; people flailing about on spectral keyboards will be too much. People are already quite agile texting on phones' number pads, and iPhone already has a reasonable improvement to that. While technology mentioned in the question will be available, it will not be ubiquitous."

"Yes, this will happen, but it will not be as good as we imagine, just better than we can do it now. Voice input to computers is over 25 years old, and it still is has a way to go. Thirteen years more will cover only one-third of the remainder but will increase its deployment tenfold."

"The need for keyboard and typing would be vastly minimized with intelligent user-interface design."

"Touch, yes. Talk, no."

"This is where the gap is today, and this is where we will see most of the changes."

"You could argue that this is already the case with cell phone technology and next-gen ATM machines."

"The public use of these devices will be regulated to avoid encroachment on the 'space' of others."

"This will be common in personal spaces—auto, home computer, etc. Don't see VR software catching up that quickly to handle noisy environments, etc."

"I agree concerning touch and gesture. But voice commands in a world that is sonically polluted won't work."

"This will take more than 12 years to achieve."

"I'm not so sure about air keyboards, but I do think more touch technologies (MS Surface, etc.) will be common, and we will interact with digital objects similar to the way we interact with physical objects."

"In 2020 innovations may have moved beyond talk and touch."

"While this scenario is highly possible, the affordability of the system is unknown."

"From the iPod forward, more touch and talk technology abounds."

"I like the idea of typing because I don't mind typing, but this just seems…I don't know."

"Maybe it's a hope, rather than anything else, but the promise of voice-recognition technology for all of us, and especially the mobility impaired and visually impaired, is phenomenal—witness the recent Kurzweil et al. mobile phone for visually impaired persons. For all of us, using voice rather than touch will allow us to do more activities at once. (Maybe not such a good thing?) The trick will be to get the voice-recognition setup to be a quick-and-easy process."

"This could have happened by now if voice recognition were the public preference—we are forcing more utility out of this technology than people want."

"I suspect the voice recognition will replace the need for such technology. The rudeness and lack of courtesy of loud-talking individuals will continue in most public spaces. Although, there may be other defined spaces (both public and private) where this technology will be enforced upon people."

"Possibly a different technology will mature to do a better job."

"The advances in Pico projectors will allow greater use of haptic technologies in mobile phones and notebooks."

"This may be the most likely of all the scenarios presented thus far."

"I'm betting that those products will have already been used and regarded as passé by 2020. Twelve years is a long time in the Internet world, and there will be something vastly more efficient and interesting than this scenario."

"While such technologies may be available and growing in popularity by 2020, it may take much longer for us to get comfortable with them."

"Voice recognition I agree with—by 2020, the keyboard will be abandoned. Interacting with computers will be a lot more accessible to all people—via touchscreens, icons, voice, and handwriting."

"This is possible, but I don't see it taking off. Current voice-recognition technology is not developed enough. Touch feedback is just starting to grow."

"Talking to software or air typing is really not far from what is occurring now."

"I just recently read about a device for blind people that optically recognizes an object and tells the person what it is. For example, it may identify the object as a dollar, not 10 dollars. We are already on this road and advancing rapidly down it."

"This is a relatively safe prediction, given the number of products that are already implementing these technologies."

"This one may be an understatement."

"Definitions of polite behavior will change as public intrusions such as talking on a cell are vetoed; technological solutions will follow."

"This technology appears to have direct, immediate consumer use, and so is likely to appear within the next 5 to 20 years."

"Such techniques will be well developed and reliable. But I doubt they will be used for more than simple surfing of the Web. It will take more time to have a reliable interface allowing, for example, to speak *Hamlet* into the micro and really get it written."

"As technology advances, I expect to see fewer people typing."

"Technology needs to evolve into something other than what it is today, and talk/touch devices will be the norm."

"This viewpoint is very optimistic. People half a century ago also thought we would be on Mars by now. Change is very slow."

"Agree on the touch feedback, less on the voice commands. There will be big steps in the voice recognition, but I don't think the interfaces will go towards full voice command. It's just less usable than the touch. I am sure that there will be huge progress in the usability of computers as well."

"I don't agree with the technological prognostications here (I think the input technology has yet to emerge), but the general idea will hold true."

"This will happen before 2020."

"This will be a reality, in part, because of federal government mandates (Section 508/Americans with Disabilities Act)."

"This scenario would require shifts in cognitive style for a significant number of people—something that there is not agreement on as a possibility."

"Can't wait. This will be one of the best innovations ever. And 'air' activity won't just be typing—it'll be touching, choosing, pointing, moving, all sorts of physical gestures."

"We no longer think someone is crazy to have a conversation with themselves because we assume an earpiece to their cell phone. However, *Minority Report*–style user experiences are just too tiring on the arm muscles to be usable. True haptic (tactile) feedback from such devices is also poor. In 2020 expect miniature thumb keyboards, not invisible ones. One-handed chording keyboards would be great, but we won't be able to wean people away from QWERTY by then."

"Touch, yes. Talk, no. We are 100 or more years away from computers that can actually converse. Speech systems are lame, and they tick people off."

"This is only the tip of the iceberg, however. Well-developed mouse devices (mobile gaming in the metaverse) are being developed (along with simple visors) and will enable one to be able to access data from anywhere at any time, not just gaming."

"Touch and talk are both already gaining in prominence and are more natural than type and mouse-based interfaces."

"The current interface with computers is on its way out. We see increased voice recognition in cars, Wi-Fi bunnies, and phones…Children faced with the barrier of learning touch typing will and are learning voice-recognition software that permits more focus on the creative process and less focus on the mechanics of keyboard typing."

"Speech will replace typing. Or a chip in our heads will enable IM by *thought*!"

"This would drive me absolutely insane. No one talking to each other, walking around air typing, loudly talking on cell phones…ugh."

"Sure, why not—as long as I can talk to it with my best Scottie [a character on the futuristic 1960s TV series *Star Trek*] imitation."

"If privacy and cultural issues are solved (remember 'talking' cars from the 1980s?), talk and touch could be common interfaces."

"We'll see more of talk and touch interfaces, but I don't think it will be as ubiquitous as described. Speech recognition has been lionized for years, and is still very early. It will be better, but the different touch methods have more opportunity."

"The air typing is far-fetched for so soon in the near future."

"Certainly so; these technologies already exist in reliable form."

"Short of technologies that read our minds, we will rely on the senses we have to interact. I don't know if smell will be integrated, but wouldn't be surprised. Holiday Inn is already using a smell to brand itself. I completely agree."

"I see this happening already and hardly think this is the future. Although it might be more widespread and increase use amongst those who are now further removed from technology because touch and talk technologies are more intuitive in their use."

"Absolutely. I think 3-D stylus devices will also be common, or sensors clipped to fingertips."

"Public commons will divide sharply in reconfiguring to either adapt or mitigate private, haptic spaces."

"If this wasn't the way of the future, Apple wouldn't have integrated the 'touch' technology into the iPhone and iPod Touch.

"It's possible that text-based input will be gestural, thereby eliminating the need for a keyboard. One could also foresee a combination of input types—touch, talk, shape, sound, etc."

"The virtual keyboard seems like something out of a sci-fi movie, and not quite as near to reality, although I can see built-in voice recognition attaining prominence prior to 2020."

"Talk and touch interfaces will become much more prevalent in our computing devices."

"To be able to interact with online materials in a more tactile way will make the technology more part of our everyday life. I would love to navigate my file structures through touch-drag interfaces and surround myself with the files on which I am working. A 3-D touch environment would be great."

"Not sure about 'air typing' (why not just text into their ubiquitous and more powerful phone?), but the rest seems reasonable."

"Blind people and people with disabilities are already using similar technologies."

"Certain things are simply a hype. Even if they become a reality, their usage will be very restricted."

"I am still not ready to interface haptically."

"Much more agreed on touch than talk."

"Most of this already occurs in at least small or prototype amounts."

"It would be wonderful—an end to carpal tunnel syndrome."

"If biometrics can protect private information and accurately verify the correct user, then this is good."

"Just as our ability to write script has diminished, I expect the keyboard to become less used."

"I think people are at the point where they almost think better when typing rather than speaking."

"This is certainly likely. Just a few years ago, people who used wireless headsets with their cell phones were looked at as obsessive-compulsive types by those tethered to the handset. Use of virtual keyboards, personal viewing devices (like the Myvu), and other methods of interaction are certainly realistic."

"I agree with this idea, but it worries me that this type of technological innovation may limit the amount of face-to-face time people spend with one another."

"It's all just technology getting easier and more convenient to use."

"Touch tech makes more sense, depending on the context. I'd really like to see us get to thought as a primary technology interface. Find ways to channel thoughts directly into interfaces. Remove a major inefficiency, which is the physical action required by today's interfaces."

"Actual speech is too slow to be used for the majority of content."

"It is very sad that some people are so disconnected to their own bodies that they will do almost anything to avoid feeling them and interacting with others by means of them."

"Voice recognition would make office life quite uncomfortable—we will still have to rely on some silent mode of conversing with the computing devices, whether this is a keyboard—or a chip implanted in the brain :-?"

"People will develop social cues and symbols to let others know when they are receiving an incoming call on their hands-free phones or taking a moment to air type or interact with the network. There will continue to be places and times where it is socially unacceptable to take a call or interact with the network."

"The military does a lot of this already."

"The details may be slightly different (e.g., less kludgey interfaces will be developed), but the general idea is correct."

"This will be driven by economics."

"People will carry around RFID tags with their own device settings that will tell the technologies we interact with how we want our interface configured."

ENDNOTES

1. See Gates and Gibson interviews in the *Rolling Stone* 40th Anniversary Issue, published November 1, 2007, online, at http://www.rollingstone.com/news/story/17168320/the_rolling_stone_40th_anniversary_issue/3.
2. See the Lightglove Web site, available at http://www.lightglove.com/.
3. See *Human Interaction Speeds*, by Bob Bailey, September 2000, available at http://webusability.com/article_human_interaction_speeds_9_2000.htm.
4. See the MIT Computer Science and Artificial Intelligence Laboratory Web site, available at http://www.csail.mit.edu/.
5. See the University of North Carolina–Chapel Hill Computer Integrated Systems for Microscopy and Manipulation Web site, available at http://cismm.cs.unc.edu/?s=nanomanipulator.
6. See *Converging Technologies for Improving Human Performance*, National Science Foundation, Department of Commerce, June 2002, available at http://www.amsoftwareservices.net/KnowledgeBase/Converging%20Technologies%20for%20Improving%20Human%20Performance.pdf.
7. See "Give Your Computer the Finger: Touch-Screen Tech Comes of Age," by Gary Anthes, published in *Computerworld*, February 1, 2008, available at http://www.computerworld.com.au/index.php/id;1847998544;fp;;fpid;.
8. See *Hello, Internet, I'm Home: Zogby/463 Internet Attitudes Poll*, by Zogby, October 24, 2007, available at http://www.zogby.com/news/ReadNews.dbm?ID=1378.

Scenario 7

The Evolution of the Architecture of the Internet

Prediction: *Next-generation research will be used to improve the current Internet; it will not replace it. In 2020 the original Internet architecture is in the continuing process of refinement—it has not been replaced by a completely new system. Research into network innovation, with help from the continued acceleration of technologies used to build, maintain, enhance, and enlarge the system, has yielded many improvements. Search, security, and reliability on the Internet are easier and more refined, but those who want to commit crimes and mischief are still able to cause trouble.*

Expert Respondents' Reactions	
Mostly Agree	78%
Mostly Disagree	6%
Did Not Respond	16%

All Respondents' Reactions	
Mostly Agree	80%
Mostly Disagree	6%
Did Not Respond	14%

Note. Since results are based on a nonrandom sample, a margin of error cannot be computed. The "prediction" was composed to elicit responses and is not a formal forecast.

Respondents were presented with a brief set of information outlining the status quo of the issue 2007 that prefaced this scenario. It read:

> Due to concerns over Internet security, reliability, and complexity, the National Science Foundation in the United States is funding research into the building of a "next-generation" or "clean-slate" Internet. The NSF initiatives include the Global Environment for Networking Innovation (GENI—building a test network on which researchers will be able to try out their ideas) and Future Internet Network Design (FIND). The European Union is funding research through its Future Internet Research and Experimentation (FIRE) program. Creating an all-new Internet might solve problems like viruses, spam, phishing, and worms. But it would cost billions of dollars, and there is a debate among experts about how long it might take. If a next-generation Internet is built, some people are concerned it will be characterized by intrinsic features that will allow governments and corporations to exercise more control over what happens online. So, the constant question remains: How do we raise barriers against spam, cybercrime, and terrorism and provide secure systems for digital transactions without infringing on civil liberties?

OVERVIEW OF RESPONDENTS' REACTIONS: Respondents clearly feel the current structure and basic architecture of the Internet will continue to underlie the technology. They believe there will be significant enhancements and updates, but a "new" system will not "replace" the current architecture; transformations will occur gradually. They pointed to two major changes that are already running in parallel with legacy systems: IPv6, the new protocol; and implementation of elements of the Semantic Web, which will

make it easier to find and link related information. Some argued, though, that by 2020 there will be more specific "walled gardens" (restricted areas of interaction and information–separate networks) that will be secure and give content control to the garden creators. Others suspected there may be split networks or partitions of the Internet. Some said they expect governments and corporations to leverage security fears to retain power over who can do what on the network. While protections are constantly added to the Internet, respondents said crime, piracy, terror, and other negatives will always be elements in an open system.

There was resounding support for the idea that the current Internet architecture will be continually refined and not completely replaced by a next-generation Internet, with four of every five responses mostly in agreement with the scenario. Just 5% mostly disagreed. "Legacy computing platforms tend to last a long time, as will the Internet," wrote **David Moschella**, global research director for the Computing Sciences Corporation's Leading Edge Forum and a *Computerworld* columnist.

"The control-oriented telco (International Telecommunication Union) next-generation network will not fully evolve, the importance of openness and enabling innovation from the edges will prevail; i.e., Internet will essentially retain the key characteristics we enjoy today, mainly because there's more money to be made," responded **Adam Peake**, executive research fellow and telecommunications policy analyst at the Center for Global Communications (GLOCOM).

Scott Brenner, consultant, technologist, and Web developer for Fortune 100 companies, commented, "The current Internet won't be replaced by a new system by 2020 any more than the highway system originally built in the 1950s has been replaced by a new system. Sure, the asphalt and concrete has long since been replaced, but no one's suggested to let the forest reclaim the land while another system of roads is built (at least not on a large scale). The Internet of 2020 will be very different from today. But it will just be a many-orders-of-magnitude improvement over what we've got now."

Alejandro Pisanty, director of computer services at the Universidad Nacional Autónoma de México, an active leader in ICANN, the Internet Society, and the Internet Governance Forum, noted, "Most of the clean-slate proposals that are being thought of in public would seem to underestimate the value of the yet-existing system, and the fact that the Internet's strong decentralization makes it incumbent on the users at the edge to apply changes they often don't master technically and for which it is difficult for them to pay. The abuse by intermediaries (from large telcos to small, local providers including small ISPs and Internet cafés) disincentivates change even further."

Some survey participants responded that there are multiple conduits in the network now, and that approach will become more formalized. "In fact," wrote **Anthony Townsend**, research director for the Technology Horizons program of The Institute for the Future, "some parts of the Internet may fragment as nations pursue their own technology trajectories."

Townsend also agreed that change will continue to be an incremental evolution, writing, "The Internet is so vastly complex, incremental upgrades seem to be the only way to get anything done. Look at how little IPv6 there is. Places like China may make big leaps and bounds because there is less legacy."

Joe McCarthy, principal instigator at MyStrands and formerly principal scientist at Nokia Research Center in Palo Alto, CA, left the door open when he commented, "Too much is already at stake on the existing Internet to build a new one. However, the recent FCC rulings that will force everyone to switch from analog to digital television shows that the [U.S.] federal government is not averse to forcing large-scale changes on its population in the conduits through which they must seek electronic information and entertainment."

Hal Varian, chief economist for Google, wrote, "The research on next-generation Internet will pay off by allowing some retrofit of the current network."

An anonymous respondent commented, "Our current approach to Internet architecture is dying already, as there is too much content to create intuitive taxonomies. Search will reign supreme."

Just one respondent who provided an elaboration jumped in with a clear statement of expectation that significant system upgrading is needed. **Ian Peter** of Ian Peter and Associates and the Internet Mark 2 Project, a pioneer who helped develop the Internet in Australia and the Asia-Pacific region in the 1980s, responded, "It is unlikely that TCP/IP (be it v4 or v6) will survive much beyond 2020. Current Internet standards bodies and core Internet protocols are ossifying to such an extent that security and performance requirements for next-generation applications will require a totally new base platform. If current Internet base protocols survive, it will be as a substrata paved over by new-generation smarter ways of connecting."

Although respondents expressed many fears, over the politics that could be built into its architecture, and the Internet's deepening complexities, they also indicated there is reason for optimism. "The Web must still be a messy, fabulous, exciting, dangerous, poetic, depressing, elating place…akin to life; which is not a bad thing," noted **Luis Santos**, Universidade do Minho–Braga, Portugal.

Incremental Change Will Continue; There Will Not Be a Clean-Slate Internet

A number of survey participants noted that change is rarely delivered in a wholesale way in complex systems. "Successful solutions are almost always built on existing infrastructure, rather than starting from a clean sheet—simple economics," noted **Jason Stoddard**, managing partner for strategy at Centric/Agency of Change, an interactive strategies company.

Walt Dickie, executive vice president and CTO for C&R Research, was aggressive in his support for the scenario. "I don't 'mostly' agree, I agree completely," he wrote. "The utopian vision of a next-generation Internet birthed by wise and benevolent leaders will be preceded by flocks of flying pigs, peace in our time, and the Easter Bunny."

Jeff Jarvis, blogger at Buzzmachine.com and a professor at City University of New York, commented, "Any media company that has tried to build the ultimate content management system has learned this lesson: It's never done, far from perfect, too expensive, and always behind. We will build on what we have."

"The Internet is too distributed to undergo a clean-slate facelift," wrote **Susan Thomas** of S2 Enterprises LLC.

"Incremental innovation will reign, based on short-term pressure to monetize," noted **Peter Kim**, a senior analyst specializing in e-strategy and management for Forrester Research.

Steve Goldstein, an ICANN board member whose job with the U.S. National Science Foundation in the 1990s was to help diffuse the Internet globally, commented, "Depending on where in time one reckons the start of the Internet (~1970 or ~1980), it took about 25 or 15 years for a truly commercial Internet to develop (~1994), and another 10 years at least for it to become as feature-rich as we now experience it to be (recall Mosaic, first browser in 1993; fully functional browsers on phones in early 2000s). So, even if NSF's and the EU's experimental network technologies were to be successful in developing a revolutionary next-gen Internet, I would not expect it to displace the legacy Internet until after 2020. And, I am not a real fan of either the NSF's or the EU's ability to re-create another disruptive technology to displace the Internet as we know it. There is likely to be too little funding and too much cronyism for that to happen. On the other hand, I would expect to see some developments feed into incremental improvements in today's Internet."

James Jay Horning, chief scientist for information systems security at SPARTA, Inc. and a former fellow at Xerox's Palo Alto Research Center, wrote, "Telephone managed to eventually supersede the legacy infrastructure of telegraph wires, but I don't see any correspondingly disruptive technological advantage that will cause the clean-slate Internets to replace, rather than supplement, the current one. I see a rolling transition, rather than a clean break."

Christine Boese, a researcher and analyst for Avenue A-Razorfish and Microsoft, responded, "The groups funding and building these so-called 'new' platforms are delusional. Not that there never will be new platforms, but they won't come from any of those groups. If such a new platform should magically appear, it will arise from inside the current

Internet, and it will be fully backward compatible and inclusive. There will never be a 'clean-slate Internet,' unless our culture does an Atlantis and dumps our beautiful Alexandria on the Ethers into the sea."

Hamish MacEwen, a consultant for Open ICT in New Zealand, wrote, "Looking at fundamentals such as the calendar, after lunar/solar, there was Julian, after Julian there was Gregorian. Will there be a replacement? No. Some basics reach a state of 'good enough,' and we move on to other things. So it is with the Internet. So it was with Ethernet. So it was with SMTP [simple mail transfer protocol]. Could it be better, yes. Is it good enough, yes. IPv6, yes, but there'll be a lot of IPv4 for a long time to come, probably still in 2020. 'Those who want to commit crimes and mischief are still able to cause trouble.' Now there's an eternal verity."

A number of respondents noted the need for additional bandwidth and speed. One anonymous participant commented, "If in real space, it's location, location, location; in cyberspace, it will be speed, speed, speed. Internet capacity and speed must go up several orders of magnitude if new social and psychological phenomena—the result of complexity—are to emerge."

Some who mentioned specific system upgrades noted that incremental improvements can gradually build a new architecture, piece by piece. An anonymous respondent wrote, "Applications and usage patterns will appear to evolve, but the underlying technology will probably change radically. Compare telephony. People still answer ringing phones and talk into them, just as they did in most of the 20th century, but the underlying technology is quite different."

And some respondents said users will pay for additional services—the Internet system will be tiered. "The most sophisticated technologies, interfaces, and protocols will be shifted to the fastest part of the Internet (fiber-to-the-computer, quantum computing), and older technologies will poke along on the current infrastructure; and of course, your ability to pay will predict how much time you get to spend on each system," commented **Peter Eckart**, director of health information technology at the Illinois Public Health Institute.

An anonymous survey participant remarked, "Paul Baran's principles are based on a diversity of routes over which packets can be transmitted. If the communications market consolidates, and if the topology of the networks consolidate, that diversity may be reduced, creating points of failure that have greater impact on the Net."

Another anonymous respondent noted, "I think we adopt the biodiversity metaphor for our information systems, and approach the problems of spam, viruses, etc., as an immune system problem and develop appropriate response mechanisms to handle it. I don't believe that such a system will require or necessarily allow greater control by governmental or corporate entities."

And another anonymous respondent wrote that a simple evolution of the current system "isn't what most governments would want, more what security agencies would want, nor is it what technocrats would want, but it seems a likely outcome."

Ed Steinmueller, a professor at the University of Sussex who researches the structure of high-technology industries, responded, "Although I expect a major push for the next-generation Internet, I expect it to shipwreck on the shoals of competing interests and visions."

THE MOVE TO IPv6 AND THE SEMANTIC WEB WILL CREATE NEW ONLINE OPPORTUNITIES

Some survey participants noted that the Internet is a system of networks (including the research networks Internet2 and National LambdaRail) that is already undergoing the most major overhaul since its beginnings, as improvements in the technologies of architecture are introduced and it transitions to Internet Protocol version 6 from IPv4 and as it also begins to weave in the added features of the Semantic Web, a longtime project of Web innovator Tim Berners-Lee and the World Wide Web Consortium (W3C).

"Internet2 is providing today the promise for advanced networks of tomorrow; unexpected jumps in optical networks will permit new types of access to rich media data and HD-based imaging," wrote

Don Kasprzak, chief executive officer of Panaround.com and a former system engineer at Apple Computer.

Paul Jones, director of ibiblio.org at the University of North Carolina–Chapel Hill, noted, "The work already underway on National LambdaRail and Internet2 is showing evolutionary improvements."

An anonymous respondent wrote, "Semantic technologies will advance to build the next-generation Internet." Another wrote, "There will be new meshes, new systems, new layers, but nothing will 'replace' the Net; merely upgrading protocols."

Paul Greenberg, president of The 56 Group LLC, commented, "There is no reason to create a new ether out of whole cloth. With the implementation of the address protocol IPv6, which provides an infinitely large number of Internet addresses, we don't have to worry about it running out of 'space,' so to speak. The new forms of the Web, like Web 3.0—the Semantic Web—will begin to show us how to interact with the Web in context, ways we can hardly imagine now will provide us with new directions. The idea of specialized search will unlock much of the so-called 'dark Web'—that portion of the Internet that isn't really being searched with Google or any other engine for that matter. Yet, there is always the possibility with something that covers as much ground as the Internet for breaching it. If it is secure, given the old problems it has, there will be someone who will creatively find a way to commit criminal mischief. So, problems will continue, but there really is no reason to create a whole new Web."

Micheál Ó Foghlú, the research director for the Telecommunications Software & Systems Group (TSSG) at the Waterford Institute of Technology, Ireland, and a member of W3C and an active participant in next-generation research, wrote, "In the short term, we need to put effort into migrating from IPv4 to IPv6 to respond to the looming crisis in IPv4 address space...My research group, the TSSG, plans to participate in the research efforts of GENI/FIND and FIRE and already has to some degree; these are medium- to long-term and cannot come to fruition by 2020. Most of this work will involve various overlay networks (over the IPv4 or IPv6 Internet) but some will take a clean-slate approach, and any clean-slate approach is very unlikely to be widely deployed in the next 12 years. So I hope to see a healthy IPv6 Internet with a legacy IPv4 Internet both operating in 2012, and lots of interesting ideas from research being deployed as overlay networks over that basic infrastructure. I do not see the private telecommunications infrastructure adopting the open-Internet model, though it may use IP technologies, so there will still be a number of interesting networks in 2012, most using forms of IP. One other interesting trend in network infrastructure development is the use of carrier-grade Ethernet, pushing previous LAN technologies into use within a wider remit, such as metropolitan networks. The promise is that these are cheaper to deploy and manage even compared to IP networks. IP will still be needed to interconnect these networks, and IPv6 will be needed."

Todd Spraggins, strategic architect for Nortel Carrier Networks and president and chairman of the board of directors of the Communications Platforms Trade Association, responded, "The Internet can never be 'replaced,' as the next best thing will not overlay it but be integrated, thus always having the appearance of being extended by the uninitiated outsider."

An anonymous survey participant commented, "The transition from IPv4 to IPv6 is uncertain enough, demonstrating the enormous cost and resistance to any global changes to the Internet. As the Internet grows, the task only gets harder."

Developments Will Be Driven by Security; There Will Be Increased Privacy Concerns

Many respondents said a further surrender of privacy in exchange for security will play out in a big way before 2020. "The arms race between the good guys and the bad guys doesn't slow or stop—it goes on hyperoverdrive," predicted **Sean Steele**, CEO and senior security consultant for infoLock Technologies. "Average business users and consumers will have more, not less, security in transactions and communications, but will be required to use more invasive technologies and techniques, such as biometric authentication (e.g., fingerprint recognition, voiceprinting, iris/retina scanning, etc.)."

Robert Eller of Concept Omega, a marketing and communication company, agreed that identification will be based on genetic information. "We will eventually only be able to interact with the Web with a personal biometric/genetic code which will imprint on any interaction we provide," he responded. "This should remove all forms of fraud or spam. To allow for privacy in 2020, laws are required for government access to this data when reason for fraud/misuse are evident."

Bertil Hatt, an Internet researcher employed by France Telecom and Orange who is completing a PhD, predicted that in 2020 "most piracy has been solved through licensing, although corporate-secret appropriation (CSA) has taken the lead. Most malware used to come from rogue countries who have been so ostracized for harboring spam, virus, or worm editors that they finally took part in global agreement on extraterritoriality of digital crime and e-terrorism. Phishing is still rampant, perpetrated by very small actors, but widespread knowledge and Bayesian filtering considerably limits its impact."

Thomas Lenzo, a business and technology consultant with Thomas Lenzo Consulting, wrote, "By 2020, beyond technology, there must be multinational initiatives to coordinate efforts to fight cybercriminals; laws must change to combat evolving cybercrimes; nations must cooperate in their arrest and prosecution. There must be a unified global effort to deal with those countries that encourage or employ cybercriminals."

Chris Miller, senior vice president for digital operations for Element 79, commented, "The how, when, and what we use to access the Internet will change (smartphones vs. computer; anywhere vs. home/work becomes the norm). Hackers will continue to be a part of society, but their mischief also drives innovation as it does today… Cybercrime or cyberterrorism takes on more priority. Look at the Middle East and Asia shutdown and delays due to the FALCON cable cut. At this time, we don't know what caused it."

Leonard Witt, an associate professor of communications at Kennesaw State University and author of the weblog PJNet.org, remarked, "When have we ever stopped crime? If it is a choice between having some criminals around and having a repressive government, I will take the former; they are much easier to deal with."

Respondents held out hope that privacy protections can somehow be preserved. "If enough people demand privacy protection, that will improve, too," noted **Peter W. Van Ness**, president of the Van Ness Group, a Web-development company. "If we do not demand it, privacy will be traded away for increased security and reliability; that is not a good trade."

An anonymous respondent separated security concerns from medium used, writing, "Spam should simply be illegal, with recipients rather than government having the primary right of enforcement. This is just like junk fax laws today. Financial institutions need to be held responsible for security breaches leading to financial losses for customers or to identity theft. It is their dud security that makes cybercrime possible. Terrorism is an international criminal problem, not a military problem, as in the misconceived metaphor of the War on Terror. Balancing numbers of deaths against the rights of defendants is difficult, but no harder than it ever was before." And another echoed the sentiment, commenting, "Nothing is wrong with the technology involved in the Internet. How it is being used is wrong when it comes to spam, phishing, etc. We need new policies and education, not new technologies. We also need a paradigm shift so that profit will not trample human rights."

CORPORATE AND GOVERNMENT CONTROL ARE AMONG THE PRIMARY CONCERNS ABOUT NETWORK CHANGE

Many respondents' negative remarks about the diffusion of a "clean-slate" Internet were prompted because they saw it as a threat to civil liberties. "The Internet is not magical; it will be utterly overmanaged by commercial concerns, hobbled with 'security' micromanagement, and turned into money-shaped traffic for business, the rest 90% paid-for content download and the rest of the bandwidth used for market feedback," wrote **Tom Jennings** of the University of California-Irvine, creator of FidoNet, the first message and file-networking system online, and the builder of *Wired* magazine's first online presence. "Notice that ARPAnet was handed to commercial interests; it wasn't turned into a national/international resource for citizens (and don't tell me that megacorporations are citizens)."

An anonymous survey participant commented, "Next-generation networks are being largely designed and implemented by commercial companies pursuing marketplace opportunities and governmental infrastructure requirements; the existing, stable, public network infrastructure will increasingly support IP functionality until 'the Internet' disappears."

Nick Dearden, campaigns manager for Amnesty International, the human-rights organization, responded, "All I would like to do is point out the risks. The Internet has, in many ways, grown up from the grassroots; it wasn't controlled by governments or corporations. That fact has led to it being a useful space, beyond normal social controls that we see, for instance, in the broadcast and print media. As governments and companies extend their control—sometimes to near-monopolies—over sections of the Internet, this space has closed down. On the surface, controlling spam seems like something few people would argue with. But in China, the war against spam has actually been used to crack down on all matter of political activity. The only way to protect free space is to ensure that any systems created to deal with real problems on the Net—e.g., child pornography—are grounded in human rights and protect fundamental freedoms like freedom of speech. To date, these rights have taken a backseat in discussions of Internet governance, and I am therefore fearful of how new-generation research will be utilized."

Howard Rheingold, Internet sociologist and author, noted, "The Internet's end-to-end architecture is being compromised when the Great Firewall of China filters packets and blocks data for political reasons, and the architecture of participation that made the Web possible is under attack when broadband providers break 'network neutrality' for commercial reasons. But the problems with replacing something as widespread and flexible as the present Internet—with all its problems, which may indeed necessitate radical redesign—are economic, political, and formidable. Who is going to design, govern, deploy, pay for the new system, and how are the world's major political and economic players to agree? Starting the Internet was simple back when everybody trusted Jon Postel. The world lacks that technopolitical simplicity today."

"There will be strong resistance to a replacement Internet with more government controls—it won't be adopted universally and requires that to occur in order to succeed," wrote **Tim Grafton**, market research director for UMR Research Ltd., based in New Zealand.

Theresa Maddix, a research analyst for ForeSee Results, responded, "NSF initiatives, GENI, FIND, and FIRE are all well-intentioned and led by very bright individuals. However, the information wants to be free. It was the release of the Internet from government hands and agencies that allowed it to explode. Google and others are always building better spam filters. Cybercrime still is much lower than noncybercrime."

Don Heath, a former leader of the Internet Society and member of the U.S. State Department Advisory Committee on International Communication and Information Policy, noted, "The Internet has achieved its remarkable success because it was not controlled by any one entity or government. As soon as governments attempt to exercise control or otherwise regulate the Internet, its usefulness will greatly diminish."

Giulio Prisco, CEO of Metafuturing Second Life and a former department head at the European Satellite Centre, commented, "Seeing slow, dinosaur governments trying to compete with fast P2P innovation is pathetic. Governments should just lower taxes and stay out of innovation's way, and I say this as a former government officer."

The delicate balance of all interests was pointed out by **Jerry Michalski**, founder and president of Sociate, a technology consulting firm and former managing editor of *Release 1.0* and cohost of the PC Forum. "The Internet is what it is because commercial interests and government agencies didn't know what it was (DARPA aside)," he responded. "There is no way to build anything like that anymore, so I have no hope that something better can be built, or that everyone can be migrated to it. That said, I'm worried about Net neutrality, and

I see many ways in which today's Internet could be hobbled significantly or improved greatly over time, with no big disjunction."

An anonymous respondent agreed, commenting, "The real question with regard to architecture is policy based, not technology based… Even without a completely new Internet, governments will play a larger role in 2020 than in 2007."

There Will Be Other Networks or Partitions; "Walled Gardens" Will Be Leveraged for Control

Some respondents expect that various motivations will cause more separation of networks. "Those with resources and security concerns will have access to 'better' and more secure channels. Speed and security will increase for everyone, but someone will figure out how to partition off areas of the network for elites," wrote **Ted Coopman**, a communications technology lecturer at San Jose State University.

Michael Zimmer, resident fellow at the Information Society Project at Yale Law School, responded, "The most likely scenario: A secure architecture to complement the existing Internet backbone for those who want to use it. One alternative view might be a new Internet-like infrastructure emerging tailored specifically for secure mobile-data transfer, capitalizing on the rise of mobile telephony."

Cambria Ravenhill, manager of national channel planning at TELUS Communications, wrote, "The Internet will split into the 'official' Internet, where most civic life and corporate and government transactions occur, and an 'underground' Internet fueled by scarcity economics."

Jay Neely, founder of News Armada, a Boston-based online news and community company, commented, "If government encouragement does not occur within the next 5 years, while there will still be refinements made to existing infrastructure, the process will be too slow for some organizations, and we will see development of separate networks, like Internet2 for universities. While unlikely, it is possible that a future technological megacorporation could build an Internet-like infrastructure that competes with publicly available

Internet; concerns about civil liberties and tracking are even more valid in this scenario, but may be overlooked by the general public due to the convenience of the advanced infrastructure."

"There will be two Internets," predicted **Garland T. McCoy**, founder of the Technology Policy Institute, a think tank focused on the economics of innovation, "one for 'us' and one for the financial institutions, security folks, spooks, government agencies, major corporations, etc. That almost exists today."

Mark Youman, principal at ICF International, a Washington, DC, consulting-services company, wrote, "The current Internet will be improved rather than replaced wholesale, but it will be one of MANY global networks. Institutions, industries, and other groups will construct independent networks when the Internet becomes too overrun or corrupt to serve their needs. Access to these networks will be part of what defines the 'haves' from the 'have-nots.'"

Could a division of networks lead to another type of visible divide and possibly even an Internet class war? Some respondents mentioned the possibility.

John Jordan, an associate professor of communications at the University of Wisconsin–Milwaukee, wrote, "Like a highway, significant parts of the Internet likely would need to be shut down and closed off in order to receive a major upgrade. Unlike a highway, the public will not stand for this, necessitating that instead of a completely new Internet infrastructure, people get slightly better service over time as patches and upgrades are made, but this leaves open potential problems familiar today. At the same time, private business ventures and new housing developments in exclusive neighborhoods will experiment with and implement new Internet architecture, leading to a point in the more distant future where there may be two Internets, creating a true Internet class war."

Benjamin Ben-Baruch, senior market intelligence consultant and applied sociologist for Aquent, predicted that there will be two Internets, the original and the next generation. "Those with the resources to move much of their communications and functions to this new architecture will do so—and early adopters will have to pay hefty costs to do so. But along with this high-cost barrier will come control of this new environment. Part of the digital gap in the future will be between those who operate on both the current and next-generation platforms and those who are limited to the current Internet...Security and privacy on the current Internet will be increasingly compromised. There will be two reasons for this slowly but steadily decreasing security and privacy: (1) Hackers and pirates will develop security-breaking technologies faster than security technologies can be developed and rolled out. (2) As the secure, next-generation platform is developed, hackers and spammers and pirates and other Internet criminals will focus on the much easier but very lucrative prey on the current Internet."

It Is Possible—Even Before 2020—That Some Revolutionary Idea Could Shake Things Up

A few respondents noted that breakthroughs incorporating influences from biology, nanotechnology, and other sciences could push Internet evolution in new directions. "By 2020, two major advances will have significant impact. The first is bioengineering and nanotechnology, allowing the Net to be 'embedded' into individual humans (scary, eh?); the second is quantum computing that will significantly alter the current electrically loaded computing engines," predicted **David Hakken**, a professor of anthropology at the Indiana University School of Informatics who studies social change and the use of automated information and communication technologies.

Roberto Gaetano, an ICANN board member who also works for the International Atomic Energy Agency, commented, "I concur in seeing the development of the 'next-generation Internet' as evolutionary rather than revolutionary. However, I am wondering whether by then we would not start seeing something that is started based on some new concept that we can't even figure out today. And I wonder whether this is not likely to come from a sociocultural environment that is completely different from ours."

An anonymous survey participant remarked, "I can't help but feel the speed of technology change may surprise us with something totally new. I remember when the Internet was not the World Wide Web, but a series of numbered addresses and text based. While the concept of connecting people to people, places, and things will remain, how it is done will probably look different."

Worldwide Series of Packet-Switching Networks Known as "the Internet" Has Always Been in a State of Change

What most people refer to as "the Internet" is actually a network of networks that is under full-time development by an international group of engineers, scientists, technology developers, and other volunteers, many of them members of the Internet Society's suborganizations, including the Internet Engineering Task Force (IETF). These men and women have built an ever-more efficient Internet, advancing it continuously and adding improved capabilities as it has grown from a network of several computers that allowed text exchanges among a few people in the early 1970s to a network of networks including a billion computers and mobile devices connecting to allow 1.4 billion people to send and receive text, graphics, and full video.[1]

The term *World Wide Web* is not interchangeable with *Internet*. The World Wide Web is the system of hyperlinked documents that are accessible on the Internet through the use of Web browsers. The people working to develop the Web are members of the World Wide Web Consortium (W3C). The Web is an Internet use that can be exercised through Internet protocols; separate protocols have been established for people to use e-mail, voice telephony (VoIP), streaming media, file sharing, remote access, and collaboration.

Some networks—including the research networks Internet2, National LambdaRail, and JANET—are not publicly accessible, and various corporations and governments have *intranets*—closed systems—so private networks already exist and they have been available since the earliest days of internetworking. Internet2, for instance, is a high-performance, high-speed alternative network used and improved by research institutions, and its members have also been doing innovative work toward future enhancements of the Internet.

Human networks built and continue to build the network of networks. The architecture of the Internet is a complex structure of layers imagined, developed, and improved over the decades to operate under dozens of protocols and standards that are suggested, debated, tested, authorized, challenged, updated, and tended to by specialized worldwide volunteer organizations like IETF and its working groups.

In his response to the *Future of the Internet III* survey, Jonne Soininen, manager of Internet affairs for Nokia Siemens Networks, gave an Internet engineer's view of the reasons for going beyond the work of standing Internet-architecture organizations and starting up next-generation research projects like GENI, FIND, and FIRE. "It will not be possible to replace the current Internet with a completely new Internet with a different paradigm," he wrote. "...The different national and international research programs are important, however, as they allow the researchers to think 'outside the box' and create new, innovative solutions, which one day may become features of the Internet."

GENI, the Global Environment for Network Innovations, is an initiative geared to inspire researchers to experiment with new approaches to future Internet designs. BBN Technologies, one of the firms participating in the original research from which the Internet was developed, is hosting the offices for the project. For these researchers, "clean slate" simply means the approach should be to look beyond the status quo and imagine a wide-open future. Some people have expressed doubts that this U.S. National Science Foundation–funded work will make a positive contribution. Regardless, the plan is to build a GENI substrate on which network experiments—possibly hundreds or thousands of them—can be layered, all running simultaneously, and including design concepts that are radically different from the current structure. The goal is for researchers to look ahead 10 to 15 years and imagine how to meet the needs for improved security, more functional mobility, and the ability to allow the networking of trillions of ubiquitous digital devices. The popular press in the United States announced in 2007 that products generated by GENI might possibly replace the Internet or be operated on a parallel network. NSF has been funding a number of other future-Internet projects through its Future Internet Network Design (FIND) program. The European Union is grouping some of its future-Internet initiatives under the umbrella program known as Future Internet Research and Experimentation (FIRE). Japan's program is named AKARI ("small light" in Japanese). All generally plan to continue to retain particular design principles of the original Internet, including the end-to-end concept and the maintenance of a common layer and added layers with independent design and independent functions.[2]

The GENI substrate may not be ready for experiments to begin until between 2010 and 2015. Some of the scientists behind the GENI plan, which was first discussed in 2005, say it could eventually inspire applications that might start out on a parallel network that people migrate to over time—in a way similar to how people are now switching from using land-based phone lines to mobile phones as the two networks run simultaneously. Members of civil-society groups are casting a suspicious eye toward next-generation efforts because governments and telecommunications companies might leverage them in some way to reassert some of the control they have lost in the Internet age.

The engineering adjustments and advances in each of the layers of today's Internet are addressed by working groups led by the people for whom that area of expertise is a specialty—for instance: the W3C's latest major innovation is the Semantic Web; the biggest project IETF is in the process of launching today is the transition from IPv4 to IPv6; a subgroup of the Internet Research Task Force (IRTF) is assessing the potential for new routing architecture that might improve support for the billions of new users likely to come in the world's developing countries over the next 20 years. An organization of the Internet Society, like IETF, IRTF is analyzing the Internet's backbone routers, which are owned by government agencies, large corporations, and Internet carriers. The core routers may be stressed by explosive growth in the routing table, driving up carrier costs, so engineers are rethinking routing.[3]

Researchers like the people on IRTF and IETF modify and add to the legacy Internet, anticipating future challenges and working to keep pace with demand. One of the major driving forces behind change is the need to adjust to handle more people and more content being moved on the Internet. Precise measures of Internet load are not possible, but former AT&T Labs researcher Andrew Odlyzko, director of the University of Minnesota–based Minnesota Internet Traffic Studies (MINTS), estimated that worldwide traffic on the Internet in 2007 measured 3,000 to 5,000 petabytes per month. A petabyte is one quadrillion bytes—that is a thousand trillion, a 1 followed by 15 zeros. So, MINTS estimated that 3,000 to 5,000 × 1,000,000,000,000,000 bytes of information are being transferred from place to place globally monthly online.[4]

To the thousands of people who worked in their development, the next-generation Internet is already arriving in the form of IPv6 and the Semantic Web. These updates in the Internet protocol and in the Web are being accomplished to address the challenges presented in a world with ramped-up demands tied to security, mobility, and ubiquity.

IPv6 is already being used in parallel to IPv4, but its adoption is moving slowly. Observers expect the dual-stack system to continue for a number of years, possibly through 2020. Some governments, including those in China, Japan, and Korea, are offering incentives or setting requirements for IPv6 implementation. In the United States, the government required that the network backbones of all federal agencies must be capable of implementing the new protocol by July 2008. European Union countries have set a goal—not a requirement—of 25% deployment of IPv6 in commercial, government, and residential networking by 2010. Every networked device has to have an IP address in the current Internet system. Under IPv4, there is a possibility for 4.2 billion unique IP addresses; IPv6 allows for nearly 340 undecillion—or 340 trillion trillion trillion unique addresses. It will also allow networks to support video, voice, and data in new ways. It is designed to enhance security, mobility, quality of service, multicasting, peer-to-peer support, and autoconfiguration. The gradual switch from IPv4 to IPv6 is expected to remain transparent to end users.

The Semantic Web, the key initiative to improve Web functionality being developed by W3C for the past decade, reached a landmark in January 2008 when SPARQL, its query language, became a standard. It makes it possible to query data from across the Web and from multiple data sources; thus, as people adopt the tagging system of the Semantic Web, the entire content of the Web can become a vast database. Just as some are referring to IPv6 as the "next-generation" Internet, people are referring to the Semantic Web as the next generation of the World Wide Web. Over the next decade, it may improve data aggregation to such an extent that an Internet search that yields hundreds of responses today (many not associated with the searcher's needs) will usually deliver only the information you need. This is incredibly useful in a world with more than 172 billion Web sites and 3 to 4 million more being created each month.[5]

The Sematic Web offers even more possibilities—it is not all about annotation of the current Web, as Rensselaer Polytechnic Institute Professor James Hendler likes to point out. As described by one of its developers, OpenLink Software CEO Kingsley Idehen, it allows Web users to look at data from a range of perspectives, peeling back the layers of information and setting them free. The concept is so revolutionary that people have difficulty describing it in just so many words. Web innovator and MIT research scientist Tim Berners-Lee described it as Web 3.0. "Linked, open data is a movement," he told participants at the Linked Data Planet conference in New York. "It's a grassroots movement, and it's about information that is free to use in the Linked Data format." He explained that the movement is about "getting one format across applications" so the Semantic Web standards can enable people to gain access to the information they want and use it any way they want, for instance, being able to mesh data from a personal bank statement and a personal calendar. Berners-Lee said in June 2008 interviews that he would like to see a future Web that allows people to connect their ideas with the ideas of others, building a system for people to share parts of ideas in a way that can make them whole.[6]

Success for the Semantic Web will depend upon people working together to accept its standards (GRDDL, RDFa, OWL, SPARQL, and others) and its naming and tagging ontologies. Sites already implementing Semantic Web elements include DBpedia, Twine, Garlik, GeoNames, RealTravel, and MetaWeb. Like IPv6, it is already being implemented to a small degree running in parallel with previous standards, and its influence, while expected to be significant, cannot yet be predicted. It is changing search, data optimization, and the way Web applications are built.

The Semantic Web has some vocal opponents and critics. One is blogger, author, and speaker Cory Doctorow who described in a 2001 essay titled "Metacrap: Putting the Torch to Seven Straw-Men of the Meta-topia" what he considered to be the seven obstacles to getting reliable data: People are lazy; people cannot accurately observe themselves; people lie; people are stupid; schema are not neutral; metrics influence results; and there's more than one way to describe something.[7]

Another critic is author, blogger, and professor Clay Shirky. In one of his columns, he wrote, "The Semantic Web is a machine for creating syllogisms…despite their appealing simplicity, syllogisms don't work well in the real world because most of the data we use is not amenable to such effortless recombination…There is a list of technologies that are actually political philosophy masquerading as code, a list that includes Xanadu, Freenet, and now the Semantic Web…like many visions that project future benefits but ignore present costs, it requires too much coordination and too much energy to effect in the real world, where deductive logic is less effective and shared worldview is harder to create than we often want to admit." Shirky did note that some positive results will come from the Semantic Web.[8]

Accompanying the hopes for the Semantic Web, IPv6, and other Internet development are fears about the future of privacy and identity and the need for security in an architecture in which growing amounts of information are shared in a worldwide data cloud. For instance, security is vital for companies and individuals who employ Semantic Web products and technologies. When search for aggregated data becomes more specific, a simple search for a person's name can yield health records, parking tickets, mortgages, signatures, travel records, video-viewing habits, and any other recordable, data-based information. People's Web-search histories, financial transactions, mailing lists, and surveillance photography of them and their homes are being collected and can be accessed, forming a "digital shadow" for every individual and group on record.

The architecture of the Internet is a result of practical and political decisions, and down-the-line ramifications of each modification or addition can alter the landscape in significant ways. The escalating struggle for political and economic power in the Internet age is making itself manifest. There are too many examples to cover here, but following are a few.

When the politics of Internet architecture are discussed, ICANN is often mentioned. The nonprofit Internet Corporation for Assigned Names and Numbers is in charge of the domain-naming system and routers of the Internet. ICANN is a work in progress itself, much like the Internet, growing massively in scope and the number of participants in its deliberations since its start in 1998. Its administrators and subgroups make global and local decisions of major economic and social consequence. Because it operates under the Joint Project Agreement (JPA) with the U.S. Commerce Department (due to the Internet's inception in the United States), many global Internet stakeholders said the United States is playing too big a role in oversight of ICANN. The JPA is set to expire in September 2009; it may be renewed, or the United States may decide to extend it, as it has previously.[9]

Internet service providers are starting to test methods of control over data flow; motivations are security/trust issues, load control, and profit. Advances in routers and network-monitoring equipment now allow Internet service providers to monitor the content of packets in real time and decide how they should be handled. This is called *deep-packet inspection*. For instance, ISPs may offer to filter spam from e-mail, exercising censorship. Acts like this erode the Internet's end-to-end principle; a router makes a judgment based on message contents. ISPs are also under pressure to filter the flow of copyrighted content. They can exercise the ability to slow down certain content and allocate bandwidth differently to different customers. In 2007 and 2008 the governing bodies of France and the United Kingdom and the European Parliament have been in various stages of discussion of legislation to require ISPs to monitor content, report any suspected abuses, and remove users. No such legislation is being considered in the United States at this point, but ISPs are taking some action on their own.

In 2007 and 2008 Comcast, a U.S. ISP, used traffic-shaping measures, packet forging, to disrupt its heaviest users—the move was aimed at people who consume significant bandwidth sharing materials through sites like BitTorrent. A study by the Max Planck Institute found that both Comcast and another U.S. ISP, Cox, were "throttling" peer-to-peer uploads.[10] A lawsuit was filed against Comcast, and hackers expressed their opposition to the corporation's tactics by attacking its site.

The debate over Internet capacity was reintroduced in 2007 and 2008 when reports—many funded by telecommunications companies—indicated the likelihood of what came to be referred to as an *exabyte flood* of data that might congest and slow the Internet. The report to get the most attention was *The Internet Singularity Delayed*, a study put together by Nemertes that predicted "brownouts" due to overload by 2010 or 2011 if there is not more investment in infrastructure. Nemertes anticipated demand will grow by 100% or more annually.[11] Both Cisco and MINTS estimated growth will be at 50% a year.

Researchers noted that data is growing by a factor of 10 every 5 years. Internet users created 281 exabytes of new data in 2007; one exabyte is about 1.1 billion gigabytes. George Gilder's Discovery Institute released a January 2008 report that estimated that by 2015, IP traffic in the United States alone could reach an annual total of one zettabyte—one million million billion bytes—by 2015. Discovery began using the term *exaflood* in 2001 when total traffic was about 15 exabytes per month.[12]

However technology-industry engineers say service providers are staying ahead of the curve, backbone providers are keeping up with increased loads, Internet capacity is adequate, and it can be scaled up to meet needs; upgrades are easy once fiber is in place; improvements can be made on the ends of the system—in fact, new compression technology is expected to help handle loads; AT&T and Verizon, the two biggest service providers in the United States, have made major investments in upgrades. Members of the Higher Speed Study Group of the Institute of Electrical and Electronics Engineers are discussing the development of the next-generation 100-gigabit Ethernet; the fastest existing standard today is 10-gigabit Ethernet. The technology employed by the computers used to interface with the Internet continues to get faster, cheaper, and more efficient. While they are not expected to come into consumer use for some time, in February 2008 Intel introduced a quad-core chip that has more than 2 billion transistors. It is designed for high-end servers. Moore's law states that the number of transistors that can be placed on an integrated circuit will double every 2 years. Intel released the first chip to contain a billion transistors in 2006, and in 2004 the leading chip contained 592 million transistors.

In the United States, arguments over ISPs' and telecoms' desire to perform deep-packet inspection and exercise methods of priority-based quality of service became one of the latest skirmishes in the "network neutrality" debate. Proponents of Net neutrality believe all Internet content, platforms, and sites should be treated equally and that all people should have equal access—no cable television–like tiered prices for access, and everyone should be able to use as much bandwidth as they would like when they would like to use it.

As convergence continues and television-programming distribution and telephone services move online, broadcast and telecommunications companies are working to protect their interests and define their position in a new-media world, and people who are suspicious of these corporations' motives are tracking the action and using the Internet as a platform for advocacy of end-to-end openness.

Security and trust issues generally drive the changes in architecture that interfere with the openness of the Internet and principles like end-to-end. Security trumps anonymity; anonymity trumps security. The Internet's swift evolution to a big, bold network people everywhere depend upon has made security of the essence; but people want it to retain its openness while adding more security, and that is hard to reconcile. Over the past 30 years, Internet engineers have been patching security problems each time a vulnerability appears; some researchers say the most important aspect of next-generation Internet experimentation will be its enhancement of security. Advocates for openness fear the outcome.

It is difficult to quantify the actual security threat levels due to the fact that most groups doing research and publishing figures describing the level of Internet crime (security firms, IP and copyright owners, police agencies, governments) are likely to benefit in some way from hyping the statistics. The U.S. president's Information Technology Advisory Committee reported in 2005 that the frequency, impact, scope, and cost of cybersecurity incidents are continuously increasing. A 2005 Pew Internet report noted that 43% of Internet users reported having adware or spyware illicitly installed on their computers.[13]

There is no doubt that the proliferation of networked devices, software with various vulnerabilities, and the growing number of users are putting the system under more stress all the time. Akamai Technologies Inc., a provider with servers in more than 3,000 locations in 750 cities, issues "State of the Internet" reports based on data from its global server network. In the first quarter of 2008 denial-of-service attacks, Web site hacks, and network events originated from 125 countries.[14] Arbor Networks released a study identifying malware-infected botnet PCs and denial-of-service attacks as the most serious threats facing ISPs, hosting companies, and other providers in 2007. The survey of 75 service providers estimated that tens of millions of PCs worldwide are infected with botnet programs.

Engineers and researchers say the Internet's complexities lead to more incidents, glitches, and hitches than hackers and crime. Things occasionally break down and need repair, or parts of the system fail; an example is the failure of the electrical grid in Canada and the eastern United States in 2003. Corrupt data, temporary system overloads, hardware failure, and cascading situational difficulties can cause problems. Considering the complexity of today's Internet, it is amazingly reliable.

One of the most delicate features of the Internet architecture is the undersea cables used to transfer data from continent to continent. In December 2006 an earthquake in Taiwan stopped transmissions on eight submarine cables that carry traffic from all over the world to Asia. Traffic was temporarily rerouted to satellites and landlines. During a span of about 2 weeks in early 2008, a number of undersea cables joining countries in the Middle East and Asia were cut. FALCON, SEA-ME-WE 4, FLAG, and DOHA-HALOUL suffered damage; the incidents disrupted Internet communications temporarily until workarounds were accomplished and repairs were made. While some people suspect the acts were intentional, experts in the field note that cable cuts occur fairly frequently for innocent reasons such as the dragging of a ship's anchor. Expansion of the network of networks is continuing to build through underwater linkages. Examples: The Eastern African Submarine Cable System is expected to make a big difference in connections on the east coast of the continent of Africa when it is finally ready for action in 2010 (the World Bank is investing $32.5 million of the $235 million cost); joint planning began in June 2008 for a new high-speed, fiber-optic line between the United States and Japan. The $300 million undersea link, called Unity, will eventually account for 20% of the capacity available across the Pacific. It is being funded by Google, India's Bharti Airtel, Malaysia's Global Transit, Japan's KDDI, and Singapore's Pacnet and Sing Tel.

In any discussion of wired architecture, the "last-mile problem" or fiber to the home (FTTH) is an important element. Canada, Japan, and South Korea are among the countries with a reputation for investment in delivering high-speed Internet capabilities to individual end users, and they have been lauded for their diffusion models, which combine municipal systems with privately deployed networks to wire everyone who wants online access in highly populated areas. The average advertised bandwidth in Japan is nearly 1Gbit/sec. In France and Korea, the rate is about 40Mbps. In the United States, the advertised bandwidth for most "high-speed" users is 2 to 5Mbps, with a few extremely wired metro areas receiving better service, but a New Yorker who wants the same quality of broadband as a Parisian has to pay about $150 more per month. The politics over the differences in speeds and prices are contentious in the countries that are lagging; in the United States, many observers blame telecommunications companies and a lack of government oversight.

An OECD study found that broadband diffusion is best effected when there is a mix of competition and active regulation, to ensure an open challenge to vendors that employs an element of oversight. The June 2008 OECD broadband report noted, "The use of broadband in education, for telework, for e-government services, energy, health (telemedicine), and transport (intelligent transportation systems) is still in its infancy. Organisational and institutional barriers hamper the necessary innovation and structural changes needed and leave many OECD countries struggling to move beyond pilot projects. The notions of ubiquitous networks, broadband-based home management, and other new forms of broadband use have yet to develop and diffuse. A number of broadband-related security threats have emerged in OECD markets over the last three years. The transition to fibre connections and symmetric bandwidth will make these threats more virulent. New or more pronounced consumer and privacy issues are transpiring with broadband's 'always-on' connections and its participatory features."[15]

It is estimated that there are more than 1.4 billion people using the Internet today. In 1996, 66% of all people online lived in the United States; in 2007, 77% of Internet users lived in other parts of the world, with just 23% in the United States. Sites in China, including TenCent, Sina, and Baidu, reach more native online users than

Microsoft, Yahoo, or Google. Internet accessibility is now described as a basic human right by advocacy groups; this has major implications for the continued development of wired and wireless Internet architecture.

In a keynote speech at the Association of National Advertisers in October 2007, Microsoft CEO Steve Ballmer said, "Within 10 years, the consumption of anything we think of as media today, whether it is print, TV, or the Internet, will in fact be delivered on IP and will all be digital. Everything will be delivered digitally." While Ballmer's remarks were directed toward his vision of reality in the world's most developed countries, it is true worldwide that primary access to the vast majority of information is increasingly found to be online. Lack of access and connection costs are still an issue in many parts of the world. NGOs that work for human rights report that common hurdles to connection and the development of rural, remote, and dispersed populations' capacity for use of the Internet include a lack of infrastructure, a lack of investment, obsolete telecommunications regulations, and, in some cases, mercenary tactics employed by corporations and/or governments in regions that lack appropriate public policies.

Even the leaders of governments and corporations in countries located in the highly developed global North are being asked to consider Internet access to be a basic human necessity. Jim Kohlenberger wrote the September 2007 report *Universal, Affordable Broadband for All Americans* for the Benton Foundation. It calls for "a new vision for Universal Service—to make broadband as universal as telephone service is today," and it outlines 12 steps for advancing a "Universal Service Program for the Information Age," adding that "broadband is now becoming an essential utility, and no longer can be seen as a discretionary luxury." He added, "Soon broadband will offer the most affordable conduit for making phone calls to anywhere in the world, deliver the video and audio programming we want where and when we want it, and allow us to remain connected to friends, family, and co-workers...All of our basic communications, be they text, audio, or video, will soon require a single broadband connection."[16]

Additional Responses

Many other participants in the survey shared valuable input on the evolution of Internet architecture. Among them:

> "Refinement and layer improvements as a development and complement to the current public Internet is definitely the way of the future...More of an evolution than a revolution :-) And like any other piece of life, bad things will be attempted by bad people, but they should be a recognized and risk-managed minority..." —**Cheryl Langdon-Orr**, a leader in the Internet Society, ICANN and ALAC (At-Large); a board member of AUDA, the group in charge of Australian domain name registration

> "Second only to the telephone system, which it is already replacing, the Net has the largest installed base in the world. Furthermore, even if a better system exists in the lab, it would have to *increase* the freedom of the individual, relative to today's Net (a trivial restatement of 'better'), and now that the world's governments and IP-based businesses understand the Net, they would never allow that to happen. The Net we have today isn't the best Net imaginable, in other words, but conserved behavior of both users and opponents both make it the best Net possible." —**Clay Shirky**, consultant and professor in the Interactive Telecommunications Program at New York University; an expert on the social and economic effects of Internet technologies

> "Privacy is vanishing and will continue to vanish with the next-generation Internet. Even today, governments and private companies know much about us and our every transaction—speaking, writing, spending, saving, etc. As we refine technology to reduce abuse by cyber criminals, we also will have to be more vigilant about misuse of our private information by governments and large corporations. I am hopeful that we can increase personal integrity and ethics such that abuse will not occur because of personal moral ethics, even as ability to abuse becomes easier." —**Ed Lyell**, professor of business and economics, Adams State College, Regis University, San Luis Valley Board of Educational Services; pioneer in issues regarding Internet and education

> "The Internet of 2020 will not be uniform, but graded in terms of capability, oversight, privacy, etc., and these grades

of service will be priced accordingly." —**Oscar Gandy**, author, activist, retired emeritus professor of communication, University of Pennsylvania

"The Bible may have been the first printed book, but Rabelais' ribald fictions established the publishing market." —**Michael Botein**, professor of law and founding director of the Media Center at New York University Law School; consultant to the FCC

"The legacy/inertia of the current model seems sufficient to slow any shift over to 'Internet 2.'" —**Jamais Cascio**, originator of Open the Future, also works with the Institute for the Future, Institute for Ethics and Emerging Technologies, and Worldchanging

"Criminals are more intelligent than the innovators." —**Lutfor Rahman**, chairman of the department of computer science, Stamford University, Bangladesh; leader of the Association for Advancement of Information Technology

"As technical capabilities increase, new formats, standards, and functions will be layered into the existing Internet. Vast swaths of functionality will be layered on top of existing aspects of the framework, similar to the way that the Web was deployed on top of the existing network, alongside Gopherspace. There will be a 'new Internet' and 'old Internet' running on the same network, similar to the way a city has an old quarter and a new quarter, all sitting on top of a shiny, new subway system and a 100-year-old sewage system." —**Ross Rader**, director of retail services, Tucows Inc.; works with ICANN in the Registrars Constituency, part of the Generic Names Supporting Organization (GNSO)

"At all times, the legacy investment in existing technology forces a slow pace of change—the world cannot afford a quantum leap across the globe." —**Adrian Schofield**, manager of the applied research unit, Johannesburg Centre for Software Engineering, South Africa; leader in the World Information Technology and Services Alliance

"If there is a radical departure from the current Internet, it will come from an unexpected source and most certainly not from the NSF program. History never repeats itself. TCP/IP succeeded because of a number of historical contingencies that cannot be reproduced." —**Milton Mueller**, professor, Syracuse University and Technology University of Delft; expert on Internet governance, technology policy; active in ICANN's Non-Commercial User's Constituency

"There is too much invested in the current infrastructure that brings Internet access, and it is too pervasive to be replaced. It will be upgraded, differentially, over the long haul." —**Steve Jones**, professor of communication and associate dean of liberal arts and sciences, University of Illinois–Chicago, cofounder of Association of Internet Researchers

"Cultures, people, and thereby their use of technologies are more conservative than revolutionary (for better or worse). We're still stuck with exposed power lines, despite the disasters that trees and ice storms can bring because it's too expensive, the argument goes, to move to buried lines. By the same token, at least in the U.S., I don't think people will prefer an enhanced version of the current Internet over a potentially revolutionary replacement for the important political and ethical reasons mentioned but simply for economic ones: It'll be cheaper, and certainly we can't count on a state-supported system to subsidize any changes by very much. The situation in Europe may well be different, given the much stronger role of the state as a coplayer in these sorts of initiatives (see, e.g., Niels Ole Finnemann's analysis of the importance of the state in the Nordic countries in establishing and expanding the needed infrastructure for the Net). Finally, no technology has even been proof against bad actors. While technological instrumentalism ('a tool is just a tool') is certainly false and misleading, it's nonetheless correct to say that hammers and screwdrivers can be used to build useful and beautiful artifacts: They can also be used to kill people. So, I think we'll do what we've been doing for millennia—muddle through as best we can! The optimistic part is, that usually seems to work (e.g., no nuclear wars, so far, at least)." —**Charles Ess**, a professor of philosophy and religion and researcher on online culture and ethics, Drury University, Springfield, MO, and active leader of the Association of Internet Researchers

"There will be a way to improve the actual applications but not replace them. The misuse of the Internet should be heavily penalized. The implementation of an identity-management system and an accurate global jurisdiction will help locate the harm-doer and impose predefined sanctions." —**Hanane Boujemi**, ICT researcher for DiploFoundation, working on educating people about Internet policy and Internet governance, Malta

"The need by communities and individuals to use the Internet to solve local problems will generate resistance to any efforts to privatize the high-bandwidth Net." —**Cliff Figallo**, social innovator and original member of the first online community—The WELL, now of AdaptLocal.org; expert in fitting and implementing social Web applications to groups

"The Internet itself was built on earlier work, so it's not likely something radical happens to erase the Internet of today to replace it with something new. Even efforts at a new architecture will build on what we have today, and even if the temptation to announce a 'new Internet' comes, people will see beyond the hype and know that new things are often a recreation of old platforms. Fortunately, if a new Internet is created, this comment will go with the old one, and I won't be caught saying this :)." —**Gbenga Sesan**, Internet for development consultant, Paradigm Initiative, Nigeria; his work is tied to the use of ICTs in socioeconomic transformation, focusing on underserved groups

"In the same way that transportation has many varied paths (rail, road, boat, air, etc.), the Internet will have many channels or 'styles' for network architecture—context will mean everything." —**Michele Perras**, artist, consultant, researcher, and futurist with Interactive Ontario

"This seems obviously true, with the caveat that the most sophisticated technologies, interfaces, and protocols will be shifted to the fastest part of the Internet (fiber to the computer, quantum computing), and older technologies will poke along on the current infrastructure. And of course, your ability to pay will predict how much time you get to spend on each system." —**Peter Eckart**, director of health information technology, Illinois Public Health Institute

"The idea that anyone could create a network that no one could hack, phish, spam, or otherwise vandalize seems rather silly. If one person can figure out how to create a network, another sufficiently motivated person can figure out how to hack it. The current Internet will be improved and refined, and we will see development of security protocols that are harder to hack. For instance, I foresee greater use of biometric technology in transactions involving the transfer of money from one party to another such that having someone's credit card number, for instance, would be virtually worthless." —**Janet Snowhill**, automated system coordinator, Chemeketa Cooperative Regional Library Service

"No matter how good the security and reliability of the Internet improves in 2020, there will always be vulnerabilities, and cybercriminals will find a way to exploit them, cause technical damage to the architecture, as well as economic harm to individuals." —**Cristos Velasco**, director general of North American Consumer Project on E-Commerce; Mexico-based attorney and active in the Internet Governance Forum and Internet Society

"The Internet cannot be 'replaced' by something different (which would also involve disruption of a vital context of life)—it will be developed..." —**Norbert Klein**, member of ICANN's GNSO Council and Internet Society leader who works with Open Institute Cambodia, a company whose primary focus is on information

"This sounds about right, a balancing act between security and search on reliable platforms and applications with open development of tools; and some will be bent on creating viruses and causing trouble." —**Joanna Sharpe**, senior marketing manager, Microsoft

"Unfortunately, there are just as many great scientific minds on the wrong side of the law." —**Janie Graziani**, manager of new media and technology for the American Automobile Association

"[The Internet will be] more of the same with bigger stakes and more complexity." —**Richard Hall**, professor of information science and technology and codirector of the Laboratory for Information Technology Evaluation, Missouri University of Science and Technology

"The Internet will continue to wrestle with these problems, and we will see the rise of more walled gardens, and privately owned and controlled networks, in part as a way of governing against spam and other security issues. Like AOL, Compuserve, and other early ISPs, these will run in tandem with the existing Internet, possibly built on its infrastructure and possibly on privately held wireless infrastructure." —**Alexander Halavais**, professor and social informatics researcher, Quinnipiac University; explores the ways in which social computing influences society

"It is almost impossible to stop governments and powerful corporations from eventually controlling the Internet in ways that consolidate their power and/or improve their profits. Viruses, spam, and the like will provide the perfect excuse for these organizations to successfully convince most people that a more controlled Internet is in their best interest despite the opposite being true." —**Christian Ferris**, Washington University

"One word: IPv6 (i.e., the replacement for the addressing system, which hasn't replaced the old system, and is only slowly coming into use)." —**Seth Finkelstein**, anticensorship activist and programmer, author of the Infothought blog and an EFF Pioneer Award winner

"The current Internet architecture is so pervasive that incrementalism is the only possible solution." —**Bill St. Arnaud**, chief research officer, CANARIE Inc., an industry-government consortium that promotes and develops information highway technologies in Canada; active in Internet2

"Another scenario (the second so far) that comes quite close to my own prediction." —**Rollie Cole**, director of technology policy, Sagamore Institute for Policy Research, a nonprofit, nonpartisan think tank headquartered in Indianapolis, IN

"The only thing higher walls do is make committed people who want to get over them climb higher." —**Sam Smith**, Web interface developer, University of Manchester, U.K.

"Any 'new' network technology will have to be incrementally deployable and yield clear benefits as it's incrementally deployed." —**Brough Turner**, chief technology officer and cofounder of NMS Communications; oversees evolution of technology and product architectures

"There will be complementary Internets that will increase the digital gap. The Internet2 will have the latest technologies and improvements, and entities that are only connected to the historical network will be excluded." —**Rafik Dammak**, software engineer, STMicroelectronics, Tunisia; DiploFoundation participant in the study of Internet

"It is very rare for new information technologies to completely replace older technologies. Typically, the newer technologies build upon and supplement older technologies. The same should be true for the next-generation Internet building upon the current Internet." —**Gary Kreps**, chair of the department of communication, George Mason University; formerly founding chief of the health communication and informatics branch of the National Cancer Institute

"Depends on what you want to accomplish. Systems will have to evolve to distinguish which kind of 'actor' is involved in each transaction." —**Greg Laudeman**, utilization catalyst and facilitator, community technology specialist, Georgia Tech Enterprise Innovation Institute

"Replacing the Internet would require too much capital. Also, governments are not creativity organizations. This needs to be left to industry, and industry will be hesitant to change." —**Todd Wagner**, health economist, Health Economics Resource Center, Palo Alto, CA, part of the U.S. Veterans Administration; also involved with the Center for Health Care Evaluation

"Those of us whose livelihood depends on the Net are annoyed, nay angered, at the time wasted on deleting spam, creating spam filters, etc. However, I am a realist. There are too many people out there with too much time on their hands, and they will always figure a way to hack the Net faster and better than our experts can design ways to stop them." —**Judith Siess**, president of Information Bridges International, Inc. and publisher and editor of the *One-Person Library* newsletter, author and blogger

"2020 is only 12 years away. We still may be coming out of this recession and the payback for the Iraq War. So who's paying for this? Nice idea, no surprise; programmers always argue for reprogramming from the bottom up. Good luck with that." —**Karen G. Schneider**, research and development, College Center for Library Automation, Tallahassee, FL; expert and thought leader in the library and technology community

"It is impossible to say today if we will see an evolution of Internet or a next Internet. We need to work on the two issues." —**Sebastien Bachollet**, president of the Internet Society of France, operates the European Global Event on Domain Names and Address Systems, known as EGENI, active participant in ICANN

"The Internet we have will be extended, but not replaced. Think of the roads in our country. They rarely get replaced, simply improved. There's not enough money, will, or need to make a new Internet. Plus, there are strong special interests that *wouldn't want* that, since their business is built around providing security and responding to threats. And the concept that some governmental/business body that could create an unhackable, unexploitable Web is laughable. Even if a site was somehow unhackable, that simply doesn't address the corruptibility of humans. We're suffering from the fallout of governments like Russia, with nuclear secrets leaking to nations with money and a desire to be nuclear superpowers." —**Buddy Scalera**, vice president for interactive content and market research for CommonHealth Qi, in charge of interactive online strategies, including social and viral marketing

"The more technology changes, the more it stays, in concept, the same." —**Stan Felder**, president and chief executive officer, Felder Communications, a marketing and advertising firm in Grand Rapids, MI

"It would be more costly to try to improve current infrastructure than to build a new one, much in the same way that it is less costly to build a new physical structure than to try to remodel an existing structure while you are trying to live and/or work in the present one. It only creates more expense and waste as you try to accommodate the users." —**Debbie Murray**, associate director, health education through extension, University of Kentucky

"The scenario pictured here seems very plausible. The very assumption that the Internet could be 'cleaned of terrorism' is inadequate because the Internet is primarily a communications medium. Just as telephone lines cannot be cleaned of terrorism, neither can the Internet. As for the other issues ranging from spam to phishing to viruses, the development will continue as it has so far: New challenges will arise all the time, defense mechanisms will evolve to counter and master those challenges. In a continuous process, the Web infrastructure will mature and evolve." —**Peter Bihr**, freelance consultant on Web strategies, communities, blogging, and social media; based in Berlin, Germany

"Corporations and governments still will be wrestling with how to adapt to the common man's technology. Corporations mostly will benefit by internalizing technology to create efficiencies of production and service. They will not have mastered how to adapt technology to a customer interface. If anything, a corporation's customer will become more independent and elusive for the corporation. Any corporate attempt to benefit by the collection of personal data will remain flawed. Government, however, will still be trying to figure out what all the x's and o's mean. Like the soon-to-be divorce, government will be the last to know." —**Eric James**, a respondent who chose not to share any other personal information

"An evolution scenario is much more credible than a 'revolution' scenario." —**João Miguel Rocha Filho**, director, DataOne, a provider of software for connecting to Linux; based in Brazil

"We add new highways to the current system of interstate travel because it makes economic sense or due to safety issues—same for the Internet. And just like our highways, some will still use the Internet (any Internet, current or future versions) to traffic illegal goods or activities. I would hope that there is a multinational patrol force to help police the Internet in the future." —**Teresa Hartman**, associate professor and head of education, University of Nebraska Medical Center

"In the history of communication technologies, new technologies provide new opportunities to innovate, but they also create new opportunities for fraud. Writing leads to genres like history and the essay, as well as the letter and the contract, but it also permits forgery. Printing, too, allowed the recreation of traditional documents and the creation of new kinds of writing: from the novel to the newspaper to paper money, but it, too, provided space for criminals and jokesters to trick or swindle us. So it's no surprise that the same thing happens with digitizing and the Internet. As we develop new ways to authentic online docs, cheaters develop new ways to circumvent our controls. The spiral will continue, and who's to say that a new, from-the-ground-up Internet will be unhackable? As in, remember the Titanic?" —**Dennis Baron**, professor of English and linguistics, University of Illinois, runs the Web of Language site and researches the technologies of communication

"Enhancement of present infrastructure is more likely than replacement, but it will only occur if a government administration is willing to support telecommunications companies by partially funding the overhaul and, more importantly, forcing them to follow through, as they've already received $200 billion in public funding to provide fiber optic to every home and have never done so." —**Jay Neely**, social strategist in the process of founding News Armada, a Boston-based company working to advance Internet-based news and commentary and community online

"The current Internet will become so full of spam and so infested with bots and malware that it will be almost unusable. Efforts to clean things up will fail and a new architecture will have to be developed." —**John Jobst**, IT specialist, U.S. Army Corps of Engineers

"The current-generation Internet is still being deployed. IPv6 has yet to take off. Bandwidth is still limited. It seems that the spam problem can be addressed, however, by 2020 (barring the use of quantum computing for spam)." —**Brian T. Nakamoto**, cofounder of MrJoy Inc. and product-line manager for Everyone.net (a leading provider of outsourced e-mail solutions for individuals and companies around the world)

"A complete replacement would seem to be a power grab, and the opposing outcry would be too loud and far-reaching to ignore." —**Woody Degan**, chief executive officer and operations director, Memphis Sound Entertainment; Consumer First Consulting, IT consulting

"Just as the width of modern railroads are based on the separation between the wheels of Roman ox carts, certain aspects of the basic structure of the original Internet will continue to underlie the Internet of the future: The need to maintain backward compatibility will ensure that this is the case." —**David F. Salisbury**, associate director for science and research communications, Vanderbilt University; formerly science and technology reporter for *Christian Science Monitor*

"By 2020, the next generations of search, security, and reliability will be much stronger, and we'll look back at the ways things are now in 2007 like it's the old black telephone with a megaphone mounted on an oak box on the wall with a town gossip switchboard operator connecting us…not quite tin cans, but almost!" —**Christopher Brown**, strategist and managing editor of new media for the U.S. television program *America's Most Wanted* on FOX

"Whether or not there is a next-generation Internet, there will always be ways of causing trouble…and, for this disruptive factor, we should be grateful. It pushes humans to grow in important directions." —**Mary Ann Allison**, principal, The Allison Group, has worked with Microsoft, Glasgow's Urban Learning Space, and other businesses, governments, and NGOs

"This is prediction of nonprogress. But for someone who recalls Gopher, Archie, and BBS [bulletin board system] posts, the Internet is not all that different today. Graphics and GUI [graphical user interface] aside, the underlying architecture is relatively constant." —**Richard Fowler**, auditor specialist, Northrop Grumman

"The underbelly will continue to grow, and the scum and villainy found there will expand to fill the available space. At the same time, the speeds will be increasing and the ability of constructive people to produce usable product will continue to increase. Replacing the Internet is not a good idea. Anything put up in its place WILL suffer from regulation and corporate interference and kill any innovation that might grow from the improved technology." —**Eric Kreider**, director of Web services, the University of Akron, OH

"While I am not in favor of cybercrime per se, I am also not in favor of someone else defining it for me and labeling agitators as criminals." —**Alex Don**, linguist and educator

"This continual process of refinement is in the best interests of the Internet. Crimes and mischief will cause their troubles but will not thwart the revolution in connected intelligence that the Internet has become." —**Barry K. Chudakov**, principal, the Chudakov Company, a marketing and advertising strategies creative consultant who has worked with many major corporations, including Microsoft and Disney

"Older people will be joking about the Internet connectivity like they do know about party lines and three-digit phone numbers." —**Robert Grant**, chief executive officer, VoyaCare, Inc.

"Like ancient cities built on top of older ones in the same location, it's not always an immediate transition." —**Hank Dearden**, director of business development, Digital Industry, Inc., a provider of technology services in the Washington, DC, area

"The rate of organic evolution in the Internet will continue at a high rate. Wholesale replacement will only occur under government fiat (think China), and will probably be killed by its extraordinarily and unavoidably high development and install cost." —**Jim Lucas**, Web manager, CACI, a provider of national security, defense, and intelligence-related solutions in the interests of the U.S.

"As the move towards more online commerce continues, there will be more motivation for hacking and cyber crimes." —**Kathryn K. Goldfarb**, president, KG Communications, an independent consultancy

"The issue is backward compatibility. Better technology causes disruptions. Government and corporations will have to learn that they can't legislate backward compatibility. Here's a quote: 'Standards should be discovered, not decreed,' from a computer scientist explaining why TCP/IP crushed OSI [Open Systems Interconnection], as described in the book *Where Wizards Stay Up Late: The Origins of the Internet*." —**Dick Davies**, partner, Project Management and Control, Inc.; past president of the Association of Information Technology Professionals

"One simple factoid, attributed to FCC member Kevin Martin: YouTube in 2008 takes up as much bandwidth as was used by the entire Internet in 2000. Net packing and P2P sharing and technology will only do so much before the pipes become clogged. Alternatives, such as broadband over power lines and Internet2 and over-the-air digital signals, will be needed to satisfy demand. Just as the Interstate highway system was created to supplement clogged state and federal roads even though the state and federal roads continue to be able to handle much of the traffic." —**Michael Castengera**, senior lecturer at the University of Georgia's Grady College and president of Media Strategies and Tactics Inc., a media consulting firm

"Like the switchover from analog TV to digital TV, the migration from the current Internet to a 'new-generation' version will take a very long time because it has to satisfy (and not harm) so many different sectors. The vested interests in the current system will stand for gradual improvements to it, but it is quite difficult to imagine a radically different system which would take a quantum leap into a whole new world of communication. That day will come, surely; but not likely by 2020." —**Fredric M. Litto**, consultant for Pearson Education Global e-Learning, president, Brazil Distance Learning Association

Anonymous Comments

The following section includes a selection of contributions to the discussion of the evolution of the architecture of the Internet from survey participants who chose to remain anonymous. Among them:

"Current Internet will solve its own problems through evolution. Governments, by definition, are always second generation and, to that degree, self-defeating. Security is a trade-off. Having antivirus, antispyware, firewalled computers is a trade-off for living in a Microsoft world. Trade some of your touchy-feeling conveniences for a more robust (and community, not government-evolved) system like Linux/BSD, and you have more robust security."

"There will be two Internets—one for government, libraries, education, where you have to have a valid username linked to your 'real-life' identity, and a second, Wild West Internet for the viruses, phishing, pr0n, recreation. [*pr0n* is a slang term for "porn" that originated when spam filters began searching for and destroying messages with "porn" in the subject line.]"

"Internet2 will always be academic. Actual Internet and its technologies (TCP/IP) will still be with us for a long, long time."

"Totally agree. There's way too much investment needed to do an overhaul, so next-gen Internet will be more of an overlay. I could see the key backbones being upgraded, but all of the other 'pipes' require so much more. WiMAX (wireless broadband) infrastructure will increase and help, but that is also an 'overlay' technology, not new infrastructure."

"It's hard to imagine how a next-generation Internet will eliminate viruses, spam, phishing, and worms."

"The Internet is now too large and ubiquitous for a 'flag day.' The results of 'next-generation' or 'clean-slate' projects are therefore likely to have their primary impact on the current Internet (IPv4 + IPv6) than on a greenfield replacement for it."

"GENI's purpose is certainly to improve communications—its goal is not to replace the Internet (per se). If that happens, great, but it isn't GENI's purpose."

"Even if a brand-new system was invented, it wouldn't be a substitute for the current one by 2020."

"The roads may be improved, but they'll follow the same routes."

"We'll see a lot of spectrum open up, and governments will begin using those in different ways. Will this be 'one Internet'? What counts as 'the Internet'? Arguably, there are many Internets right now. Regardless, a lot of social divisions result from people not experiencing the same Internet. Still, things will improve on all fronts with regard to security and reliability and search."

"I see a two- or three-tiered system evolving."

"Virtual worlds and 3-D presence will create another layer of identity and trust to our online interactions."

"GENI is not unlike the DARTNET, CAIRN, and PlanetLab of the past. They will be used to prove out ideas, but will not become the fundamental basis. That said, yes, there will have been great improvements in Internet service."

"The Internet has, from the beginning, grown by evolving rather than through revolutionary change, and I don't see this changing. Previous attempts to 'replace' the Internet have failed miserably."

"I doubt that anyone would support having to move to an entirely new system of Internet just because of the huge hassle."

"Thirteen years is not enough time to replace the Internet and access networks. IPv6 will perhaps be well deployed, and we will have new technology from this R&D to make the Internet more secure, etc. But a completely new infrastructure will be unlikely."

"The Internet is too big and is too decentralized to be upgraded to a new system. We'll be lucky if we're at IPv6 by then."

"The complexity and dynamism of the current Internet makes it beyond centralised top-down control or straightforward replacement."

"The ones causing the troubles are going to refine their efforts at computer crime, making it all the worse."

"The biggest error here is thinking the National Science Foundation is doing anything. *China!* Or perhaps a China vs. U.S. virtual war, but then it wouldn't be NSF, it would be the Department of Defense running the Internet, which isn't that different than how it got started. The Internet will always be hackable and open to crime...That it might be the battle-ground between countries or insurgencies is totally believable. So not just individual criminals but collectives are acting as criminals."

"A completely new architecture will be built, but it will be more controlled and the current Internet will be the underground."

"A global transition to a completely different Internet architecture will not happen in the foreseeable future."

"I suspect that there will be investment in dedicated safe networks driven by commercial demand, but the cost will delay introduction. And the benefits are key. Incidentally, I am reminded of one of Parkinson's laws: When [the Internet] is perfect for [what people use it for], it's dead (i.e., people are using something else, innovation and change are no longer happening). So almost in introduction, the next-generation, perfect Internet will have to cope with things it was not designed to do."

"New research will yield better designs but not better by enough to mobilize sufficiently enough to permit these new designs to get much beyond the labs before some patch to the existing system sucks the wind out of them. There is too much invested in the current infrastructure to allow any change that is not hugely beneficial."

"There will be technologies to augment the existing Internet, not replace it."

"Well, yeah. After all, you can still plug a 1930 phone into the current network, and it'll still work."

"There will be new ways to prevent and detect online crimes and mischief."

"There will always be bad people, as well as government officials wanting too much control."

"The analogy I would use is the human body. We replace our parts in a continuous stream, with cells dying and new ones taking their place. The management of the Internet will function similarly—we will never flip 'the switch' to a completely new Internet, but rather will keep changing out parts as technology advances."

"History teaches that installed processes and investments are terrifically difficult to replace in the absence of radical changes in functionality…Incrementalism will win out."

"Due to costs and potential civil liberties challenges, the next-generation Internet architecture will likely not appear by 2020."

"Couldn't agree more. We now have a billion embedded users."

"Open system a must—this is a major threat to governments in general, and especially communist and controlled governments. I see a huge effort to develop a new network with much more controls—and in this case, the horse is out of the barn, and it will be very difficult and expensive for any alliance to change the basic structure."

"The disruption from replacement would be catastrophic."

"2020 is really not that far away in terms of revolutionizing the architecture of the Internet. Evolution is more likely."

"The current Internet was not designed to fulfill the needs of the future. We need a new one."

"Developing a second Internet would cost too much time and money, and would ultimately fall victim to the same problems the current one does."

"Those who want to commit crimes and mischief are *always* going to be able to cause trouble if they try hard enough, no matter if it's a new system or an old system. New systems simply have new bugs to be discovered and worked around."

"It will change for sure. But will it still be the Internet? Could be something else entirely. After all, in 1982 nobody could have predicted the Internet (as it has evolved today) at all!"

"It is hard to see a complete replacement of a system in which there has been significant investment; particularly with the current economy. Current investments in cyberinfrastructure do not seem to be going that way in the U.S."

"Social tagging and other social research applications will play a huge role here."

"Crime has always been a feature of human life and will most likely continue in some form or another."

"Though 95% of the population will use the Internet for good, the 5% who choose to use it for evil will make security a continued issue. Those who have not been able to connect meaningfully with others may also find themselves so disenfranchised that they behave in Columbine kinds of ways."

"I don't think we can even imagine what the format will be 12 years from now. I did not think in 1982 that I would ever be able to afford a computer, yet within 3 years I was paying for one on time."

"Intelligent crime will increase with technological advancements."

"Present IPv4 will be replaced by IPv6 by 2011. After that, research into a new network will start, and by 2018, a new network will be started up."

"This sounds like the same chance as me having a flying DeLorean."

"It really is beyond government control—or at least beyond any particular government to reform."

"It's human nature…someone always tries to cause mischief in most systems, from the errant school kid to the sophisticated hacker."

"Consider what the Internet was only 5 years ago."

"Too much can be achieved by replacing the existing Net not to have built the future Net."

"Historically, we tend to improve what works for infrastructure rather than replace—and the Internet works just fine."

"We will continue to improve the current technology, not replace it. However, over time, the improvements—due to their complexity—will appear as though the entire system had been replaced."

"There will always be miscreants, but efforts to improve the current Internet are ongoing. (Highly irritating to think of all the resources wasted on combating spam!)"

"Replacing the Internet is impractical. The power of the network is its diversity. The new tools will just have to integrate to the current network."

"The technology may be there, but the 'powers that be' will have difficulty cooperating and power sharing."

"If a 'next-generation' Internet is developed, it may be used for some types of interaction, but people will still prefer the original."

"I can't see that the funds would be available for this by 2020."

"TCP/IP may be replaced by something far more efficient. The GUI layer will change. But, I don't think we're all going to go to bed one night and wake up the next morning with a new Internet."

"A complete overhaul/change in the system would require more funding and disruption than the public could handle."

"Enhancements are good as long as they don't infringe on civil liberties. There really is not a way to completely squelch mischief, thankfully."

"The Internet of today is nothing like the Internet of 10 years ago. In 10 more years, I can't imagine that it would be the same as now."

"Of course it will. The Internet has become the primary source of research in the last 10 to 15 years. Though that still does not guarantee substance of content or its validity."

"I think it possible that something new will evolve."

"There could be multiple systems."

"Technological evolution will follow the model of biological evolution, building more complex structures on top of the simpler base without eliminating it."

"Because the original Internet infrastructure only reaches the remotest areas in 2018, it will be a while before the NGI (next-generation Internet) replaces it."

"For a minority, a new system may be being used, but most people are still trying to come to grips with the current system and will not be open to a new system by 2020."

"It will evolve, not replace. We have come a long way, but more research is needed."

"There will always be a need for someone to build a better mousetrap."

"I hope everyone will fight against the build of a whole new network with whole new rules."

"First, they have to agree on what to do."

"If you build a better mousetrap, humans will still find a way to outsmart it."

"'Those who want to commit crimes and mischief' will *always* be able to cause trouble. And most people would prefer the existing infrastructure to one where 'the trains run on time.'"

"The Internet routes around censorship; it also routes around security. People are lazy and gullible. That won't have changed by 2020, so security threats will still exist."

"It is too difficult to put the genie back in the bottle."

"New technologies; new hacks. A totally new Internet will require new technological standards, and it takes time."

"*Plus ca change, plus c'est la même chose*. Hopefully, we will have better controls in place to combat issues like spam and viruses. But there will always be evil ones seeking to exploit weaknesses."

"Participation culture—like Google, for example—will create a seamless interface for ongoing development."

"A second secure network will surely be developed as a response to hackers."

"It is human nature to attempt to disrupt systems such as the Internet. If humans can invent it, then humans can disrupt it."

"Yes, the 'last mile' may be WiMAX or fiber, but the essential structure will remain."

"I see the current Internet following Milton Friedman, libertarian economics, and eventually we will be able to effectively police the community and create a world where there is more opportunity through the legitimate channels. As Internet giants become bigger and bigger, they will stay on top of research trends, and any innovative technologies will immediately be introduced through our current Internet."

"There is a move afoot to examine popularly held distinctions between crime and creative rights."

"Since there is so much dependency on the Internet (phones, medical awareness, communication with others, conducting business), having a new Internet seems unlikely."

"Just like the health care system that is so engrained in our society, the fundamental architecture of the Internet will not be changed. There would be too much pushback."

"The idea of a 'clean' Web is a long way off. Although there are certain aspects of the Internet that almost everyone would like to see disappear, I doubt that it will happen soon."

"It will take a very long time before there will be some massive change to the current architecture—largely because it is not only the U.S. and Europe that are online."

"Most clean-slate design efforts are fundamentally flawed in that the present set of problems of the Internet are the other side of the very strengths of the Internet, openness, intelligence on the edge, flexibility, low cost, etc. You can have all both ways. There are problems with the computer-host environment as well. Thus, addressing the problem at the right place and striking a good balance of the up- and downsides of the same important character of the Internet should be the focus. The Internet is no different from any other way to communicate and share information."

"The government has time and again demonstrated its total incompetence in the construction and management of technology; we should want to keep them as far from it as possible."

"With the work in virtual firewalls advancing, I don't see the need to create a whole new Internet. It's always been a dynamic entity."

"Spy vs. spy."

"Governments will build the next Internet just as they built the one we have and, this time, having had a glimpse of what it can do, they will look to develop it to their advantage."

"I doubt the world's governments and commercial entities will ever be sufficiently coordinated to enable the generation of a new Internet. The changes are far more likely to be patchworks as entities attempt to fit acquisitions into their existing structures."

"Create a new Internet? The current one is moving too quickly, and I don't foresee anyone catching up and creating something people would use, not to mention the fact that this would require international governments working together. More likely, the current Internet will continue to exist as some sort of cool underground where you can do what you want."

"2020 isn't that far off. Things don't change that quickly. 1996 is as far off as 2020, and the network isn't substantially different."

"There is no fence tall enough for a man who wants to cross it. It will be too costly and impractical to build a new Internet from scratch since it is impossible to ignore the importance of the current Internet, as imperfect as it is."

"Physically, the Internet will continue to be a network of networks. However, there will be some safe tunnels for corporations and other groups through the current networks that provide better speed and assured authentication. These safe zones will blend seamlessly with the rest of the Internet. New forms of viruses and security breaches will be problematic. Law enforcement will have special units to police online crime."

ENDNOTES

1. From *Usage and Population Statistics, World Internet Users*, by Internet World Stats, published by Miniwatts Marketing Group, June 2008, available at http://www.internetworldstats.com/stats.htm.
2. See two reports led by David Clark: *New Arch: Future-Generation Internet Architecture*, a 2003 DARPA report, available at http://www.isi.edu/newarch/iDOCS/final.finalreport.pdf; and *Tussle in Cyberspace: Defining Tomorrow's Internet*, a 2002 SIGCOMM conference

paper, available at http://conferences.sigcomm.org/sigcomm/2002/papers/tussle.pdf. Also see GENI information, available at http://www.geni.net/documents.html.

3. See "Paving the Way for IPv6: Community Meets to Discuss the Challenges," from the *Internet Engineering Task Force Journal*, volume 3, issue 2 (October 2007), available at http://www.isoc.org/tools/blogs/ietfjournal/?p=151#more-151.
4. See *Year-End 2007 Monthly Internet Traffic Estimate*, by Minnesota Internet Traffic Studies, available at http://www.dtc.umn.edu/mints/home.html. In making the estimate, extrapolations were leveraged based on snippets of information, including input from 100 of the largest public Internet exchanges in the world.
5. From the Netcraft June 2008 Web Server Survey, available at http://news.netcraft.com/archives/web_server_survey.html. Reports are filed monthly. The April 2008 survey report indicated that Google's blogger service alone added 1.1 million extra sites.
6. See *Tetherless World Research Constellation*, published by the Rensselaer Polytechnic Institute, June 11, 2008, available at http://tw.rpi.edu/launch/; and the main Web site for Tetherless World, a wiki page, available at http://tw.rpi.edu/wiki/index.php/Main_Page.
7. See *Metacrap: Putting the Torch to Seven Straw-Men of the Meta-Utopia*, by Cory Doctorow, published August 26, 2001, available at http://www.well.com/~doctorow/metacrap.htm.
8. See *The Semantic Web, Syllogism, and Worldview*, by Clay Shirky, first published November 7, 2003, on the Networks, Economics, and Culture mailing list, available at http://www.shirky.com/writings/semantic_syllogism.html.
9. See *Joint Project Agreement Between the U.S. Department of Commerce and the Internet Corporation for Assigned Names and Numbers*, 2006, available at http://www.icann.org/en/general/JPA-29sep06.pdf.
10. See *Glasnost: Results From Tests for BitTorrent Traffic Blocking*, published by the Max Planck Institute for Software Systems, May 2008, available at http://broadband.mpi-sws.mpg.de/transparency/results/.
11. See *The Internet Singularity Delayed: Why Limits in Internet Capacity Will Stifle Innovation on the Web*, published by Nemertes Research, November 2007, available at http://www.nemertes.com/internet_singularity_delayed_why_limits_internet_capacity_will_stifle_innovation_web. The 70-page report indicated that by 2010 the Internet's capacity will not accommodate user demand, resulting in "brownouts" or interruptions to applications and longer download times.

12. See *Estimating the Exaflood: The Impact of Video and Rich Media on the Internet—A Zettabyte by 2015?* by George Gilder and Bret Swanson of the Discovery Institute, January 29, 2008, available at http://www.discovery.org/a/4428. Estimates include that by 2015 video calling and virtual windows could generate 400 exabytes, video gaming and virtual worlds could total 200 exabytes, movie downloads and P2P file sharing could total 100 exabytes, and growth will make the U.S. Internet 50 times larger than in 2006.
13. See *Spyware: The Threat of Unwanted Software Programs Is Changing the Way People Use the Internet*, by Susannah Fox for the Pew Internet & American Life Project, published July 6, 2005, available at http://www.pewinternet.org/ppf/r/160/report_display.asp.
14. See *Report: The State of the Internet, by Akamai*, April 2008, available at http://www.akamai.com/stateoftheinternet/.
15. See the Organisation for Economic Cooperation and Development (OECD) Broadband Portal, http://www.oecd.org/document/54/0,3343,en_2649_33703_38690102_1_1_1_1,00.html. New reports are continually added to this site. OECD emphasizes the importance of competition in markets. The number of broadband connections in rural and dispersed locations is improving, but the quality and consistency of the connections varies.
16. See *Universal, Affordable Broadband for All Americans*, by Jim Kohlenberger for the Benton Foundation, published in September 2007, available at http://www.benton.org/benton_files/BroadbandForAll.pdf.

Scenario 8

The Evolving Concept of Time for Work, Leisure

Prediction: *Few lines divide professional time from personal time, and that is OK. In 2020 well-connected knowledge workers in more developed nations have willingly eliminated the industrial-age boundaries between work hours and personal time. Outside of formally scheduled activities, work and play are seamlessly integrated in most of these workers' lives. This is a net positive for people. They blend personal/professional duties wherever they happen to be when they are called upon to perform them—from their homes, the gym, the mall, a library, and possibly even their company's communal meeting space, which may exist in a new virtual-reality format.*

Expert Respondents' Reactions	
Mostly Agree	56%
Mostly Disagree	29%
Did Not Respond	15%

All Respondents' Reactions	
Mostly Agree	57%
Mostly Disagree	29%
Did Not Respond	14%

Note. Since results are based on a nonrandom sample, a margin of error cannot be computed. The "prediction" was composed to elicit responses and is not a formal forecast.

OVERVIEW OF RESPONDENTS' REACTIONS: The vast majority of respondents agreed with every aspect of the scenario except for the "net-positive" outcome; this is where the debate was centered in the written elaborations. Some people are hopeful about a hyper-connected future that they said will offer more freedom, flexibility, better mental health, and positive life improvement. Others expressed fears that mobility and ubiquity will be a burden in an always-on world that causes stress and could lead to the disintegration of family and social life. They said hyperconnectivity might lead to oppressive surveillance by bosses and government. Other observations from respondents: People will rebel against corporate control of their lives. Workers and institutions will have to draw boundaries. Successful employers will adjust by taking holistic approaches that might focus more on work output (projects completed) than input (amount of time in the cubicle). Because work infiltrates every corner of life, people will be motivated to pursue satisfying employment, rather than settling for a "job." Deepened personal networks will strengthen professional outcomes. The workforce will be more dispersed. There will be an increase in divorce. People will not take the time to enjoy nurture or nature.

While 29% disagreed, most respondents mostly agreed that by 2020 the formalized delineation of social, personal, and work time will be eliminated for knowledge workers in the world's most developed areas, and this will generally be a positive change. There were varied responses about the pluses and minuses of the "always-on" environment. Most of the people who wrote elaborations exhibited concerns about the potential negatives of hyperconnectivity.

The following anonymous responses are a sampling of typical attitudes and commonly held views:

- "What a nightmare! It's bad enough already, with 24-hour e-mail responses expected."
- "What's going to happen to focus?"
- "Agree…You can be enjoying deep-sea fishing as you do your stock quotes."
- "It will increase the number of people involved in freelance employment."
- "If this takes place, you'll find me in a less developed country where my time is MINE."
- "People will work more from home and remotely instead of wasting time commuting to cubicle hell."
- "It will not be a net positive for anybody but type As and geeks—people who didn't have a social life in the first place."
- "As corporations expand their demands and intrusions into employees' personal time, workers will eventually rebel."
- "It's already happened, for better or worse. Get over it."

Respondents noted that work and play evolve as humans and their tools do, and they pointed out that set "workdays" are a recent human concept. "This was basically life for my great grandfather in northern Sweden and my grandfather in Nebraska," wrote an anonymous survey participant. "Work and life were not separate events."

"The 9-to-5 approach will disappear completely, with few exceptions," responded **Roberto Gaetano**, ICANN board member. "The current separation between 'work time' and 'free time' is a by-product of the industrial revolution, and is bound to disappear with it. Whether this is positive or negative, I don't know, because the pressure of being 'always at work' just because you have the ability to be 'always connected' will be high. But we would need to build a new way of life that has to cope with this."

"The boundaries between work and home, or private life, will have been transformed," wrote **Oscar Gandy**, author, activist, and emeritus professor of communication at the University of Pennsylvania, warning, "The sorts of stress-related illnesses that we see will be astounding."

Nicholas Carr, author of *The Big Switch: Rewiring the World, From Edison to Google*, noted that hyperconnectivity is already reality for some people, writing that it is a net positive for corporations and will cause "the expansion of the work to encompass all time and all space."

Gbenga Sesan, an Internet-for-development consultant for Paradigm Initiative in Nigeria, saw positives, responding, "Even those who live in developing (or underdeveloped) nations will be able to overcome the barrier of geography through Internet access and other connected devices. It may be 'plug-and-pray' and not 'plug-and-play,' but it plugs anyway! It's now 4:05 a.m. in Lagos, Nigeria, and I'm asking myself if everything I've done in the last 5 hours will count as work, rest, play, or sleep-mode tasks. In 2020 professional and personal time will be as far from each other as fingers from the keys on a mobile phone. Multitasking will no longer mean driving and talking alone, but it will include work and play at the same time."

Jerry Michalski, founder and president of Sociate, formerly of *Release 1.0* and cohost of the PC Forum, wrote, "It's healthy to have flextime and other ways to work when you're sharpest and avoid temporal hassles like rush hour. Some people like to keep their work and private lives very separate; they will find this new world hostile. I'm on the other side—seldom not thinking about the things I care about professionally. One big caveat: We have to have a better 'Do Not Disturb' function. Without it, we are all at the end of electronic leashes, and a major backlash will be much more likely."

And **Charles Kenny**, senior economist for the World Bank, expert on technology and economics, noted, "I hope 200 years' worth of social progress towards the paid holiday doesn't end like this."

Tom Jennings, creator of FidoNet and builder of *Wired* magazine's first online presence, reflected the feelings of many survey respondents who disagreed with the scenario, writing, "We *work more*, *work more* from home, take *more work* home, and are, overall, paid less for it. Notice that all cars have cup holders now; the extreme technology for such exotic things existed in 1960, only no one wanted them! We drank coffee at home or in a café or restaurant!"

A high percentage of the respondents who wrote explanatory elaborations to this scenario used the phrase "this is already happening" or something along similar lines. Of course, the people invited to participate in the *Future of the Internet III* survey are well-informed, technology-savvy knowledge workers, so many are living hyperconnected lives. An example of one of many dozens of responses in this vein comes from **Louis Naugès**, president of Revevol, an enterprise 2.0 company with offices in France, Spain, the United Kingdom, and the United States, who wrote, "Already there! This is the way 100% of our employees work at Revevol, our company; 1Gbit/sec.-minimum networks, wired and mobile, available anywhere, anytime on any device will make this one a no-brainer."

Susan Crawford, founder of OneWebDay and ICANN board member, commented, "It's just how our lives work. Somehow we've got to figure out how to fit in 8–9 hours of sleep a night as well, just so we won't hurl our ever-present handsets against the wall."

And **Christine Boese**, a researcher and analyst for Avenue A-Razorfish and Microsoft, wrote, "While *I* have few lines dividing my professional from personal time, and I love my life that way, everyone I know has clearly and emphatically communicated to me that they *strongly* demark their personal and professional time, and only allow the professional to intrude with the greatest reluctance. They are not embracing this world I live in, and when I think about it, I have always been this way, long before technology ever came to dominate my life, when I lived as deeply inside books and personal projects that consumed my life then, just as they do now. I could surely do with a little less 'helpful' discipline from them telling me to 'get a life,' however."

HYPERCONNECTIVITY ALLOWS PEOPLE TO BE PRODUCTIVE ACROSS MANY ASPECTS OF THEIR LIVES

Respondents who are already integrating work and personal time in jobs they enjoy and as members of families who do not mind the integration happily responded that such connectedness will be a net positive. "We are enjoying the benefits even when I am at my birth place, a remote village—Ikrail in Bangladesh," wrote **Professor Lutfor Rahman**, chairman of the department of computer science at Stamford University, Bangladesh, and a leader of the Association for Advancement of Information Technology.

An anonymous respondent noted, "Remote teleworking will become commonplace and second nature, giving the average American the equivalent of 3 weeks of vacation time no longer spent commuting."

Havi Hoffman, senior editor for product development at Yahoo and blogger, responded, "Perhaps this is the latest and most distributed version of an aristocracy human civilization has developed. Bestsellers like *The Four-Hour Work Week* are bellwethers of this trend. This rulers' club, though, is widening and becoming more diverse. I bet if a person had access to a database of Davos World Economic Forum attendees over the last 10–12 years, one would see a pattern of greater diversity, greater inclusion of people more removed from seats of power, but still connected and influential in part because of their significance in the social OS that is growing like a social commons of metadata about our relationships, our expertise, our causes and passions. The value of weak ties and the portability of connectedness make this work/play continuum possible if not probable. It won't be true for everybody and the divide between the elite and the poor/the 'underclass' could continue to grow."

Cliff Figallo, founding member of the first online community, The WELL, now of AdaptLocal.org, wrote, "The world is increasingly characterized by uncertainty, so people refuse to divide their lives into professional and personal. Staying connected and informed is the security blanket that people demand." An anonymous survey respondent wrote that in 2020 "life happens, with graceful sliding between aspects as the need arises."

People on both the pro and con sides of hyperconnectivity said it will influence people's health. While those who fear it said it will cause stress-related illnesses, those who welcome it said the flexibility it offers may improve mental health.

Christine Satchell, a senior researcher at the Institute for Creative Industries and Innovation at Queensland University of Technology, responded, "People can work when they are at their best, and by allowing them to mix professional and personal duties, they can spend longer periods of time in front of their machines, actually accomplish more work and get less burnt out." An anonymous respondent commented, "This is a great vision for knowledge workers, and can cause reduced stress and improved health."

Micheál Ó Foghlú, research director, Telecommunications Software & Systems Group, Waterford Institute of Technology, noted, "It would be better to think in terms of more people having more professional attitudes to work where more emphasis is on outputs and less on just turning up and signing in. This does not mean that private time disappears." An anonymous respondent commented, "Companies have found that people are more efficient when the company enables personal time and, in fact, becomes part of their social environment. People are social, after all."

Michael Castengera, a senior lecturer at the University of Georgia's Grady College and president of Media Strategies and Tactics Inc., noted, "Many, if not most, people derive their identity from what they do. It defines who they are. The blending of personal and professional existence will be heightened by the Internet connections."

Some survey participants predicted that technological change will help people find more satisfying work that they will then enjoy interlacing throughout their lives. "People will begin to do more 'work' that suits them rather than just having jobs," wrote an anonymous respondent. And another noted, "This will take time, but I believe the real-world interaction of the Internet and the 'long tail' of the Semantic Web will enable everyone to find the perfect job for them, the right opportunity, so that people in general are happier, healthier, and more productive."

Some respondents predicted that the future workforce will prefer a blur of work and personal life. "Flexible, technology-based work environments will be attractive to next-gen workers," wrote **Michael Stephens**, an assistant professor at Dominican University in River Forest, IL. "The benefit of this is improved productivity, happy workers, and increased return on investment."

Peter Bihr, a freelance consultant on Web strategies, communities, blogging, and social media, based in Berlin, responded, "An optimistic estimate could even go so far as to assume that those well-connected knowledge workers—who are privileged by definition—will be able to work less as their contributions are so efficiently managed that, say, 6 hours of net work will be the norm instead of, say, 9 hours."

Jamie Richard Wilson, a journalist and freelance Web developer, predicted, "Nearly every knowledge worker will have at least one virtual personal assistant, a computer that is completely capable of performing time-consuming and mundane tasks. A lot of work will be voice dictated, allowing workers to 'work' from virtually anywhere. Handheld computer technology will be portable enough that it can be carried in a pocket, and most data storage and software will be kept online, leaving workers free to literally take their work with them wherever they go."

And **Hamish MacEwan**, a consultant for Open ICT in New Zealand, commented, "The 9 to 5 of the industrial era was required so worker units, generated by homogeneous 'education' that set strict times for functions, would be available to manipulate tangible products. Where we seek ideas and thought, there is no schedule."

Hyperconnectivity Will Create Unrealistic Work Expectations and Stress, and Intrude on Lives

Many people saw hyperconnectivy as a threat. Among the hundreds of elaborations provided by the respondents who chose to explain why they agreed or disagreed with the scenario, only a few people perceived that blending work and personal time would tilt people's lives toward more time for family, friends, and personal pursuits. One of the only examples of a respondent foreseeing employers losing time to people's personal pursuits: "Are you kidding me?" asked an anonymous respondent, "Work is work, and play is play. Can you be happy when your employees are integrating PLAY with work seamlessly?"

The vast majority of respondents who wrote elaborations equated hyperconnectivity with more work, not more play. Those who agreed with the scenario and saw it as a net positive tended to be people who also noted that they enjoy their work lives and find connection to be valuable. Those who predicted that work will impinge on personal time primarily perceived employers as profit oriented, not people oriented.

"Most employees are essentially being exploited to earn shareholders huge amounts of money, and, unless cooperatives and profit-sharing all of the sudden become the norm, employees are not likely to feel that it's 'OK' to work on their own time," noted an anonymous respondent.

Another anonymous survey participant wrote, "Seems like a capitalist wet dream and a Marxist's nightmare. I pray it doesn't happen." And another anonymous respondent wrote, "This structure will dominate, benefiting younger and geekier workers but devastating older workers, especially women and family-centered communities. There will be a significant cost for this shift, primarily in the form of cultural coherence and social norms. It will fuel divisions in the U.S. and abroad, increase illness and disease, and otherwise make people miserable. It's coming, and the cost will be tremendous. (None of it will be about virtual reality.)"

Another anonymous respondent explained, "People died in the 1800s in the fight against an 80-hour workweek to get a 40-hour workweek. When people in the 1980s decided it was 'cool' to consult or telecommute, the workweek grew back to Victorian-factory size. An 80-hour workweek is now common again. Individuals do best when they have collective bargaining like labor unions to fight back against the demands of big business."

Benjamin Ben-Baruch, senior market intelligence consultant and applied sociologist for Aquent, commented, "In 2020…a myth will develop that outside of formally scheduled activities, work and play can be seamlessly integrated in most of these workers' lives. Employers will attempt to convince us that this is a net positive for people because we will be able to blend personal/professional duties…However, the reality will be quite different. Because we can be surveilled whenever we are 'connected,' and especially because we can be surveilled whenever we are connected using our employer-provided devices, we can and will be controlled. Our employers will gain even more control over work-time discipline and over our lives and will be able to force even more productive working hours from us. Our lives will, in fact, be increasingly controlled by those who provide us with the devices that will have become increasingly necessary for us in both our work and personal lives, as well as those who own and control the networks and network sites that we use and visit. Some companies will try to distinguish themselves as companies that do not actually use their power to watch and control us—but most companies will do the 'fiscally responsible' thing of using available technology to assert control."

Concern over surveillance was also the central concept in the elaboration from **Steve Sawyer**, an associate professor in the college of information sciences and technology at Penn State University. Sawyer's research includes the uptake and uses of computing by knowledge workers. In his 2020 scenario, "corporate control of workers' time—in the guise of work/ family balance—now extends to detailed monitoring of when people are on and off work. The company town is replaced by 'company time management,' and it is work time that drives all other time uses. This dystopia challenges the concept of white-collar work, and unionism is increasingly an issue."

Brian Dunbar, an Internet manager for NASA, commented, "It will be a means of requiring workers to give up ultimate control of their lives to employers; to always be on call and connected. The social fabric of family and friends will be further disrupted."

Charles Ess, a professor of philosophy and religion and research on online culture and ethics at Drury University, responded, "This might be a positive scenario for some in the U.S. and, perhaps, Japan. But, for example, in Europe and Scandinavia, there is considerable resistance to what is seen as the American model of working more and more and having less and less of a life. People may be forced into blurring the boundaries between the personal and the professional for economic reasons, but they're not happy about it and do not see it as a positive. An alternative scenario is to see the well-connected knowledge worker described here as simply a drone in the Borg hive: always connected, never free to be/do anything other than contribute to the collective. Upper managers who keep their BlackBerrys and Treos by their side for the 4:30 a.m. phone calls, even during 'vacation,' already come close to this depiction. Those on the outside who enjoy at least an occasional freedom from the Net would see such a drone as a slave, not as a free human being."

Joanna Sharpe, senior marketing manager for Microsoft, commented, "When people are too blended in the mashup between work and play, they are missing valuable time and experiences that probably shouldn't be preempted by a work need, i.e., an important event being with your family or friends and working at the same time, so both groups suffer due to lack of focused attention."

Victoria Nash, director of graduate studies and policy and research officer at the Oxford Internet Institute, responded, "The result may be longer, less efficient working hours, and more stressful home life."

Scott Smith, principal at Changeist LLC and a consultant, futurist, and writer, noted, "Evidence is mounting that blended work/play scenarios enabled by pervasive connectivity aren't a net positive for many able to experience this blend today. Access opens the door to time pressure, the need to respond, and expectations of 24/7 productivity. It isn't clear how this will change for the better in 13 years' time." **Hal Varian**, chief economist at Google, shared a similar sentiment and added, "Institutions will have to be proactive in drawing some boundaries; burnout is real."

A number of respondents noted cultural differences and said they expect more negative effects in the United States than in other places.

"This is perhaps the scariest development of all for workers in the U.S.," wrote an anonymous survey respondent. "In other cultures, where leisure and relaxation are more valued and respected than they are here, the blurring of work and playtimes will be a net positive, but in American culture, this will make work even more of a beast that cannot be sufficiently fed, and we will see increasing problems in our society with mental and physical health related to overwork."

Alejandro Pisanty, director of computer services at the Universidad Nacional Autónoma de México, an active leader in ICANN, the Internet Society, and the Internet Governance Forum, presented this 2020 scenario: "Those who have not been absorbed by the madness will forcefully try to make their cofamilials escape the vertigo and take walks, swim, what have you, with a clear separation of work and fun. Memories of 19th-century analyses of the relation between worker and owner will fade but not die. Payment will not be proportional to commitment. This always-on commitment to work will have become a standard, and those who are even conscious of the division between work and rest will carry holograms of Winston Smith and place them on their all-pervasive ID card-weblog-Facebook page, albeit steganographically in order not to fall afoul of the powers that be."

WE WILL ADJUST, DEVISING NEW WAYS TO BALANCE LIFE AS WELL AS WE CAN

Many who expect the future depicted in the scenario commented that social adjustments will be made to deal with the new realities. **Brad Templeton**, chairman of the Electronic Frontier Foundation, wrote that he expects people will "develop tools to isolate personal time more effectively, and only have it pierced when truly urgent—people will come to accept that."

Howard Rheingold, Internet sociologist, university professor, and author, noted, "We're beginning to see people finally erecting personal and social boundaries around the use of mobile technologies because the colonization of every sphere of our lives—homes, cars, family life, social events, toilets, movie theaters, concert halls, subways, classrooms—of these devices is beginning to make people angry."

An anonymous respondent explained that we are already in a transition toward a healthy assimilation of new realities. "In this honeymoon, individuals are so pleased with the technology, they tend to overuse it, creating a blur between private and public. As maturity sets in, so does the desire for, perhaps it could be called, private time…The excitement of the honeymoon era will transform into an alienation era or backlash era where people will find themselves unsettled, striving to regain quality in life—and this will lead to a mature era where a greater sense of ethic is developed, creating a health personal space that is entered when appropriate and not at the yank of a chain."

Ross Rader, a director at Tucows Inc. and member of ICANN's Registrar's Constituency, part of the Generic Names Supporting Organization, responded, "Work flow, organizational structures, and personal space will all need to be reconfigured to properly deal with the implications of removing this division, and this is the activity that will take place over the next 12 years."

Mary Ann Allison, principal of The Allison Group, predicted, "We will have adapted to this blurring—which might otherwise be termed integration…and, at the same time, will have many widely used and 'approved' time-out activities ranging from 'no-contact' vacations to official 'no-schedule' times in organizations' workday structures."

An anonymous respondent expects technology to help people keep things in balance, explaining, "While work or play will be accessible from any location in the future, stronger delineations between work/play will appear in 2020 and will be held in place by automated tools that reroute or turn off one's accessibility."

Rollie Cole, director of technology policy for the Sagamore Institute for Policy Research, suggested that the scenario will "not be an unmixed blessing," adding, "I could see a backlash leading to regulations about 'no-employer-contact' hours."

An anonymous respondent commented that employers will come to measure employee contributions in new ways, explaining this vision of 2020: "Employees will be rewarded based on their contribution to the corporate knowledge and information space either through the creation of knowledge content or through their personal connections. Those people who are good 'people people' will continue to maintain and grow their personal networks through a mix of online and offline interactions. Because virtual meetings have become so easy, people will place a premium on taking time to meet in person."

A number of survey participants suggested that the nature of work is going to change for the positive.

"I hope that future work activities will become more creative and fun for people," wrote **Gary Kreps**, chair of the department of communication at George Mason University, formerly founding chief of the health communication and informatics branch of the National Cancer Institute.

An anonymous respondent commented, "The trend over the last few hundred years will continue…less and less work will be for survival, and more will be for entertainment purposes. People will enjoy their work more."

Dan Larson, CEO of PKD Foundation, a nonprofit organization working for patient advocacy and education, responded that young workers today are ready to take a healthy approach to a 24/7 work/leisure mix. "Anyone who has hired younger-generation employees knows they are generally unwilling to work the long hours their grandparents did," he explained. "They don't sell their soul to the company store. Rather, they value, greatly—their own personal, non-work time and space. With the accelerated pace of everyday life, the importance and value of rest, relaxation, renewal, and diversion from the work world…will only become greater."

Chris Miller, senior vice president for digital operations for Element 79, noted, "As the millennials enter the workforce, this is how they have always lived their life. The boundaries between work, school, play, etc. have never been there, so why should it start when they formally enter the work force? I do think there will be some forced off-the-grid time, which can be from a conscious choice or at the other end as a 'punishment.'"

Paul Greenberg, president of The 56 Group LLC, also noted that young employees are already happily blending personal time and work time, writing, "Forrester and Gartner Groups are showing that 'consumer behavior' is being brought into the workplace; everyone uses their BlackBerrys at work for personal things and at home for work things; there is a whole class of activity sometimes called 'prosumer' that accepts that people are doing work at home and home stuff at work. This is an easy trend, as younger generations get more comfortable with the idea of no real separation. With the increasing smartness of mobile devices, and the creation of the New Desktop—a unified communications and collaboration hub that is increasingly prevalent in the workplaces adopting the social behaviors associated with Enterprise 2.0 (see P&G, Samsung, salesforce.com, IBM, etc.), this one should be a no-brainer by that time."

John Jordan, an associate professor of communications at the University of Wisconsin–Milwaukee, responded, "Blending of work/personal time has been going on for years, and the final removal of the seam is all but inevitable. Most of the talk about this right now is focused on how this will result in a loss of personal time: more stress, less time for family, etc. But the other side is just as important and shows how this likely will balance. Rather than having employers spend time and energy trying to keep employees 'on task' and halting them from using company resources for personal use (e.g., browsing Amazon while at work), this barrier will also fall. The focus will be on accomplishing a task, not logging hours. This will make time more flexible for employees, and will allow sufficient management by employers who switch to compensation plans based on work accomplished rather than time spent. This will be a radical new model of employment, but it will happen."

Ivor Tossell, blogging journalist and technology columnist for the *Toronto Globe and Mail*, noted that modern tools are just evolving to fit the evolution of human desires. "The rhetoric of employment has shifted from conceptualizing it as a means of sustenance to a vehicle for personal fulfillment. More and more people are saying they'd rather work than retire, even if they could afford it. Technology will not drive this change, but it will enable it."

CONNECTIVITY INFILTRATES NATURE AND ARCHITECTURE; EXISTING HUMAN SYSTEMS WILL BE TRANSFORMED

Some respondents looked ahead and imagined how human systems might change as hyperconnectivity becomes more prevalent between now and 2020, with its positives and negatives. "The notion of time will change," responded **Rafik Dammak**, a software engineer for STMicroelectronics in Tunisia.

"Large existing bureaucracies will increasingly be challenged by this trend," commented **Ed Lyell**, an Internet pioneer in issues regarding education. "Schools, which I study, are already way behind the opportunity presented by even our current Internet world. Children know that learning can take place, anytime, anywhere, and in multiple modalities. Yet we only acknowledge or seem to respect the learning that takes place in a top-down, time-dependent school system. I first said this 30 years ago, but it becomes more ubiquitous in the future. Formal schooling is often a barrier to an individual's learning."

"One of the things I have predicted as a futurist for the last 5 years," wrote **Robin Gunston**, consulting futurist for Mariri Consulting, "is a major change in employment contracts as a result of this type of scenario. For effective utilization of scarce human resources, we have to free people to work on an outcome basis irrespective of location or time. Many of us already do this as consultants, but the vast majority of information and knowledge workers are hidebound to a desk, a fixed location, and fairly inflexible working hours."

Jeremy Yuille, digital media coordinator at RMIT Communication Design and former director at IXDA, wrote, "Many Generation Xers will not deal well with this collapse of work/play/home, whereas Generation Y will find it quite normal. The current 7- to 20-year-olds will be even more aligned to multitasking, but this will have an effect on 'normal' society, e.g., the concept and role of the family and what it means to work will fundamentally change."

Adrian Schofield, manager of the applied research unit at the Johannesburg Centre for Software Engineering in South Africa, predicted, "There will be an increase in divorces caused by the intrusion of work time into family time."

Kathryn Greenhill, an emerging technologies specialist at Murdoch University, commented, "The integration of personal and professional time, however, will result in far fewer children being born to people in professions, as they realize that being 'always on' is not compatible with children's concepts of time and development. Lives will be lived too fast for people to slow down sufficiently to gently nurture."

Utopia and dystopia are represented in the ideas of the next two respondents' remarks.

Joe McCarthy, principal instigator at MyStrands and formerly principal scientist at Nokia Research Center in Palo Alto, CA, saw positive outcomes in this realm in years to come. "Concurrent with this shift will be a tendency for people's professional lives to reflect their personal values—work will become meaningful and, thus, will seem less like work because 'workers' will be fully engaged in the missions, goals, and activities of their organizations...many of which will increasingly be organizations of size one."

Mary McFadden, a respondent who chose not to share any other personal identification, predicted the following 2020 scenario: "The individual disappears into the corporation. Work rules and regulates lives, and every place is a company town. Resorts holidays become popular by advertizing disconnection. The poor have computers; the rich have teachers. We live longer, but only with the use of drugs and technological upgrades. Our emotions are not our own, but part of pharmaceutical biochemistry created to prevent us from being unhappy or able to recognize that we are out of touch."

Hyperconnectivity, Lifestreaming... Where Is "Always On" Going to Take Us?

People of the world are becoming ever-more connected through an evolving set of mobile-computing tools networked in the global communication system.

Barry Wellman and Anabel Quan-Haase were the first Internet researchers to apply the term *hyperconnectivity* to the always-on state of engagement experienced by humans as they express themselves to others through an expanding universe of Internet-based tools, now including e-mail, SMS, instant messaging, VoIP, wikis, and social networks. Microblogging through the use of Twitter and associated tools is another of many ways people are beginning to do more "lifestreaming" online, sharing nearly constant updates throughout the day with friends, and even with strangers at times, through Internet tools.

Internet pioneer David D. Clark, a leader in the next-generation Internet initiative from MIT and a former chief protocol architect, has predicted the need to accommodate a trillion connected devices online in the next 13 to 18 years. Billions of cell phones will be in use. Chrysler, announcing in-car Wi-Fi availability beginning in 2009, and more automakers are likely to offer connections. Appliances are expected to follow.

Businesses are attuned to and beginning to finance the study of the evolution of the always-on world, seeking ways to leverage it in their favor. Nortel, a Canada-based technology equipment vendor, has made hyperconnectivity a core of its business strategy, and it has a site centered on it at http://www.hyperconnectivity.com. Nortel sponsored a spring 2008 study by Interactive Data Corporation (IDC) that assessed the work/life balance of 2,400 men and women in 17 countries and found that 16% already fit the criteria for being "hyperconnected" tech users—defined in the study as people who report they are "reasonably happy with their work/life balance…[and] willing to communicate with work on vacation, in restaurants, from bed, and even in places of worship." The hyperconnected people in this study reported they use a minimum of seven communications devices and nine communications applications each day, including e-mail, mobile phones, IM, Web conferencing, and social networks. The highest concentrations of hyperconnected people are found in the United States and China. Most are working in the banking and technology industries, primarily in management positions, and 60% are under the age of 35. Those identified as "hyperconnected" reported they use their Internet connections for both personal and professional communications, with little notice of boundaries between the two. The report classified approximately 36% of respondents as "increasingly connected"—defined as people who are using as many as four devices to access six types of communications applications daily to suit their personal- and work-communication needs. Most of the people who are classified as increasingly connected are seen as likely to be joining the hyperconnected group in the next few years. IDC also reported that there will be 1 billion mobile workers by 2011. The largest percentage of hyperconnected people were found to live in the Asia Pacific region, and Latin America is seen to have the most significant growth potential. More than one in three people surveyed also reported that they use online social communities for business communications.[1]

AOL and Opinion Research Corporation surveyed more than 4,000 U.S. residents in 20 major cities in 2007 and found that "e-mail use on portable devices has nearly doubled since 2004, and as a result people are checking e-mail around the clock." The study found that the average person checked e-mail five times a day; portable-device users checked e-mail every time a message was delivered, no matter where or when; and 43% reported keeping the portable device nearby at night to listen for incoming messages. In addition, 83% of e-mail users reported they check mail every day while on vacation, 40% said they figure the availability of e-mail access into their vacation planning, and 37% said they check their e-mail in the bathroom. Only 15% of the respondents said they consider themselves to be "addicted" to e-mail.[2]

BlackBerry maker Research In Motion (RIM) reported in 2007 that the average BlackBerry user processes 1,200 time-sensitive phone calls and 2,500 time-sensitive e-mails annually. All types of these mobile devices have become known as "crackberries" due to the tendency for some highly connected people to be addicted to checking them 85 times a day or more. Research by Ipsos Reid that was sponsored by RIM indicated that the average BlackBerry user converts

60 minutes of "downtime" into "productive" time daily. A Digital Life America consumer-trend survey of 2,600 Americans conducted by Solutions Research Group in 2006 found that 33% of respondents strongly agreed that "devices like the BlackBerry chain you to work more than they liberate you," 34% of respondents strongly disagreed, and 34% fell in the middle. In the same survey, 46% percent of respondents strongly agreed with the statement "technology like e-mail and cell phones give people more time to do the things they like"; 24% strongly disagreed, with 31% neutral in response. Nineteen percent of BlackBerry owners reported working 50 hours or more a week, and 11% of the overall group surveyed said they worked 50 hours or more.[3]

A 2007 Ipsos survey commissioned by Lexmark found that 83% of respondents found wireless devices allow them to be more productive, 80% said the devices allow greater work flexibility, 64% reported they make their work more rewarding, 92% said they use the devices to work in nonwork situations, 73% said they keep the devices on during weekends, 55% said they used the devices to work while spending time with friends or family.[4]

While the people who are using the technology may consider it reasonable to allow wireless devices to interfere in nontraditional settings, those who are not wielding smartphones or PDAs may disagree. A Yahoo HotJobs survey found that the top five "unacceptable behaviors" most often seen by offensive technology users in April 2008 were: accepting a personal call while in a meeting or presentation; answering the phone or e-mails while at dinner; talking on the phone in the bathroom; talking on the phone while in close quarters (plane, train, bus); and answering a work call or e-mail during personal time after work hours. Eighteen percent of respondents admitted that they have been reprimanded themselves for bad manners related to their use of wireless devices.[5]

An argument over whether employees should be paid overtime for working on their BlackBerrys outside office hours made the news in June 2008. Writers at ABC News in the United States were not paid overtime for their after-hours work on the devices, under a long-standing contract waiver. When three new writers were hired, the Writers Guild of America–East challenged the waiver. The company took away all writers' BlackBerrys in response; the protest was quickly withdrawn and the devices—key tools for the newswriters' work—were returned. In an interview with *USA Today*, Lowell Peterson, Writers Guild–East executive director, said in reference to writing articles, posting blogs, drafting documents, doing online research, and the many other tasks performed outside work hours on the small devices, "We don't want this to grow into a major work commitment that people don't get paid for…It was important to us to make it clear that here's where we stand. This is not going to become a 24/7 workplace."[6]

In 2008 most employers who equipped workers with mobile devices had a tacit or explicit agreement with them that while they may expect them to occasionally do work outside their regular shifts on the devices, the workers may also conduct some personal business during work hours as well.

Cognitive scientists who study how the brain handles multitasking have found that people do not perform well when they are interrupted while performing a task or when they try to do more than one thing at a time. Decades of research have found that the quality and depth of output is restricted as a person attends to more tasks. When people try to complete two or more related tasks at the same time or in alternating order, it takes twice as long to complete them and the rate of error goes up. In many studies, distractions have been found to take a toll on concentration.[7] One example: A study of workers at Microsoft showed that they took an average of about 15 minutes to return to challenging tasks such as writing computer code after responding to vocal face-to-face conversations with coworkers, IMs, or e-mail; the interruption would usually cause them to stray to other tasks before getting back to the work they had been doing before being interrupted.[8]

Psychiatric research distinguishes few differences between addicted online game players and wired workforce addicts. Jerald Block wrote a March 2008 editorial for the *American Journal of Psychiatry* arguing that Internet addiction—involving people who are hooked on viewing online pornography, working online, e-mailing, text messaging, social networking, and doing excessive gaming—is now a serious public health issue that should be recognized as a clinical disorder in the *Diagnostic and Statistical Manual of Mental Disorders*. He noted four primary components: excessive use, often associated with a loss of the sense of time or a neglect of basic drives; withdrawal, including feelings of anger, tension, or depression when the computer is inaccessible; the drive to get a better computer, more software, more hours of use; negative repercussions, including argumentative behavior, lying, poor achievement, social isolation, and fatigue. In his argument, Block reported that in China as many as 10 million teen Internet users could be considered addicts, and South Korea's government has estimated that 210,000 children are affected and in need of treatment.[9]

Microsoft, Hewlett-Packard, Johnson & Johnson, and other corporations now offer research funding into issues tied to human-computer interaction through groups such as the Institute for Innovation and Information Productivity and the Information Overload Research Group.

Intel engineers and managers looking into making adjustments to ease work stress ran a 7-month pilot during which a group of workers was required to set aside a 4-hour stretch each Tuesday morning for a "quiet time," during which e-mail, IM, and phones were turned off, there were no meetings, and visitors were discouraged. Forty-five percent of post–pilot survey respondents found it improved their effectiveness, efficiency, and quality of life, and 71% suggested that the idea be extended to other Intel work groups. Interruptions were permitted on 'Quiet-Time Tuesdays' when urgent situations required them.[10] Some employers are coaching employees to set boundaries, and many are offering their employees time-management seminars. The deputy minister of Canada's Citizenship and Immigration office sent out a memo in February 2008 asking employees to implement a BlackBerry blackout between 7 p.m. and 7 a.m. on weekends and holidays.

It is not just adults who work who are hyperconnected. Children are running circles around them. Pew Internet reported in December 2007 that 93% of U.S. teens use the Internet, and 64% of them have created online content of some sort; in many families, the parents wish their *children* would spend more time offline with them.[11] The recent work of MacArthur "genius" grant recipient Elinor Ochs, an anthropologist and director of UCLA's Center on Everyday Lives of Families, is concentrated on a 4-year study of modern family life. She told *Time* magazine in 2006 that one of the most dramatic areas of change she has noted between her current work and a similar study she did 20 years previous is the impact of multitasking. "I'm not certain how the children can monitor all those things at the same time," she said, "but I think it is pretty consequential for the structure of the family relationship."

As a result of information overload, people who are highly connected are often immersed in a state of "continuous partial attention," a phrase popularized by Linda Stone. She described this as a state in which most of the individual's attention is focused on a primary task while also monitoring one or more background tasks and shifting attention between them as necessary. David Weinberger has applied Stone's idea to describe how people use social media, labeling it as a sort of "continuous partial attention to continuous partial friendship."[12]

Due to the flood of information online and the ability to access it from anywhere, attention has become a rare commodity; people are predicting that an "attention economy" will emerge, and some say it already has. Other catchphrases tied to information overload include information pollution, interruption overload, and the tyranny of e-mail (tyranny of IM, etc.).

Research firm Basex chose "information overload" as its 2008 Problem of the Year. The company's chief analyst Jonathan Spira likes to point out the fact that there is no such thing as multitasking—he said we cannot do more than one thing at a time and we are really simply switching between tasks, and added that the interruptions suffered in the process result in a need for recovery and time for refocusing on the primary task at hand. He and others participated in a March 2008 expert panel discussion titled "Silicon Valley Fights Back Against the (Information) Monster It Created." During the discussion, Spira said he estimates that knowledge workers lose about 2 hours a day to interruptions. But John Poisson, CEO of Tiny Pictures, responded, "Saying there's too much information is like saying there's too much food at the buffet; just stop and manage it."[13]

The Economist hosted a February 2008 online debate series titled "Freedom and Its Digital Discontents" that included a segment on information overload. The point of argument was this statement: "If the promise of technology is to simplify our lives, it is failing." Participants were invited to debate the pros and cons, and a vote was taken; 47% of voters agreed with the statement, and 53% disagreed. In his support of the statement, Richard Szafranski, partner in Toffler Associates, wrote, "We love—to the point of gluttony—to communicate, play, invent, learn, imagine and acquire. Information technology has given us tools to do all of those anywhere and round the clock. We are awash in the benefits that high-bandwidth fixed and mobile wireless communications, email, text messages, pictures, games, data and information give us, including instant access to thousands of products. The seductive ease with which we can engage in any and all of those activities, or quests or endeavours makes it difficult and stressful to not be overwhelmed by choices. Choosing takes time and our time is not unlimited. Devices and applications that save us labour in one area may merely allow us, and sometimes seem to compel us, to invest labour in other areas...Technology has imposed the encumbrance of overchoice on us." Jon Maeda, president of the Rhode Island School of Design, argued, "We voluntarily let technology enter our lives in the infantile state that it currently exists, and the challenge is to wait for it to mature to something we can all be proud of. Patience is a virtue I am told, and I await the many improvements that lie ahead. To say that technology is failing to simplify our lives misses the point that in the past decade we have lived in an era of breakneck innovation. This pace is fortunately slowing and industries are retrenching so that design-led approaches can take command to give root to more meaningful technology experiences. There are advanced developments underway at MIT, CMU, and Stanford for improving user interfaces, data visualization, network reliability, and energy management that will reduce the 10% of downsides we feel today compared with the 90% of upsides brought on by both life-saving and life-styling technologies…Technology will unite with design and the arts in unprecedented harmony such that not only will our lives be simplified, but more importantly satisfying."[14]

People have recognized the drawbacks of the always-on life, and have been sharing information about coping and making adjustments. Gina Trapani started Lifehacker.com, a coping blog, in 2005. Life hacks are productivity tips and tricks to help people cut through information overload and streamline their lives. The site became so popular that it now employs three additional writers, and it frequently carries guest posts from celebrity bloggers, researchers, and others. Another popular life-management site is 43 Folders.

One of the most read self-help books of 2007 was *The 4-Hour Work Week: Escape 9–5, Live Anywhere, and Join the New Rich* by Tim Ferriss. It has been compared to Henry David Thoreau's *Walden*. Part of the fairly obvious but successful Ferriss approach is to cut down on or eliminate e-mail, IMs, and wireless devices, which complicate life rather than simplify it. His methods include the application of "selective ignorance," and he suggested outsourcing work.

Software plug-ins are now available to help Internet users monitor the sites they visit and how long they stay on them; 8aweek and RescueTime are two such applications. They can be set up to restrict the time you spend on a site or to block certain sites and are aimed at helping users help themselves to stop wasting time.

The opportunity to leverage ubiquitous, mobile connections has magnified the ability of individual computer users to network their knowledge to collaborate and create projects, databases, and other material of value through shared or collective intelligence. People are using their laptops and smartphones to add value to their work lives, their personal lives, and sometimes both simultaneously. Scientists who study neurobiology are saying there is an underlying therapeutic effect in the sharing of information through blogging and other online writing formats. Research has shown it improves immune cell activity, aids memory and sleep, speeds healing after surgery, and may trigger dopamine release. Some hospitals have adopted programs that include the option of blogging as part of preoperative and post-operative care. The California Health Care Foundation, a nonprofit that publishes original health policy research, published a report by Jane Sarasohn-Kahn on the increasing importance of social media to patients and health care consumers, *The Wisdom of Patients*, in April 2008.[15]

When people connect online outside of the workplace, they are not always substituting work for family or social time; a lot of the impact has been a transfer of time once spent watching television or reading traditional publications. At the Web 2.0 Expo in April 2008, a conference centered on adding to the momentum already gathered behind this phase of Internet development, Clay Shirky, author of *Here Comes Everybody: The Power of Organizing Without Organizations*, noted that the content on Wikipedia today represents 100 million hours of human thought, and he added that, given the fact that Americans watch 200 billion hours of television every year, if that cognitive surplus is converted to development of collective-intelligence projects, it could create thousands of projects with the potential of Wikipedia. "This isn't the sort of thing society grows out of," he said. "It is something that society grows into."[16]

ADDITIONAL RESPONSES

Many other participants in the survey shared valuable input on the concept of time for work and leisure. Among them:

> "In an increasingly connected society, distinctions between work and leisure will still exist, but certainly not in the 8-hour chunks of the 20th century. Individuals will find themselves most productive when using the same tools for both work and personal purposes, rapidly switching between the two."
> —**Peter Kim**, senior analyst, Forrester Research, Boston; specializes

in e-strategy and management, social marketing, blogs; recent reports include *Microblogging for Marketers*

"I do not want to live in an 'on-call' society, tied to a device. As it is now, face-to-face communication is interrupted, by the cell phone, which people 'instinctively' respond to. The face-to-face interaction is on hold. The devices are calling the shots." —**Charles Hill**, a respondent who chose not to share any other identifying information

"The idea of separating work and home life is a distinctly Western, Victorian one. It basically doesn't exist as a concept before the late 19th century. It won't last much longer, but not just because of technology—because the way we organized cities for the last 100 years was something of an anomaly." —**Anthony Townsend**, research director for the Technology Horizons Program of the Institute for the Future, providing long-range forecasts on technology; he is also a cofounder of NYCwireless

"People with high Internet connectivity in future will likely be dependent for this on their engagement with their companies, but by the same token, their very existence will be dictated by their belonging to that company or corporation, such that they may as well be labeled as 'slaves,' with the proviso that, should they wish to be freed from the company, they will also be freed from any connection with their fellow man via the Internet. In this scenario, employees will have very little choice as to where and when they devote their time to the profit-making of the company. Of course, I may see this as a negative thing, but perhaps the generation of today are already being trained to recognize that being a company employee first means they can be recognized as an identity/a person only as a function of that." —**Alex Don**, linguist and educator

"While I agree that fewer lines now divide professional from personal time, I don't think this is A-OK. We are seeing now an entire self-help industry growing up around the withering effects of the so-called attention crash. While we will continue to have ubiquitous access—and others will have ubiquitous access to us—this is today inspiring many to seek release in yoga, meditation, tai chi, Pilates, etc. This tendency will not abate but grow stronger. The fewer the lines that divide professional from personal time, the more we will dive inside; we will drive ourselves away from the attention crash to places of peace and quiet." —**Barry K. Chudakov**, principal, the Chudakov Company, a marketing and advertising strategies creative consultant who has worked with many major corporations, including Microsoft and Disney

"Sounds like my life/work balance, right now! Great if and when the rest of the connected world catches up :-)." —**Cheryl Langdon-Orr**, a leader in the Internet Society, ICANN and ALAC (At-Large); a board member of AUDA, the group in charge of Australian Domain Name registration

"This is already taking place today in less developed nations as well, with more employers tolerating the use of Facebook and MSN on office PCs." —**Fadi Salem**, research associate, Dubai School of Government; research focuses on e-government and development in the Middle East and North Africa

"Technology will liberate well-connected knowledge workers from the 'commute.' It may also liberate them from a traditional sense of job security." —**Mark Youman**, principal, ICF International, a Washington, DC, consulting-services company that works with government and commercial clients

"Most people fall into careers that complement their personal interests, so they will always be the drivers of this kind of work/play/work innovation. My passion is my work, so when I do what I love, I excel at work. And what we cannot experience at work—such as fulfillment and glory—we'll enhance in secondary channels, including video games, virtual environments, and social networks." —**Buddy Scalera**, vice president for interactive content and market research for CommonHealth Qi, in charge of interactive online strategies, including social and viral marketing

"In this picture, life seems perfect and ideal. There is no war, no conflicts, no fights within relationships, no environmental disasters, no storms, and no droughts. In short: There are no accidents. The world is a place where nothing happens. A bit silly, no?" —**Geert Lovink**, professor and expert on culture, sociology and the Internet; based in Amsterdam; author of *Dark Fiber* and *Uncanny Networks*; responsible for the Institute of Network Cultures

"I agree, however, by 2020 I think people will be beginning to see this as a net negative and be looking for a return to late 20th-century split of work/social time (they will want their lives back)." —**Adam Peake**, executive research fellow and telecommunications policy analyst, the Center for Global Communications (GLOCOM); studies the intersection of public policy and the Internet

"2020 seems a little early for this." —**David Brin**, futurist and author whose 1998 nonfiction book *The Transparent Society: Will Technology Force U.S. to Choose Between Privacy and Freedom?* identified key issues of concern

"Likely, but as always, the cause of much complaining." —**Jamais Cascio**, originator of Open the Future, also works with the Institute for the Future, Institute for Ethics and Emerging Technologies, and Worldchanging

"Sad but true. Connectivity is hard to get away from, and it'll become more so. Also, since few of us in developed nations by 2020 will place faith in standard forms of employment, we'll be relying on taking work (in the form of projects) as it flies at us. That means we'll need to be ready to catch it at any time." —**Dian Schaffhauser**, writer and editor for CampusTechnology.com, THEJournal.com, *Redmond Magazine*, *Computerworld*, and Web Worker Daily; founder of Sourcingmag.com

"This seems to be the case now, minus the virtual reality part, oh, and the seamlessly part!" —**Steve Jones**, professor of communication and associate dean of liberal arts and sciences, University of Illinois–Chicago, cofounder of Association of Internet Researchers

"That sounds like my life right now! And yes, it's net positive." —**Paul Miller**, technology evangelist on the senior management team at Talis, a company delivering human-centric Web applications, based in Birmingham, U.K.; a blogger for ZDNet

"I always tend to agree with the approach of creating a divine line between work time and playtime. There should be a balance, otherwise it can be described as an obsession with one or the other." —**Hanane Boujemi**, ICT researcher for DiploFoundation, working on educating people about Internet policy and Internet governance, Malta

"This scenario will not be for the majority of citizens." —**Janet D. Cohen**, blogger, futurist, and trend analyst

"'Blending personal and professional time' is well described as *workaholic*. I suspect that for their personal health, people will, if anything, learn to impose better boundaries." —**Fred Baker**, Cisco Systems Fellow, former Internet Society (ISOC) chairman of the board; Internet Engineering Task Force (IETF) leader; an architect of the Internet

"The work/play diffusion will result in a harsher result for those on the other side of the divide." —**Michele Perras**, artist, consultant, researcher, and futurist with Interactive Ontario

"Communication-free zones are established on resort, beachfront, and tropical properties and become popular tourist destinations. Digital communications are not functional at these locations, and visitors notify friends and coworkers before departing." —**DJ Strouse**, international relations and computer science student, University of Southern California

"This is already happening." —**Richard Hall**, professor of information science and technology and codirector of the Laboratory for Information Technology Evaluation, Missouri University of Science and Technology

"This concept assumes that people don't have families and friends and that the only thing they enjoy is work. There are, of course, people like that. But most normal people will be very unwilling to stop reading a bedside story to their children, or enjoying the view as they shoot down the piste, just because they are connected. There are boundaries between areas of life for a reason—work, play, children, eating, sex, learning…" —**Jeremy Swinfen Green**, Telecom Express, an interactive marketing company

"This is already true in 2007 for a large number of knowledge workers. Even for those who have clear lines between work and home, there is a cultural understanding that this division is less prevalent for professionals. It's not a question of 'if' but, rather, of what proportion of the population will become a part of this group." —**Alexander Halavais**, professor and social informatics researcher, Quinnipiac University; explores the ways in which social computing influences society

"Mobile devices have already spilled work hours over into personal time, and I think it is inevitable that the opposite trend begins! I already see this beginning to happen in my own workplace." —**Christian Ferris**, associate director of career advising at Washington University

"Yes, though I wish I could honestly tick 'mostly disagree.' As an academic, I do tend to blur professional and personal life. But as an academic, I can choose which bits to do at home (i.e., the ones that I like). I would dislike being in the situation of being able to be contacted 24/7 by a superior." —**Emma Duke-Williams**, lecturer in the School of Computing and researcher, University of Portsmouth, U.K.; education blogger

"This sounds like a description of 2008. (Also not clear that there's any real basis for calling this a net positive, as the change strikes me as incommensurable with the earlier era)." —**Clay Shirky**, consultant and professor in the Interactive Telecommunications Program at New York University; an expert on the social and economic effects of Internet technologies

"That's how I live now!" —**Robert J. Berger**, CTO for Cinch; expert on backbone networks, access networks, wireless networks, and innovative Web applications, a frequent speaker on next-generation networks

"I agree it's likely to happen. I don't think it's a good thing, especially not a 'net positive.' Having to be on-call all day, all week is *not* a sign of progress. Per the saying, 'The Union Movement: The People Who Brought You the Weekend.'" —**Seth Finkelstein**, anticensorship activist and programmer, author of the Infothought blog and an EFF Pioneer Award winner

"You don't have to wait until 2020; this is already happening with anybody who owns a BlackBerry or iTouch." —**Bill St. Arnaud**, chief research officer, CANARIE Inc., an industry-government consortium that promotes and develops information highway technologies in Canada; active in Internet2

"The fact that few lines divide professional from personal time is very true. However, I'm not certain that this is OK. I rather think it's beyond our control, considering the integration of digital medium into daily business and personal activities." —**Sam Ozay**, e-learning and e-communication specialist and solutions architect at Postmodern (Asia/Pacific); formerly general manager at European Language Centre

"Agree but for the virtual reality bit. Much of this is already the case (who knows where you are when your office number diverted to your cell phone rings)." —**Sam Smith**, Web interface developer, University of Manchester, U.K.

"It has already begun!" —**Don Heath**, Internet pioneer; former president and CEO of the Internet Society; member of U.S. State Department Advisory Committee on International Communication and Information Policy

"I agree for two interrelated reasons. First, technology already enables this, and second, more and more professional endeavors are global/multinational, meaning that at least some collaboration will require live sessions that will require participants to join in at all hours of the day and night (I already have to rise in the wee hours of the morning to take part in committee work of just such an international organization)." —**Steve Goldstein**, ICANN board member, retired from the National Science Foundation, where his job in the 1990s was to help diffuse the Internet globally

"Personal time will be one of the most valuable assets 10 years from now." —**Sebastian Ricciardi**, associate with Jauregui & Associates, a law firm in Buenos Aires; leader in the Argentina chapter of the Internet Society, formerly of ICANN's At-Large Advisory Committee

"This is already the case for some of us." —**Itir Akdogan**, PhD candidate and lecturer, University of Helsinki, expertise in ICT in empowering women and girls

"It's that way for me already." —**Dan Lynch**, founder of CyberCash, Inc. and Interop Company and an Internet pioneer; board member of Santa Fe Institute; director of computing for SRI International in the late 1970s

"For some workers, the blending of personal/professional time and 'always-available' connectivity will be a good fit, but good questions are being asked about the degree to which such a lifestyle inhibits reflection, creativity, and innovative thinking." —**Michael Edson**, director for Web and new-media strategy, Smithsonian Institution

"This will be true for younger individuals who see these lines blurring already. Like everything, however, I suspect these things will change when people have kids and then, as parents, they value time with the family and opt to protect this time." —**Todd Wagner**, health economist, Health Economics Resource Center, Palo Alto, CA, part of the U.S. Veterans Administration; also involved with the Center for Health Care Evaluation

"Many professional women with families have already tried this. Peter Drucker understood the problem when he talked about 'widow-maker' jobs. Sometimes we all need to get away from the constant pressure or interacting with everyone in our lives at once. I do not think that can be changed with technology." —**Ramona Nelson**, PhD, codirector of the health care informatics program, Slippery Rock University

"Pressures on mass-transit systems, on fuel supply, and global time shifting will make this inevitable. The trade-off is that we will have to blend flexible working hours with personal time while still striking a balance." —**Jill O'Neill**, director of planning and communication, National Federation of Abstracting and Information Services; author of the Infotoday blog; based in the Philadelphia area

"As more and more people leave the conventional workforce (by choice or layoff) and become entrepreneurs, it will only become more ubiquitous. For one who owns his or her own business, evening is no different than 9 to 5, weekends no different than weekdays, and holidays don't really exist. If there's work to be done, clients to serve, then you work—if not, THEN you can play." —**Judith Siess**, president of Information Bridges International, Inc. and publisher and editor of the *One-Person Library* newsletter, author and blogger

"This already occurs for many people, and will only increase. The Internet and e-mail and virtual reality have made the always-on personal/work spaces merge." —**Catherine A. Fitzpatrick**, Open Society Institute, Physicians for Human Rights; lecturer on humanitarian issues; formerly UN representative for International League of Human Rights

"Yes, I agree, but as someone who already does this to a certain degree—I wonder if it's healthy. Will it result in a work/life balance that is really off?" —**Tiffany Shackelford**, consultant who works with clients such as Phase 2 Technology, Stateline.org, Foneshow, WebbMedia, and Daily Me

"A big driver will be environmental. Why again are we producing all that smog to sit in an office and send e-mail to the worker in the next cubicle? There will be some 'correction' here, as workers learn how not to get sucked dry by employers who 'empower' them to work 24/7." —**Karen G. Schneider**, research and development, College Center for Library Automation, Tallahassee, FL; expert and thought leader in the library and technology community

"Yes, but keep 'real' holidays!" —**Sebastien Bachollet**, president of the Internet Society of France, operates the European Global Event on Domain Names and Address Systems, known as EGENI, active participant in ICANN

"I currently work 60 hours a week, with computers assisting me at home, the office, and on my belt. I look forward to the day when my office is virtual, and I'm confident it will happen before I retire in 10 years." —**Bill Warren**, vice president of government relations, Walt Disney World; founding editor of the *Orlando Business Journal*

"I tend to work far more hours because when I have down time at home, I usually am sitting with my laptop working on my job. At the same time, I feel if I need to do something personal while on the job, I can." —**Debbie Murray**, associate director, health education through extension, University of Kentucky

"I agree this will happen but don't agree it will be a net positive. Examples of the refreshing power of a change of scene are too common to believe that this is an overall improvement in our mental state." —**Jerry McCann**, vice president and director of the social marketing group at Carton Donofrio Partners Inc., marketing and advertising firm in the Washington, DC, area

"This is already happening, but I do not believe it is 'positive.' It is part of an ideology that forces the brain worker to run behind a business target that is always moving ahead at high speed." —**João Miguel Rocha Filho**, director, DataOne, a provider of software for connecting to Linux; based in Brazil

"Things will change more fundamentally. The scenario described here is a repositioning of the idea of personal/professional from the 20th century into the 21st. Perhaps machines will do all of the 'work,' and humans will only do things they perceive as 'creative.' We do not know yet." —**Judy Breck**, blogger at GoldenSwamp.com—aimed at "watching the global golden age of learning emerge from the open Internet"

"Why are you asking 'if' work and play will be seamlessly integrated in most of these workers' lives? It already has happened—certainly it has for me." —**Leonard Witt**, associate professor in communication, Kennesaw State University, GA; research interest is citizen journalism and user-generated content; author of weblog PJNet

"For many of us, that's already the case. It will be even more so in 2020." —**Joan Connell**, online editor, *The Nation* magazine, formerly an executive producer for MSNBC.com, senior editor for MSN and a Pulitzer finalist for her reporting

"This is dystopian but not really a vision. Most of the processes you mention in your statement are already going on. NOW." —**Oliver Quiring**, PhD, Institute for Communication Science and Media Research, LMU, Munich, Germany

"As a current knowledge worker, I have had blurred lines between personal and professional time for many years, and find it works well in my life. I do not mind 'working' on my off time—I love my career, and it defines me. This is possibly due to having lived a military life, where one never really sheds the military 'on-duty' status. This has also led to considering my life to be more transparent than my parents' lives—everything is known and will be known about you in the military, even as a family member." —**Teresa Hartman**, associate professor and head of education, University of Nebraska Medical Center

"The line between work and play is bound to be broken. This is good for people who are able to focus on the work that pays the bills. Not so good for many others. There should be a high school course that teaches how to relax, get away from work at times, and yet take responsibility for what needs to be done." —**Mike Samson**, interactive media writer and producer

"Why wait for 2020? All we need to do is look at the waves of people using mobile digital devices to conduct business from their cars, at their dining room table, and in every way possible. We haven't embraced the seamless integration of digital connectivity, it has swallowed us." —**Stephen Schur**, director of online communications, Ramapo College of New Jersey

"A possibility indeed (though teleworking has been around as a 'future certainty' for decades now) for some very specific professions. It is an impossibility for others." —**Luis Santos**, Universidade do Minho–Braga, Portugal

"This is how I live today. It will get better when we have better means to substitute face-to-face meetings." —**Jeff Jarvis**, top blogger at Buzzmachine.com; professor at City University of New York Graduate School of Journalism; contributor and consultant to the *Guardian*; advisor to start-ups

"With greater connectivity and the ability to capture, process, send, and receive information through multiple devices anytime and anywhere, we move closer to perpetually 'plugged-in' society." —**Ed Dieterle**, Harvard Graduate School of Education; research tied to handheld devices for ubiquitous learning

"This has always been true, and the Internet simply increases connectivity options. But there is still the 'off' button, and there will still be people who, for reasons of privacy, sanity, or stealth, just switch off or let the machine get it." —**Dennis Baron**, professor of English and linguistics, University of Illinois, runs the Web of Language site and researches the technologies of communication

"Unless there's a radical shift among the younger generation (the millennials), people are going to continue to value separation of personal time and professional time unless forced into a new scenario. Even then, longing for 'the good old days' will abound. Corporations and the people who control them are not liable to give up their greed." —**Kathee Brewer**, freelance journalist and consultant, Galveston, TX; editor of *AVN* online magazine, a publication for the adult entertainment industry

"This scenario has only a single wrong note: 'and that is OK.'" —**Walt Dickie**, executive vice president and chief technology officer, C&R Research, one of the principal developers of CheckMetrix in the early '90s

"The erosion of barriers between work and personal time is an insidious trend, and it's not OK. Corporations have found new ways, largely through technology, to squeeze more labor out of their workforce by blurring the line between both time and space constraints of what is work and what is free time. Knowledge workers have not submitted to this invasion peacefully—it has been forced upon them as a precondition for participating in the current job market. You cannot get a corporate job these days without being expected to be available at home or in transit whenever the job requires your attention." —**Sam McAfee**, chief technology officer and owner, RadicalFusion, a company that offers consulting on Web design and development, server-side programming, and database development

"People go crazy when they meld their personal and professional lives. It does not matter if they are knowledge workers or factory workers. If people don't have to go to work or spend time working for others, they won't." —**Theresa Maddix**, satisfaction research analyst, ForeSee Results

"This scenario will most likely come to pass, and I think that is unfortunate. We should not live to work, but work to live. Boundaries are critical, and technology is tearing them down brick by brick." —**Lisa Carr**, director of strategy, Targetbase Interactive, health care strategist and writer

"As the pendulum swings to a more work-life scenario and it becomes commonplace, eventually, there will be a movement (by 2030?) back to the more traditional separation of work and life." —**Mike Driehorst**, messaging strategist; leads social media for Hanson Inc., an interactive communications and video production company in Ohio

"This is happening already. It is a global business world, so conference calls at odd hours are not unusual. The only thing that might stop this from going further is attitudes of the younger generations, who want more separation between their personal and professional lives." —**Susan Frede**, vice president for research and panel management, TNS, a global market research company

"I am living in that blur now, and it works well for me." —**Stephanie Geyer**, associate vice president, e-communications and Web strategy, Noel-Levitz, an operating division of Sallie Mae that consults on recruitment and marketing of higher education

"I have no way to agree or disagree with this statement. On the one hand, there are *already* few lines drawn between these two spheres. So to that end, I agree with Mariana Almeida: As a result of this, we will need to be well versed in stress management and in creating one's own rules to shut out work." —**Alexis Turner**, webmaster, Greenwood Publishing Group, New York

"There will be growing resentment of the invasion of personal time by work responsibilities." —**John Jobst**, IT specialist, U.S. Army Corps of Engineers

"It certainly has consequences, but day in and day out, most of us feel it reflects convenience more than anything. Our other ideas and reactions are the exception, not the rule." —**David Allen**, PhD, Temple University

"Maybe I'm getting old, but I still know a lot of people (even younger people) who like to completely detach from their knowledge-worker lives." —**Brian T. Nakamoto**, cofounder of MrJoy Inc. and product-line manager for Everyone.net (a leading provider of outsourced e-mail solutions for individuals and companies around the world)

"We're seeing this already, in 2008." —**Stan Tennyson**, deputy program manager, BETA A&AS support to the Office of Warfighting Integration, chief information officer, Sumaria Systems Inc., a defense contractor

"Seems to be reality right now, but for a small group only." —**Jutta Croll**, managing director, Stiftung Digitale Chancen—Digital Opportunities Foundation, Berlin, Germany, promoting and supporting access and equal opportunities for all online

"I am not so sure about the 'that is OK' element, but the rest seems inevitable." —**Ed Steinmueller**, professor, science and technology policy research, University of Sussex; researches industrial structure of high-technology industries, coevolution of technology

"The ubiquity of mobile communication technology makes this outcome inevitable." —**Tim Grafton**, market research director for UMR Research Ltd., a market research company based in New Zealand

"This is harder than you think and doesn't depend on the technology. Many want their work/leisure time separated by forces outside themselves, i.e., an employer! Blending is difficult. We will be going in this direction, for a percentage of people, and by their choice." —**Jennifer Jarratt**, principal, Leading Futurists LLC; works with formalized methodologies to assess and interpret potential futures

"Yes, it's too hard to disentangle, especially as the number of jobs per person over a lifetime increases." —**Bryan Alexander**, director of research National Institute for Technology and Liberal Education, blogger, expert on computer-mediated pedagogy; based in Ripton, VT

"Just as child labor or 12-hour/6- to 7-day workweeks seem unreasonable now, too much professional time integrated into personal time will be resisted. The companies and individuals who thrive in this type of 'merged' professional/personal scenario will be those who recognize the mutual benefits." —**Scott Brenner**, technologist, Web developer, consultant for clients ranging from Fortune 100 companies to small nonprofits

"People who love their work rarely see the divisions. This may be sad to those who are less enchanted by work-life vagueness, but it's fine by me." —**Paul Jones**, director of ibiblio.org at University of North Carolina–Chapel Hill; the original manager of SunSITE, one of the first Web sites in North America

"I can conceive of this scenario by 2040 possibly, but in the next 12 years, we will not have developed the organizational and management skills to operate complex institutions that operate in this manner." —**Christopher Jacobs**, chief operating officer, Solutions for Progress, Inc.; formerly with KnowledgeFlow Inc. and Unisys Corp.

"For many people, this is already reality today, so this is no bold prediction. But it might be that we will see a counterreaction as people feel exploited by employers who demand that they be online 24/7. Hence employers need to be prepared for this risk." —**Olav Anders Övrebö**, media blogger and former *Netzeitung* journalist, University of Bergen

"This scenario will develop even sooner than 2020 in developed economies. The more mobile our technologies become, the more likely this scenario will be." —**Jane Sarasohn-Kahn**, founder, health economist, and management consultant, THINK-Health, a strategic health consultancy; author of *iHealthBeat*, an online publication

"I think people will find that this scenario isn't healthy in the long run, and most people will find ways to make separations." —**Lynn Blumenstein**, senior editor, *Library Hotline, Reed Business Information*

"There is no doubt that this trend will intensify over the next dozen years. But whether this is a 'net positive for people' is controversial. Certainly, it has some identifiable benefits. But it has also led to people's overall lives becoming extensions of their work lives and consumer lives, and the elimination of boundaries and of silence is a cultural shift with many problematic aspects. As Thomas Friedman has said, it leads to 'over-connectedness being the social disease of the 21st century.'" —**Jeremy J. Shapiro**, a professor of critical social theory at Fielding Graduate University with a research emphasis on the social and cultural effects of information technology and systems

"Hopefully, this will develop in a way that will give individuals more freedom and flexibility instead of forcing people to work more hours per day." —**David F. Salisbury**, associate director for science and research communications, Vanderbilt University; formerly science and technology reporter for *Christian Science Monitor*

"Civilization, ethics, integrity are nurtured at shared family supper, private moments of physical intimacy, people reading aloud to each other, parents teaching children and turning them on to refinements and pleasures—these are left out of the scenario, and surely by 2020 people will see that they do not have to live in a *Lord of the Flies* ethical wasteland, which is what is evolving now. Nurturing children and nurturing love relationships and friendships will occur outside the professional, at places of worship, over meals, etc." —**Bruce Henry**, Concordia University, Montreal, Canada

"We will begin to see more of a breakdown in family relationships because it's too easy for parents to hide behind an away message and harder for the children to bring up (or get caught doing) something that needs attention. You know, those difficult subjects: pregnancy, drugs, hurt feelings from bullying, presuicide. Technology still doesn't pick up the nuances that are so crucial in personal relationships—and filling both professional/personal time into one seamless life takes the edge off of catching problems early because of the flip of the hair, the look in their eye, the catch in their voice. Texting, e-mail, etc. just doesn't catch the subtle telltale signals." —**Christopher Brown**, strategist and managing editor of new media for the U.S. television program *America's Most Wanted* on FOX

"The situation shall become more prevalent by far, but only in that small area at the top of the wage spectrum." —**Francis J. L. Osborn**, futurist and activist, philosophy department, University of Wales Lampeter (formerly St. David's University College)

"While the scenario is likely, I don't see the outcome as being 'a net positive' for people. It is more likely to be seen as a positive development for the organizations that pay them. It will be change driven by economic imperative." —**Sam Punnett,** president, FAD Research Inc.; has worked in the field of interactive digital media since the 1980s, for the last 9 years on strategy, marketing, and e-business development

"This is already beginning to be true and will increasingly be the case, but the elimination of divisions between personal and professional time will not be a good thing. For most people, this will be a net negative, though they won't have a choice. Stress levels will increase, health will decrease, and there will eventually be a backlash, where people increasingly try to draw the line between the personal and professional, with mixed success. People will increasingly bring exercise devices to the workplace and work at their desks while walking on treadmills and using exercise bikes. Entirely new gyms will be created aimed at enabling people to work and workout at the same time." —**Jonathan Dube**, president of Online News Association, director of digital media at CBC News, the Canadian Broadcasting Corporation, publisher of CyberJournalist.net

"Because of the omnipresence of communication tools, people in 2020 will find difficulty in escaping from people's reach. There will be a new concern in finding personal time away from work. People will experiment different ways to find balance between work and personal life, public and private times, and the hectic and leisure." —**Clement Chau**, research manager for the Developmental Technologies Research Group at Tufts University

"This mixing up of professional and personal time is clearly increasing, but for most people it's not OK. It can be OK for managers and executives who want to compete, but for other people, it means more and more work as a result of competition on the labor market." —**Yves Froidevaux**, Swiss Federal Statistical Office

"There is a subtext here that work becomes more what you want it to be, and you end up blending different bits of work into your life at different times and in different ways. There may be social issues here though, in that this is something that will increasingly be available for those in higher social classes, perhaps even creating a new sort of digital divide." —**Richard Osborne**, Web manager for the School of Education & Lifelong Learning, University of Exeter; research focus is e-learning

"People in information industries (don't forget about the trades—physical things still need to be built and maintained) will trade flexibility for time. We will work more, but be able to choose when and how much. Less time wasted on physical co-location unless it produces concrete advantages and benefits. Virtual meeting spaces will become better and more commonly used for interaction. This will be especially true in education, government, and administration." —**Ted M. Coopman**, lecturer, San Jose State University

"Technology allows us to meld both professional and personal, and it will be increasingly hard to separate the two. Just like workers will pay bills online during 'work' hours, they will also attend to work over the dinner table. This will actually improve productivity and job satisfaction." —**Rachel Kachur**, behavioral researcher, U.S. Centers for Disease Control & Prevention

"Many workers are borderline invested in this manner today. This blending is what is truly driving the innovation and the push toward the reality described in this scenario. It is self-fulfilling. The people who want this scenario are the people who would build this scenario. Unless you propose that it's technically not possible, and we know that's not true, then it must happen." —**Eric Kreider**, director of Web services, the University of Akron, OH

"There is already a backlash developing against the 'always-on' work culture, though the strength of this varies considerably between different countries and cultures. I believe that this will get stronger over time, rather than allowing the insidious intrusion of work into the rest of life that the scenario implies. (If it doesn't, heaven help us!)" —**Roderick White**, editor, *Admap* magazine, World Advertising Research Center

"People have always had differing boundaries between work life and personal life. For some, the two are intertwined. Business owners, along with those in the arts and politics, often feel they are always 'working.' This is not necessarily a new phenomenon. If obliterating the boundaries between work time and personal time becomes a requirement for maintaining a job as a 'knowledge worker,' we will see a shift in the knowledge-worker population toward those with a preference for this integration. I'm not so sure that's OK." —**Peter W. Van Ness**, president, Van Ness Group, a Web-development company; founded Personal Computer Solutions in 1983; cofounded StockPlan, Inc. and MyStockOptions.com

"The line between what is 'at-work behavior' and what is not will remain firm. Companies that have implemented Internet content filters to keep staff from using blogs, social networking sites, and e-mail during work hours are so, so sad. They don't realize that these tools help them. Given that attitude today, I can't see it changing drastically in the next decade." —**Sarah Houghton-Jan**, consultant for the Infopeople Project; digital futures manager, San Jose Public Library; author of the Librarian in Black technology blog

"I am experiencing that phenomena now, and while at times I am concerned with the degree to which I am 'connected,' at the same time, I enjoy the flexibility it provides me. Taking downtime will become a more learned and self-imposed process versus one that is dictated by physical working environments." —**Nathan Botts**, community informatics research associate at the Kay Center for E-Health Research; working on the national health information network (NHIN) and outreach

"Nope. Sorry. I don't buy it. I'd like to think that ongoing advancements in productivity and manufacturing will lead to an overall decrease in the number of work hours and the beginning of a de facto postscarcity economy. But, in general, people will be more jealous of their personal time by 2020, and will take great pains to safeguard it." —**Jason Stoddard**, managing partner/strategy at Centric/Agency of Change, an interactive strategies company; he is also a popular speaker on social media and virtual worlds

"The more things change, the more they stay the same. Work and life will blend, and people will complain." —**Dan Weingrod**, vice president for digital operations, Cronin and Co.; oversees creative online initiatives for integrated marketing communications company

"While not true for all, most work is not going to be purely knowledge work but will instead require a physical presence or telepresence, and for those tasks it will be necessary for people to still work some type of coordinated 'normal hours' for that group." —**Shawn Kelly Apochromantic**, configurations manager, designer, technologist, futurist, General Atomics, and volunteer builder in Second Life

"The hard part is creating the fluid organization to match the way work and personal time flow." —**Nancy W. Bauer**, chief executive officer and editor-in-chief, WomenMatter, Inc.

"I would quibble about whether that's a 'good thing.' These boundaries are clearly fusing, and it is unlikely that the trajectory will reverse itself." —**Amy Friedlander**, director of programs for the Council on Library and Information Resources, a nonprofit that services research and higher education

"Since this is already happening, the issue is whether 'net positive' means better lives or more work-centric lives." —**J. W. Huston**, president of Huston Consultancy and futurist

"Some people will opt out of those jobs that have and will continue to drift towards 24/7 availability, but for a growing generation, it will be the norm." —**Jim Lucas**, Web manager, CACI, a provider of national security, defense, and intelligence-related solutions in the interests of the United States

"I agree with the blending. Not sure I agree with the part about this being best for everyone. I think we all know that without boundaries, work tends to take over everything." —**Douglas Schulz**, managing editor for online publishing, America's Health Insurance Plans; formerly a Web team director for a biotech industry organization and manager of Internet and Web services at the Council of Better Business Bureaus

"Though I work in technology and may not exemplify the average worker, this is already true." —**Kathryn K. Goldfarb**, president, KG Communications, an independent consultancy

"There is no doubt that we are truly embracing the 'now' society with 24/7/365 connectivity devices and expectations from our employers, friends, and family. However, the pendulum will eventually swing back, and we will see a resurgence of simple living. Technology-free folk, who feel (or have been statistically convinced) that dedicated time and separation between the commitments of our lives, will abandon all the new interactivity and will find old ways to enjoy one thing at a time." —**Drew Diskin**, director of e-strategy, Johns Hopkins Medicine

"Most people will prefer to have some segregation of their professional and personal lives." —**Mike Langum**, Web developer, U.S. Office of Personnel Management

"Devices like the BlackBerry have already blurred the lines between work and personal time. There's no doubt that this trend will continue. Whether it's a net positive for people remains to be seen. The younger generation who grows up with this mindset will accept it much easier than those who have to adapt to it." —**Jay Buys**, vice president for digital development, Fleishman Hillard, an international marketing and communications company

"There will be a movement to establish firm walls between personal and professional time; indeed, the movement has already started. The American work ethic, while noble in spirit, can be, in practice, dangerous to one's health. The hypercapitalism begun during the Reagan years, exacerbated during the

Clinton and Bush years, will reach its endpoint when people discover that the work-life balance truly is just that: a balance between work and the rest your daily life." —**William Winton**, product manager, digital media, 1105 Government Information Group

"We are already seeing this with the proliferation of virtual offices and the ability to connect and communicate from any place with a clear cellular signal." —**Kent Kirschner**, media specialist, Neighborhood America, inviting companies to understand the power of community-building online

"Most people require separation of personal and work time as a de-stressor. This scenario may hold true for people that work for an Internet-related company, but not the majority of people." —**Mack B. Rhoades Jr.**, Web services product manager, Michael Baker Corp.

"They only affected a small part of the population at any time. 'Work' and 'nonwork' are artificial distinctions used by politicians, union labor organizers, bureaucrats, educators, and other fools trying to control others." —**Dick Davies**, partner, Project Management and Control, Inc.; past president of the Association of Information Technology Professionals

"People will accept this trade-off, as it will allow them to work anytime and anywhere." —**Lawrence Swiader**, chief information officer, U.S. Holocaust Memorial Museum

"It will happen even if most folks don't think it's OK." —**James Jay Horning**, chief scientist, information systems security, SPARTA, Inc.; a former fellow at Xerox's Palo Alto Research Center; fellow of the ACM

"It's been happening for some time. I'm already living/working just this way. And it works just fine for me and for those who remunerate my work. But it's not everyone's 'beach,' as we say in Brazil. Many people cannot work autonomously; and many employers, likewise, are not prepared for it. So, just as cinema did not displace theatre, and television did not displace cinema, etc., the new ways of working will not entirely eliminate the old ways. And I rather think that this is good." —**Fredric M. Litto**, consultant for Pearson Education Global e-Learning, president, Brazil Distance Learning Association

"This will only work if people have the ability to take time off. The expectation that workers are going to be available 24/7/365 runs contrary to human nature, and processes will develop that allow people to assert control of their time." —**Robert H. Rich**, PhD, strategic planning and evaluation, American Chemical Society

"This could be very workable for knowledge workers, and is already for many. I try to work from home at least twice a month but more often would be most welcome. However, it means a different kind of managing style, one more suited to ensuring tasks are done and goals are reached rather than a certain number of hours are clocked." —**Patti Nelson**, a webmaster who works on U.S. government sites

"Work will still be work. There are some who already have seamlessly integrated work and play. But if current trends continue, most people will be 'well-connected knowledge workers,' doing menial, bureaucratic tasks. No one will want to take that home 24/7." —**Marco Rivera**, Internet specialist, Vistronix, Inc., an information management, technological solutions, and managed services firm serving federal, state, and local agencies

"It's already happening, and more traditional companies will find a way to embrace this in their day-to-day management of people and clients." —**Helen Keegan**, founder of Beep Marketing, a self-employed consultant and a judge for the Webby Awards

"Agree with the permeability of work/play boundaries, but not the 'willingly.'" —**David Hakken**, Indiana University School of Informatics and a professor of anthropology who studies social change and the use of automated information and communication technologies

"There will be those who want to work this way, but even amongst knowledge workers, there will probably be those who still want to be able to switch off from time to time." —**Heath Gibson**, manager of research and market analysis, Big Pond, a competitive intelligence company and provider of broadband customer Web sites in Australia

"The line between professional and personal time is disappearing. This is, in part, due to portable connectedness through handheld devices and smartphones. It is also the result of people

> whose work is an enjoyable and intrinsic part of their life—where work is much more than a job." —**Elizabeth Talerman**, chief operating officer, Talerman+Partners LLC, a marketing-integration company; previously senior vice president of marketing at Martha Stewart Living Omnimedia

> "This is the province of the networked elite, and not of people in general." —**Paul Hyland**, executive producer, edweek.org; formerly a member of the board of directors for Computer Professionals for Social Responsibility and director at Media Matters for America

ANONYMOUS COMMENTS

The following section includes a selection of contributions to the discussion of the future of the Internet and the concept of time for work and leisure from survey participants who chose to remain anonymous. Among them:

> "As Baudrillard said, 'The revolution has already happened, and we lost.'"

> "The next generation is about personal endeavors. In contrast to the past, they will displace work for personal time, rather than displacing personal time for work."

> "If work could produce for most the pleasure and satisfaction of avocations, then the scenario would be a much more appealing one to most people."

> "The baby boomers are the last generation to understand what will be lost in this scenario. The question is whether they will rise to the challenge and try to secure a balance for future generations by educating them as to what they will sacrifice."

> "It absolutely will happen, and this is the way it should be."

> "I work all the time, and I'm tired of it :(."

> "I agree about the likely fusion of time and spaces, but personal/leisure time won't expand, on the contrary, and the outcome is not going to be net positive. The negative effects of overload will become even more pervasive."

"Family-oriented societies will resist the scenario where work and play are totally intermingled."

"People will want exclusive personal time regardless of whether they can blend professional duties in or not."

"This is already true now, isn't it? ;-)"

"Work is now being extended again by technology."

"While technology will have enabled us to be 'always on,' more time-conscious devices will allow us to not have to be."

"Although there will be many people who are happy to work and live this way, there will be many who want nothing to do with work once they get off work."

"Being connected will be the norm of a person's experience."

"People are working more now than in the past, upsetting the work/life balance."

"While there will continue to be some solid crossover (answering e-mails on free time or shopping during work time for a gift), it won't likely be OK to largely erase the lines."

"I'm not sure this fuzzy boundary is a good thing."

"I hope not—I want personal time to be personal, *not* integrated with work."

"This won't be for a net positive. The blurring lines between personal and professional lives take their tolls."

"This affects productivity and future employment and people taking blocks of personal time, formerly known as holidays."

"This will actually only be OK with two-thirds of these networked, knowledge-class people. The other one-third, who are introverted, have memory problems, etc., will have a hard time working to their potential."

"Some (white-collar, male) workers will see this is a good thing, however, people (especially women and those with young families) will find creative and innovative ways to balance or counteract this in order to carve out work-free zones and reduce stress."

"The boundaries between work and play will blur with gaming and knowledge work. However, there is still the need to do face-to-face and manual work—the mix will be in the balance."

"It's definitely not 'OK.' An end to individual and family time as we are forced to work longer and longer hours is neither desirable nor 'OK.'"

"The majority of the population continues to work on industrial-age tasks, and these will not be automated until robots become ubiquitous, not by 2020."

"This has been an issue for those tethered to pagers for many years. The backlash began a long time ago, and will spread."

"This will happen…It will continue to destroy the sense of the individual and all creative impulses in humans subject to this type of lifestyle."

"Many people (self included) are here already."

"You cannot seamlessly integrate work and play."

"As the European economy strengthens and the American economy weakens, European quality-of-life values will begin to hold more influence, not less."

"While good work requires time, so does personal leisure. Unless all work in 2020 is about sending e-mails back and forth, the attention needed for quality work will still be the same, and that means people will have to choose between work and family."

"Sooner or later, people will realize that being on a leash 24/7 is slavery, not freedom."

"We will see fewer and fewer offices, as people work from homes, gyms, malls, etc."

"Hopefully, there will be technology to let people still leave their work at work, and not be punished."

"Ever met a crackberry [BlackBerry] addict? I have one in my family, and he drives everybody else crazy."

"Maybe they can enable their computers to work on something when they are asleep."

"There will be a deep divide between occupational types (e.g., blue-collar vs. white-collar)."

"This may be the future, but it is more than 13 years away due to the major change in levels of productivity that it will require. France, for example, is headed toward more work and less leisure time."

"Call this the "crackberry syndrome." It's already happening for business people. Almost any job at any global company requires 24/7 access and global coordination. I do see a revolt coming soon—the Gen Yers are more into quality of life, and this will breed less interest into being on 24/7."

"I *partially* mostly agree. Technology allows people to expand their professional hours into their personal time. I have colleagues who bring their BlackBerrys on vacations because it feels less like 'work' than bringing their laptops. But the only times that vacations are pure vacations are when people go somewhere with no network access. On the contrary, I do not see employers loosening up their expectations that workers be on site during regular business hours. The extent of this may rely on the nature of the work. Some work requires intense collaboration among teams. Some work can be done by a single individual; workers who fall in this category are more likely to benefit in setting their own work-life schedules than are people in the former."

"Things will largely be much the same as they are today, although with higher technology and more ubiquitous 'connectedness.'"

"It is NOT OK for businesses to expect human beings to be 'on call' 24 hours a day, and it never will be OK no matter what kind of technology we have developed."

"Yes, this will happen. No, it won't be OK. We see the beginnings of it already, and it's going to create problems and not become a comfortable, new status quo."

"Not me, ever. If this happened, the incidence of stress-related illness, heart attacks, etc., would skyrocket. This is a depressing scenario that I would never want, and I am a techie and very dedicated to my job!"

"I'd like to think that greater flexibility will be brought into the workplace and that people will have greater freedom over how they blend their professional and personal lives, but I don't think the corporation is ready to allow worker drones that kind of freedom, at least not for the vast majority of workers."

"Most employees will never be happy blending their personal and professional lives as long as employers are trying to make as much money as possible on their work."

"This generation's children will almost surely reject their parent's technophilia and demand more human contact and more time free from the constraints of work."

"It is never OK for someone to not have time away from work. That's how you create giant stress balls. While I can certainly see this as an increasing expectation, it is not one we should be happy about."

"It's happening already with the millennials. Whether we use the Net or something else, this is happening."

"Too few people love what they do for this to ever hold true."

"It's just the continuation of a trend that was begun in the '60s."

"People will increasingly seek vocations to which they are called, however, so stress and work will be less of an issue and they are doing what they really love."

"This will be of great benefit to individual productivity. People will be able to work when they are the most creative—assuming that companies learn how to manage this new workforce."

"Generation X and Y have grown up with a computer/phone available at all times and are comfortable with online banking, social networking, and being always in touch. Information overload will demand more and more of their time; family life and free time will suffer."

“The knowledge worker will want flexibility of work area, but clear lines between personal and professional hours.”

“I hate this scenario; where does the family structure go? We all need time away from our jobs and need our personal time.”

“I find this prospect appalling.”

“Sounds good to me. I commute an hour and a half every day.”

“While I would love to see this happen and see elements of this in today’s world, I find it hard to believe that we will move away from traditional ‘9 to 5’ working hours for most people.”

“While corporations will push for lack of division, the increased communication between workers in different countries will create a greater drive for personal time. Increased conflicts leading to emotional problems will cause more people to seek a balance between work and play.”

“As an IT worker, my work life is already integrated into my home life. It seems to be impossible to escape. It may well be the downside of being well educated and versed in technology. Instead of working a shift and going home, you have to or get to work from home, from the train, from work. You also get to use personal e-mail from home, the train, and work. I think 11 to 7, 7 to 3, and 3 to 11 will always have meaning.”

“The stress is intense and can lead to a degradation of the quality of the work, even though the quantity continues to build. It is, to put it simply, nasty.”

“This is positive for employers but *not* necessarily employees. This is already the culture in medicine and education. Employees are expected to be available anytime, even during vacations if necessary. People who have jobs they love will easily embrace this, and people who do not get satisfaction from their jobs will stubbornly try to remain ‘unplugged.’”

“I foresee a backlash where people enforce their own parameters on ‘work’ and ‘my’ time.”

"Fewer hours spent in the office are exchanged for some time online at home going over any pressing issues that may have been missed while you were at the kid's soccer game."

"It will not necessarily be a net positive, as work is never entirely left behind. This scenario is clearly in play already."

"More and more people are working from remote locations, putting in hours between activities, and increasing productivity."

"I live this life now, and I love it. I've been weaving work and family life together for years, and it works very well for me."

"By 2020, there will be a backlash to the blurring of boundaries between personal and work time. Individuals will incorporate more ways to stop work from intruding into their private lives."

"Not in 12 years, if ever."

"It is likely that social conventions for going 'offline' will be established, so the effect will be mitigated to some extent."

"The absence of a dividing line between work and leisure will probably only really effect information workers, but the revised mentality toward constant work will likely propagate throughout the economy. Between this and your world where I have to pay a nickel every time I hear a Justin Timberlake song, I would probably like to hang myself."

"This is a horror scenario, and it could also well be that just the opposite happens, because it will socially not be tolerable and will lead to empowering enterprises too much, thus driving a counterreaction."

"My professional time overwhelms my personal time."

"It will happen. I do not think it will be a good thing from a life-work balance standpoint. Even the addition of a BlackBerry means that you spend MORE time working, not that you get more done in less time or that you are more productive. Rather, from a social standpoint, you wind up wasting work time because you feel you have lost your personal time."

"There will always be a need for some structure. There will always be a deadline for a project, and not everyone is great at controlling their time."

"Technology will have to deal with bigger requests for more spare and family time from more evolved people."

"By 2020? How about in 2008? It's already significantly true."

"The distinction between 'online' and 'real life' will go, but if anything, there'll be a backlash with people being at work 24/7."

"My greatest struggle is to *avoid* this blending, as I'm never completely relaxed as it is! This might be OK for some professions, but I don't see this by 2020."

"People will be burning out because they can't escape their e-mail and other info feeds. People will eventually declare info bankruptcy, and there will be a backlash."

"Already now, many people have difficulties in time management. They cannot separate work life and family life. This causes stress and concentration problems."

"People will mix work and play more. But if you think as a parent I want to mix work with raising my children and the time I spend with them, you have to be crazy. Personal time will still be important in 2020."

"This scenario will be true in the U.S. but not necessarily in Europe, for instance, where people still take real vacations of several weeks."

"I fear the future because people will always be working."

"We're already seeing the effects of information overload and burnout for those who are truly plugged in."

"This is just exploitation with a pretty face."

"For people who experience this now (as I do), I can't say it's necessarily a net positive. It provides more schedule flexibility, but it does not provide a discernable productivity increase."

"I would agree if employers weren't so paranoid about employees not really working. Out of sight, out of control."

"I mostly disagree because I think it will be 'expected' but not OK with a lot of people."

"The death of space and time has been predicted for a long time, but has not become reality…It is likely that play will be more integrated into work and that at work, as well as at home, the work/play/life boundaries will become vaguer."

"I already work largely in this manner and love the flexibility it brings to both my work and home life."

"We will find a population oppressed and abused by the economic giants as they eke out the last drop of profit for the corporate coffers."

"While it is easy for professional duties/work to intrude on personal time, it will be more difficult for employers to accept the blending of personal and professional time."

"It's not OK. It's a great boon to businesses so people will work extra hours for nothing."

"Utopian notions like this pop up every 100 years or so. Utopian communities form and then drift away. For all but a handful of lucky people, work is necessary to survive, and survival is hard work. People need to disconnect from work or life becomes drudgery. Most of us can't turn our minds on and off like a lightbulb. If you are in the middle of an engaging leisure activity, it is ruined when the boss calls. You never really relax. And it's hard to do good work when you are interrupted by play all the time."

"Fewer lines divide personal and professional time, but that is probably not a net positive for most, only the ruling class."

"There will be enormous pushback from the majority of the population when it comes to integrating work and leisure. The few who would adopt this kind of arrangement are those same individuals whose work already pervades their daily life outside of the office."

"People will become more stressed, have more health problems, and get even less exercise than they do now."

"I have already started to blur the lines—and it seems like as long as I get the work done, it's not a big deal. I work at home; I play around at work sometimes. I hope this becomes more acceptable."

"Our lives are, for the most, linear, and as humans we prefer that compartmentalization. Any situation that is otherwise will be tantamount to chaos."

"Again, another blow to the traditional family structure."

"Life is on a fast pace, and I sometimes wonder if that is good for children and young people. Whatever happened to time to relax and reflect?"

"Work defines a person more so now than before, so this will be common."

"People don't consider personal time 'not-working' time."

"People still won't be nearly as good at multitasking as they think they are, and there will still be biannual lamentations for declining attention spans."

"Sanity requires that people can exercise some personal control over how they manage their time."

"I say mostly disagree here based on what we know about millennials—that they want personal time away from work and want a separation between their work and nonwork lives. While they're definitely into technology, I don't think that necessarily means they will want to use it the way baby boomers and Gen Xers have to stay connected to work 24/7."

"While personal hours are definitely blending with work hours, it is unlikely that corporations will willingly relinquish the idea that they are paying for your time and expect you to spend specific hours doing what they ask, where they ask, in a way they ask."

"That's the trend line, but in 5 to 7 years, there will be mounting evidence that such a trend becomes counterproductive."

"Just like the universe, work will expand into our lives more and more—and then it will have to contract or change fundamentally."

"This is the reality today. I have 60 employees who work for me, and they don't recognize a boundary."

"This has taken a role in my life. I don't see a problem with it and, as far as I can tell, is still beneficial to me."

"It is extremely unhealthy to assume no boundary between personal and professional life. Such blurring of the lines will in fact dramatically reduce productivity, and businesses themselves will encourage some meaningful division."

"People like flexible working and fitting work into their lives, but there are limits. We need turn-off time. We need to feed screaming babies. We need to get out of breath in the swimming pool."

"The current state of continuous partial attention in which workers find themselves in 2008 will by 2020 prove to be detrimental to work productivity and detrimental to personal freedom."

"In 2020: Although most tend to complain about blurred professional status, digital workers are generally employed by an individual corporation and hired on the basis of ad hoc contracts. A segment of disenfranchised workers are still directly employed, although they resent it as a competency challenge."

"For a large percentage of the working population and for students, such a model would simply allow them to accomplish their tasks."

"Given the workaholic nature of the United States, I predict that the economic interests of corporations in squeezing even more work out of their employees will win out."

ENDNOTES

1. See *The Hyperconnected: Here They Come!* an IDC white paper sponsored by Nortel, published May 2008(a), available at http://www.nortel.com/promotions/idc_paper/index.html.
2. See *E-Mail Addiction*, by America Online and Opinion Research Corp., published July 26, 2007, available at http://corp.aol.com/press_releases/2007/07/think-you-might-be-addicted-email-youre-not-alone.

3. See *BlackBerry Backlash? Americans Split on "Always On" Culture: Digital Life America Survey*, published by Solutions Research Group, February 15, 2007, available at http://www.srgnet.com/pdf/BlackBerry%20Backlash%20(SRG%20%20Report%20Feb%2007).pdf.
4. See *Knowledge Workers Putting in Longer Workweeks*, a report issued by Ipsos and Lexmark, February 20, 2007, available at http://www.lexmark.com/lexmark/pressrelease/home/0,6930,204816596_653271419_1002010230_en,00.html.
5. See *Wireless Working: Rudeness Made Easy?* published by Yahoo HotJobs, April 2008, available at http://hotjobs.yahoo.com/career-articles-wireless_working_rudeness_made_easy-378.
6. See "BlackBerries, Blogs Create Overtime Work Disputes," written by Ellen Wulfhorst for *Reuters* and published in *USA Today*, June 25, 2008, available at http://www.usatoday.com/tech/products/2008-06-25-blackberry-blogs-overtime-pay_N.htm?csp=tech.
7. See "Isolation of a Central Bottleneck of Information Processing With Time-Resolved fMRI," by Paul Dux, Jason Ivanoff, Christopher Asplunc, and Rene Marios, published December 21, 2006, available at http://www.psy.vanderbilt.edu/faculty/marois/Publications/Dux_et_al-2006.pdf.
8. See *Conversations Amidst Computing: A Study of Interruptions and Recovery of Task Activity*, by Shamsi Iqbal and Eric Horvitz, Department of Computer Science, University of Illinois, Microsoft, 2007, available at http://research.microsoft.com/~horvitz/conversational_interruptions.pdf.
9. See "Issues for DSM-V: Internet Addiction," by Jerald J. Block, published in the *American Journal of Psychiatry*, March 2008, http://ajp.psychiatryonline.org/cgi/content/full/165/3/306.
10. See "'Quiet Time' and 'No E-mail Day' Pilot Data," by Nathan Zeldes for *IT@Intel*, June 14, 2008, available at http://communities.intel.com/openport/blogs/it/2008/06/14/-quiet-time-and-no-email-day-pilot-data-is-in.
11. See *Teens and Social Media*, by Amanda Lenhart, Mary Madden, Alexandra Rankin Macgill, and Aaron Smith for the Pew Internet & American Life Project, December 19, 2007, available at http://www.pewinternet.org/pdfs/PIP_Teens_Social_Media_Final.pdf.
12. See *Thoughts on Attention and, Specifically, Continuous Partial Attention*, by Linda Stone, June 24, 2008, available at http://continuouspartialattention.jot.com/WikiHome.

13. See "Multitasking: Getting Things Half-Done?" by David Needle, published by *Internetnews.com*, March 31, 2008, available at http://www.internetnews.com/bus-news/article.php/3737601.
14. See "Freedom and Its Digital Discontents: Information Overload," published in *The Economist* (online), February 26–March 7, 2008, available at http://www.economist.com/debate/index.cfm?action=summary&debate_id=5.
15. See *The Wisdom of Patients*, by Jane Sarasohn-Kahn, California Health Care Foundation, April 2008, available at http://www.chcf.org/topics/chronicdisease/index.cfm?itemID=133631.
16. See "'Revolutionary' Collective Intelligence of Users Touted at Web 2.0 Expo," by Heather Havenstein, published in *Computerworld*, April 23, 2008, available at http://www.computerworld.com/action/article.do?command=viewArticleBasic&articleId=9080038&source=NLT_AM&nlid=1.

Methodology

This Web-based survey, sponsored by the Pew Internet & American Life Project and Elon University conducted by Princeton Survey Research Associates International, elicited a nonrandom sample of 578 leading Internet activists, builders, and commentators. An additional 618 stakeholders also participated in the study, for a total of 1,196 participants who shared their views.

Experts were located in two ways. First, nearly a thousand were identified in an extensive canvassing of scholarly, government, and business documents from the period 1990–1995 to see who had ventured predictions about the future impact of the Internet. Several hundred of them participated in the first two surveys conducted by Pew Internet and Elon University, and they were recontacted for this survey. Second, expert participants were handpicked due to their positions as stakeholders in the development of the Internet, or they were reached through the leadership listservs of top technology organizations including the Internet Society, Association for Computing Machinery, the World Wide Web Consortium, the United Nations' Multistakeholder Group on Internet Governance, Internet2, Institute of Electrical and Electronics Engineers, Internet Corporation for Assigned Names and

Numbers, International Telecommunication Union, Computer Professionals for Social Responsibility, Association of Internet Researchers, and the American Sociological Association's Information Technology Research section. For the first time, some respondents were invited to participate through personal messages sent using a social network, Facebook. In all, 578 experts identified through these channels responded to the survey.

While many respondents are at the pinnacle of Internet leadership, some of the survey respondents are "working in the trenches" of building the Web. Most of the people in this latter segment of responders came to the survey by invitation because they are on the e-mail list of the Pew Internet & American Life Project or are otherwise known to the Project. In all, 618 additional respondents participated in this survey from these quarters. Thus, the expert results are reported as the product of 578 responses, and the lines listing "all responses" include these additional 618 participants.

In the survey, participants were asked if they mostly agreed or mostly disagreed with eight scenarios about the potential evolution of the Internet we might see by 2020. They were given the opportunity to elaborate on their answers. The scenarios—woven from material collected in recent industry and research reports and from predictive statements by leaders in science, technology, business, and politics—were layered with overlapping elements to spur discussion and elicit nuanced views of the future.

INVITATION TEXT

2008 Predictions Survey Invitation

The Pew Internet Project and Elon University are conducting an ongoing survey of stakeholders about the future of the Internet, and we would like to include your views in our research.

This Web-based survey about international concerns and the Internet follows two previous surveys of thousands of Internet stakeholders that measured the expected impact of the Internet over the next decade (to see the results, please go to http://www.elon.edu/e-web/predictions/expertsurveys/default.xhtml). This year, we are soliciting predictions from thoughtful analysts in regard to a number of additional issues tied to the future of information and communications technologies (ICTs).

We hope you'll take 10 to 15 minutes to fill out our survey (to participate, you must use Firefox, Internet Explorer, or Safari as your browser). You will find the survey at http://websurveyor.net/wsb.dll/WSGateway?surveyid=195410

The survey asks you to assess several potential scenarios tied to the future of the Internet and to contribute your own thoughts about what you believe should or will happen by the year 2020. This is a confidential survey; however, we encourage you to take credit for your thoughts. After each question, you are invited to explain or expand on your views. Each elaboration you provide will remain anonymous unless you put your name at the ***start*** *of it.*

When you begin the survey, please use this personal identification number (PIN): XXX

The Pew Internet Project will issue a report based on this survey in the spring; we expect the results to be useful to policy makers, scholars, and those in the information technology industry. Material from this survey will be added to the Elon University/Pew Internet site, Imagining the Internet (http://www.imaginingtheinternet.org). We will not use your name or e-mail address for any purpose other than this research project, and we will not share your information with outside solicitors.

We are certain we have not identified all individuals whose views would be helpful to this research, so I invite you to send an invitation to participate in this survey to any friends or colleagues whose insights would be of value. Please ask them to use PIN 9000 when taking the survey.

I hope you enjoy taking the survey and sharing your views about the future of the Internet. If you have any questions, please feel free to contact me at lrainie@pewinternet.org.

Thank you,
Lee Rainie
Director, Pew Internet & American Life Project
1615 L Street NW
Suite 700
Washington, DC 20036

TOPLINE SURVEY RESULTS

Future of the Internet III Web Survey Final Combined Topline 3/04/08

Data for December 26, 2007–March 3, 2008

Princeton Survey Research Associates for the Pew Internet & American Life Project

Sample: n = 1,196 Internet experts
Interviewing dates: 12.26.07–3.03.08

Note. The current survey results are based on a nonrandom online sample of 1,196 Internet stakeholders, recruited via e-mail invitation from the Pew Internet & American Life Project. Since the data are based on a nonrandom sample, a margin of error cannot be computed, and the results are not projectable to any population other than the experts in this sample.

General Survey

Thank you for agreeing to assess the following potential future scenarios. Most are provocative extensions of recent statements by leaders in science, technology, business, and politics. Each scenario or set of scenarios is prefaced by an extremely brief explanation of the status in 2007 regarding the issues involved. You will be asked if you "mostly agree" or "mostly disagree" that a given scenario could become a reality in the year 2020. Given that these are complex scenarios contingent on many different factors, you may agree with some aspects of the prediction while you disagree with others.

The predictions made here are offered in the spirit of testing ideas about how the future might unfold. They are *not* meant to represent our "best guess" or our preferences about the future. Neither the Pew Internet Project nor Elon University advocates any policy outcomes related to the Internet.

By the Year 2020…

The first section of the survey asks you to assess predictions about what the Internet will be like in 13 years.

Please proceed to share your views about these proposed scenarios…

Mobile Telephony, Computing, and Digital Inclusion

Status Quo 2007: According to the UN/ITU *World Information Society Report 2007*, there has been some progress in improving digital inclusion: In 1997 nearly three-quarters of the world's population who lived in low-income and lower middle-income economies accounted for just 5% of the world's population with Internet access. By 2005, they accounted for just over 30%. A number of commercial and nonprofit agencies are combining forces to bring inexpensive laptop computers to remote regions of the world to connect underserved populations. In addition, by the end of 2008, more than half the world's population is expected to have access to a mobile phone.

Please Respond to This 2020 Scenario...

Q1 The mobile phone is the primary connection tool for most people in the world. In 2020 while One Laptop Per Child and other initiatives to bring networked digital communications to everyone are successful on many levels, the mobile phone—now with significant computing power—is the primary Internet connection and the only one for a majority of the people across the world, providing information in a portable, well-connected form at a relatively low price. Telephony is offered under a set of universal standards and protocols accepted by most operators internationally, making for reasonably effortless movement from one part of the world to another. At this point, the "bottom" three-quarters of the world's population account for at least 50% of all people with Internet access—up from 30% in 2005.

	Current	
%	81	Mostly Agree
	19	Mostly Disagree
	*	Did Not Respond

Q1b Please elaborate on your answer. Remember, if you want your answer cited to you in published material about this survey, please begin by typing your name. Do not type your name if you want your comments to be anonymous. We encourage you to take credit wherever possible.

	Current	
%	50	Gave Elaboration
	50	Did Not Respond

Please Respond to This 2020 Scenario...

Q2 Social tolerance has advanced significantly due in great part to the Internet. In 2020 people are more tolerant than they are today thanks to wider exposure to others and their views that has been brought about by the Internet and other information and communication technologies. The greater tolerance shows up in several metrics, including declining levels of violence, lower levels of sectarian strife, and reduced incidence of overt acts of bigotry and hate crimes.

	Current	
%	33	Mostly Agree
	55	Mostly Disagree
	11	Did Not Respond

Q2b Please elaborate on your answer. Remember, if you want your answer cited to you in published material about this survey, please begin by typing your name. Do not type your name if you want your comments to be anonymous. We encourage you to take credit wherever possible.

	Current	
%	50	Gave Elaboration
	50	Did Not Respond

Intellectual Property/Digital Rights Management

Status Quo 2007: Major content producers such as the world's music and film businesses are lobbying governments and courts to protect their assets. *Digital rights management* (DRM) is one of the umbrella terms used to describe various technologies being developed to help copyright holders control access to digital products and prevent copying. Its intent is to assure that content creators maintain control of their work and are rewarded with appropriate compensation. Opponents of DRM say its language and approach are geared toward forcing public acceptance of intellectual monopolies. They argue that the movement toward assigning ownership of everything stifles innovation and competition, saying DRM is actually "digital restrictions management," and IP stands for "intellectual protectionism" and "intellectual poverty."

Please Respond to This 2020 Scenario...

Q3 Content control through copyright protection technology dominates. In 2020 strict content controls are in place thanks to the efforts of legislatures, courts, the technology industry, and media companies. Those who use copyrighted materials are automatically billed by content owners, and Internet service providers automatically notify authorities when they identify clients who try to subvert this system. Protesters rarely prevail when they make claims that this interferes with free speech and stifles innovation.

	Current	
%	31	Mostly Agree
	61	Mostly Disagree
	8	Did Not Respond

Q3b Please elaborate on your answer, typing your name at the beginning if you would like to be credited with your response.

	Current	
%	52	Gave Elaboration
	48	Did Not Respond

Digital Identity/Security/Transparency/Privacy

Status Quo 2007: People openly share more intimate details of their lives online every day, and they are flocking to social networks and uploading and/or viewing homemade videos by the millions. Ubiquitous computing is diffusing into everyday life. Much of what goes on in daily life is more visible—more transparent—and personal data of every variety is being put on display, tracked, tagged, and added to databases. The number of mobile camera phones in use will top 1 billion in 2007; miniaturized surveillance cameras are simultaneously becoming extremely inexpensive, sophisticated, and pervasive; clothing is being designed with technology woven into the fabric; and it is expected that most surfaces can and will be used as two-way interfaces in the future.

Please Respond to This 2020 Scenario...

Q4 Transparency heightens individual integrity and forgiveness. In 2020 people are even more open to sharing their personal information, opinions, and emotions than they are now. The public's notion of privacy has changed. People are generally comfortable exchanging the benefits of anonymity for the benefits they perceive in data being shared by other people and organizations. As people's lives have become more transparent, they have become more responsible for their own actions and more forgiving of the sometimes unethical pasts of others. Being outed for some past indiscretion in a YouTube video or other pervasive media form no longer does as much damage to a person's reputation as it did back in the first decade of the 21st century. Carefully investigated reputation corrections and clarifications are a popular daily feature of major media outlets' online sites.

	Current	
%	44	Mostly Agree
	45	Mostly Disagree
	10	Did Not Respond

Q4b Please elaborate on your answer, typing your name at the beginning if you would like to be credited with your response.

	Current	
%	46	Gave Elaboration
	54	Did Not Respond

Augmented Reality, Mirror Worlds, and the Metaverse

Status Quo 2007: While most current Internet interaction is found in the user-generated content and social networks of Web 2.0, the 3-D Web-computing ecosystem is developing quickly. Augmented reality enables the enhancement of real-world information through the use and confluence of the Internet, RFID, GPS, smart-tag networks, and portable/wearable information technology. 3-D environments, which are just beginning to be more efficient and accessible, offer ideal design spaces for social and economic experimentation, rapid prototyping, and customized and decentralized production. Every item in the physical world is being mapped, tagged, and databased, as humans build mirror worlds (data-enhanced virtual models of the "real" physical world, also known as digital Earth systems, or the geoWeb), and innovate in new, virtual worlds (Second Life, Cyworld, World of Warcraft). MIT's Fall 2007 Emerging Technologies conference had a headline session titled "Second Earth: Second Life, Google Earth, and the Future of the Metaverse," with the explanation: "Social virtual worlds such as Second Life and mapping tools such as Google Earth are beginning to overlap, perhaps foreshadowing the advent of an immersive, 3-D 'metaverse.'" A 2007 Gartner study estimated 80% of all active Internet users will have virtual selves by the end of 2011.

Please Respond to This 2020 Scenario...

Q5 Many lives are touched by the use of augmented reality or spent interacting in artificial spaces. In 2020 virtual worlds, mirror worlds, and augmented reality are popular network formats thanks to the rapid evolution of natural, intuitive technology interfaces and personalized information overlays. To be fully connected, advanced organizations and individuals must have a presence in the "metaverse" and/or the "geoWeb." Most well-equipped Internet users will spend some part of their waking hours—at work and at play—at least partially linked to augmentations of the real world or in alternate worlds. This lifestyle involves seamless transitions between artificial reality, virtual reality, and the status formerly known as "real life."

	Current	
%	56	Mostly Agree
	31	Mostly Disagree
	13	Did Not Respond

Q5b Please elaborate on your answer, typing your name at the beginning if you would like to be credited with your response.

	Current	
%	40	Gave Elaboration
	60	Did Not Respond

Q6 **Talk and touch are common technology interfaces.** In 2020 the most commonly used communications appliances prominently feature built-in voice recognition. People have adjusted to hearing individuals dictating information in public to their computing devices. In addition, "haptic" technologies based on touch feedback have been fully developed, so, for instance, a small handheld Internet appliance allows you to display and use a full-size virtual keyboard on any flat surface for those moments when you prefer not to talk aloud to your networked computer. It is common to see people "air typing" as they interface with the projection of a networked keyboard visible only to them.

	Current	
%	67	Mostly Agree
	19	Mostly Disagree
	14	Did Not Respond

Q6b Please elaborate on your answer, typing your name at the beginning if you would like to be credited with your response.

	Current	
%	33	Gave Elaboration
	67	Did Not Respond

Architecture and the Next-Generation Internet

Status Quo 2007: Due to concerns over Internet security, reliability, and complexity, the National Science Foundation in the United States is funding research into the building of a "next-generation" or "clean-slate" Internet. The NSF initiatives include the Global Environment for Networking Innovation (GENI—building a test network on which researchers will be able to try out their ideas) and Future Internet Network Design (FIND). The European Union is funding research through its Future Internet Research and Experimentation (FIRE) program. Creating an all-new Internet might solve problems like viruses, spam, phishing, and worms. But it would cost billions of dollars, and there is a debate among experts about how long it might take. If a next-generation Internet is built, some people are concerned it will be characterized by intrinsic features that will allow governments and corporations to exercise more control over what happens online. So, the constant question remains: How do we raise barriers against spam, cybercrime, and terrorism and provide secure systems for digital transactions without infringing on civil liberties?

Please Respond to This 2020 Scenario...

Q7 **Next-generation research will be used to improve the current Internet; it will not replace it.** In 2020 the original Internet architecture is in the continuing process of refinement—it has not been replaced by a completely new system. Research into network innovation, with help from the continued acceleration of technologies used to build, maintain, enhance, and enlarge the system, has yielded many improvements. Search, security, and reliability on the Internet are easier and more refined, but those who want to commit crimes and mischief are still able to cause trouble.

	Current	
%	80	Mostly Agree
	6	Mostly Disagree
	14	Did Not Respond

Q7b Please elaborate on your answer, typing your name at the beginning if you would like to be credited with your response.

	Current	
%	26	Gave Elaboration
	74	Did Not Respond

Connectedness, Social Commerce, and Social Lives

Please Respond to This 2020 Scenario...

Q8 Few lines divide professional from personal time, and that is OK. In 2020 well-connected knowledge workers in more developed nations have willingly eliminated the industrial-age boundaries between work hours and personal time. Outside of formally scheduled activities, work and play are seamlessly integrated in most of these workers' lives. This is a net positive for people. They blend personal/professional duties wherever they happen to be when they are called upon to perform them—from their homes, the gym, the mall, a library, and possibly even their company's communal meeting space, which may exist in a new virtual-reality format.

	Current	
%	57	Mostly Agree
	29	Mostly Disagree
	14	Did Not Respond

Q8b Please elaborate on your answer, typing your name at the beginning if you would like to be credited with your response.

	Current	
%	43	Gave Elaboration
	57	Did Not Respond

Q9 **What do you want to know?** We consider these surveys to be a conversation with technology policy makers and leaders; we would like to know what ***you*** would like to know about the Internet and related technologies. What key research questions about the direction of technological change and about the impact of the Internet would you like to see addressed by organizations such as the Pew Internet Project? What should we be measuring? What research questions should we address in future surveys of Internet experts and leaders? Are there some critical uncertainties about Internet evolution in your field of expertise that we should position ourselves to monitor?

Did Not Tabulate

Methodology
Future of the Internet III Web Survey

Prepared by Princeton Survey Research Associates International for the Pew Internet & American Life Project
April 2008

Summary

The *Future of the Internet III* Web Survey, sponsored by the Pew Internet & American Life Project and the Imagining the Internet Center at Elon University, obtained online interviews with a nonrandom sample of 1,196 Internet experts. The interviews were conducted online, via SPSS, from September 26, 2007 to March 3, 2008.

Pew Internet and Imagining the Internet encouraged the initial sample of experts to forward the e-mail invitation to any colleagues whose thoughts on the future of the Internet they would consider useful and important. This created an additional snowball sample of Internet experts, whose ideas are also included in the final data.

Questionnaire Development

The questionnaire was developed by PSRAI in collaboration with staff of the Pew Internet & American Life Project and their partners at Elon University.

Note About Survey Data

PSRAI recommends including the following note with any public release of the data:

Note. Results are based on a nonrandom online sample of 1,196 Internet stakeholders, recruited via e-mail notices sent to an initial sample of pre-identified experts, as well as a snowball sample of their colleagues. In addition, a limited number of additional technology stakeholders were reached through an appeal using the Web 2.0 social network, Facebook. Since the data are based on a nonrandom sample, a margin of error cannot be computed, and the results are not projectable to any population other than those people who completed the survey.

ABOUT THE SURVEY PARTICIPANTS

Many top Internet leaders, activists, and commentators participated in the survey, including Clay Shirky, Fred Baker, David Brin, Susan Crawford, Brad Templeton, Howard Rheingold, Jim Kohlenberger, Josh Quittner, Seth Finkelstein, danah boyd, Hal Varian, Jeff Jarvis, Anthony Rutkowski, Michael Botein, Steve Jones, Richard Bartle, Alejandro Pisanty, Tom Vest, Milton Mueller, Bernardo Huberman, Jonne Soininen, Don Heath, Doug Brent, Anthony Townsend, Steve Goldstein, Adam Peake, Basil Crozier, Craig Partridge, Sebastien Bachollet, Geert Lovink, James Jay Horning, Dan Lynch, Fernando Barrio, Roberto Gaetano, Christian Huitema, Susan Mernit, Jamais Cascio, Norbert Klein, Tapio Varis, Martin Boyle, Ian Peter, Todd Spraggins, Catherine Fitzpatrick, Tom Keller, Charles Kenny, Robert Cannon, Hakikur Rahman, Larry Lannom, David Farrar, John Levine, Cliff Figallo, Sebastian Ricciardi, Lea Shaver, Seth Gordon, Jim McConnaughey, Neil Mcintosh, Charles Ess, Alan Levin, David W. Maher, Jonathan Dube, Thomas Vander Wal, Adrian Schofield, Clifford Lynch, Jerry Michalski, Paul Miller, and David Moschella, to name a few.

A sampling of the workplaces of respondents includes the Internet Society, World Bank, Booz Allen Hamilton, AT&T Labs, VeriSign, Cisco, BBN Technologies, Fing, Yahoo Japan, France Telecom, the International Telecommunication Union, Alcatel-Lucent, the Electronic Frontier Foundation, GLOCOM, AfriNIC, Electronic Privacy Information Center, APNIC, Universiteit Maastricht, Amnesty International, BBC, PBS, IBM, Microsoft, Forrester Research, Harvard University's Berkman Center for Internet and Society, Open Society Institute, Open the Future, Yahoo, First Semantic, CNET, Microsoft, Universidad Nacional Autónoma de México, IDG, FCC, Institute for the Future, 1&1 Internet AG, Moody's, HP Laboratories, Amazon.com, Gannett, Lexis/Nexis, Tucows, InternetNZ, ICANN, Oxford

Internet Institute, Institute of the Information Society—Russia, The Center on Media and Society, Online News Association, Nokia, the Association for the Advancement of Information Technology, Massachusetts Institute of Technology, the Institute of Network Cultures, Nortel, Disney, DiploFoundation, Information Technology Industry Council, J-Lab, Information Society Project at Yale University, Santa Fe Institute, the London School of Economics, Google, the University of California–Berkeley, NASA, the Singapore Internet Research Center, Princeton University, the federal government of Canada, several policy divisions of the U.S. government, and many dozens of others.

Participants described their primary area of Internet interest as "research scientist" (12%); "technology developer or administrator" (11%); "entrepreneur or business leader" (10%); "author, editor, or journalist" (9%); "futurist or consultant" (7%); "advocate, voice of the people, or activist user" (5%); "legislator or politician" (1%); or "pioneer or originator" (2%); however, many participants chose "other" (24%) for this survey question or did not respond (18%).

APPENDIX

SELECT BIOGRAPHIES

The 1,196 participants in the survey were allowed to retain complete anonymity, or they could enter their names while retaining the right to keep their answers anonymous; many longtime Internet luminaries chose to remain completely anonymous, and their names are not in the following compilation, nor are they used in any aspect of the final report. The survey respondents also had the opportunity to elect on each and every question they were asked whether or not they chose to have their name tied to their answer.

The following list of more than 400 biographies is a sample of respondents. It is not all-inclusive. It is offered to give a brief illustration of the expertise and background of the participants who were willing to give their names. Please note that individuals' views do not represent the views of their employers; their work is listed for identification purposes only.

Among the *Future of the Internet III* survey respondents who were willing to forgo total anonymity were the following people:

John C. Abell, new-media project director, Committee of Concerned Journalists

Stephen Abram, vice president for innovation and chief strategist, SirsiDynix, Special Libraries Association

Stephen Aguilar-Millan, director of research, European Futures Observatory, Ipswich, U.K.

Mikko Ahonen, researcher of management and information systems and ethics, University of Tampere, Finland; blogger

Shakib Ahsan, MBA and MA in educational technology, Concordia University, Canada

Itir Akdogan, PhD candidate and lecturer, University of Helsinki, expertise in ICT in empowering women and girls

Md. Tazin Alam, managing director and CEO, The ICT Associates Ltd (MKR Group), Dhaka, Bangladesh

Bryan Alexander, director of research National Institute for Technology and Liberal Education, blogger; based in Ripton, VT

Abubakar Alhassan, researcher in communication law, policy, and the digital divide at the University of Florida

Mary Ann Allison, principal, The Allison Group, has worked with Microsoft, Glasgow's Urban Learning Space, and other businesses, governments, and NGOs

Mariana Almeida, product manager of Web products for health care, Kaiser Permanente

David Alsmeyer, information services manager for British Telecom, and an expert on semantically enabled knowledge technologies

Jacobo Alvarez, chief executive officer, Multiplica, Barcelona, Spain

Betty Anderson, bilingual content manager, General Services Administration

Ken Anderson, senior researcher and anthropologist, Intel Research; specialties are globalization, identity, and urban studies

Shawn Kelly Apochromantic, configurations manager, designer, technologist, futurist, General Atomics, and volunteer builder in Second Life

Daniel do Amaral Arbix, Brazilian Ministry of Justice

Aaron Archer, PhD, a research scientist in the Algorithms and Optimization Group AT&T Labs Research, New Jersey

Geoff Arnold, senior principal, software development engineer, Amazon.com; formerly at Sun Microsystems in distributed computing at large scale

Sudip Aryal, president, Nepal Rural Information Technology Development Society

Earl Babbie, professor emeritus and writer, BEinc, prolific author of books on social research

Alan Bachers, PhD and director of the Neurofeedback Foundation

Sebastien Bachollet, president of Internet Society of France, operates the European Global Event on Domain Names and Address Systems, known as EGENI, active participant in ICANN

Fred Baker, Cisco Systems Fellow, former Internet Society (ISOC) chairman of the board; Internet Engineering Task Force (IETF) leader; an architect of the Internet

Troy Barker, principal, ICF International, a management, technology, and policy consulting firm

Dennis Baron, professor of English and linguistics, University of Illinois, runs the Web of Language site and researches the technologies of communication

Fernando Barrio, senior lecturer in business law and program leader for the Master of Art in E-business regulation, London Metropolitan University

Richard Bartle, PhD in artificial intelligence and a pioneer of massively multiplayer online games; a contributing editor for Terra Nova, the blog that covers virtual-world issues

Reva Basch, consultant for Aubergine Information Systems (online research expert); active longtime member of The WELL, one of the earliest cyberspace communities

Nancy W. Bauer, chief executive officer and editor-in-chief, WomenMatter, Inc.

Zori Bayriamova, research manager, Time Inc., previously supervisor for global brand analysis at Young & Rubicam and an associate analyst at Jupiter Research

Mark Bell, PhD student at Ball State University; manager of the SL Researchers mailing list; coauthor of *Second Life for Dummies*; a veteran of 15 years in software development

David Belman, founding partner, Threespot Media, an interactive media consultancy, previously with Magnet Interactive

Benjamin M. Ben-Baruch, senior market intelligence consultant and applied sociologist for Aquent, working at General Motors Corporation

Robert J. Berger, CTO for Cinch; expert on backbone, access, and wireless networks, and innovative Web applications, a frequent speaker on next-generation networks

Susan Bergman, director, Knowledge Center, working with NGO management in the Washington, DC, area

David Berkowitz, director of emerging media and client strategy, interactive industry strategist for 360i, writes Search Insider for MediaPost and the blog Inside the Marketer's Studio

Esha Bhatia, senior client solutions IT consultant, the Children's Hospital of Philadelphia

Peter Bihr, freelance consultant on Web strategies, communities, blogging, and social media; based in Berlin

Rebecca Blood, author of Rebecca's Pocket blog and *The Weblog Handbook: Practical Advice on Creating and Maintaining Your Blog*

Lynn Blumenstein, senior editor, *Library Hotline*, *Reed Business Information*

Christine Boese, PhD, researcher and analyst for Avenue A-Razorfish and Microsoft

Rob Boostrom, PhD student, Southern Illinois University–Carbondale, creator of the Society for Word of Mouth, "a group for educators to share ideas about sharing ideas"

Michael Botein, founding director of the Media Center at New York University Law School; consultant to the FCC

Nathan Botts, community informatics research associate at the Kay Center for E-Health Research; working on the national health information network (NHIN) and outreach

Hanane Boujemi, ICT researcher for DiploFoundation, educating people about Internet policy and Internet governance, Malta

Jak Boumans, principal director, Electronic Media Reporting; formerly editor in chief at Broadcast; general secretary of the European Academy of Digital Media

danah boyd, expert social media researcher, Harvard University's Berkman Center for Internet and Society

Michael Boyd, futurist and independent consultant

Martin Boyle, head of international ICT Policy, U.S. government department for business, enterprise, and regulatory reform; a Governmental Advisory Committee representative for ICANN

Judy Breck, blogger at GoldenSwamp.com—aimed at "watching the global golden age of learning emerge from the open Internet"

Scott Brenner, technologist, Web developer, consultant for clients ranging from Fortune 100 companies to small nonprofits

Doug Brent, COO, ICANN, with a 25-year record of management and leadership in technology companies; responsible administrator for Internet Assigned Numbers Authority

David Brin, futurist and author whose 1998 nonfiction book *The Transparent Society: Will Technology Force Us to Choose Between Privacy and Freedom?* identified key issues

Christine Brown, president, Marketing Resources and Results Inc., a blogger on branding and marketing

Erica Brown, British Broadcasting Corporation World Service News and Current Affairs

Lorelei Brown, director of user experience, expert on information architecture, interface, and design, Matrix Group

Richard Bruesehoff, director for lifelong learning, Evangelical Lutheran Church in America

Jean Burgess, PhD, Queensland University of Technology, Australia, researcher of user-generated content

Andrew Burnette, Web manager, Worldwatch Institute, and consultant on user experience, usability, strategy, and development

Jade Burnside, CEO, Ahead of the Web Marketing Services, New Westminster, British Columbia, Canada

Jay Buys, vice president for digital development, Fleishman Hillard, international marketing and communications company

Robert Cannon, senior counsel for Internet law at the Federal Communications Commission's Office of Strategic Planning and Policy Analysis and founder and director of Cybertelecom

Lisa Carr, director of strategy, Targetbase Interactive, health care strategist and writer

Nicholas Carr, former editor of *Harvard Business Review*; author of *The Big Switch: Rewiring the World, From Edison to Google* and *Does IT Matter?* and the Rough Type blog

Juan Carreón, professor at the Universidad Nacional Autónoma de México, and a columnist at CompuChannel

Tiago Casagrande, works with social communications and new technologies for verbeat

Jamais Cascio, originator of Open the Future, also works with the Institute for the Future, Institute for Ethics and Emerging Technologies, and Worldchanging

Michael Castengera, senior lecturer at the University of Georgia's Grady College and president of Media Strategies and Tactics Inc., a media consulting firm

Edward A. Cavazos, principal, Fish & Richardson, PC, a technology firm; expert on e-commerce and IP law, author of *Cyberspace and the Law*; formerly director at Bridges.org

Martha Chaconas, director for Web management, U.S. Department of State, responsible for management of operations of more than 200 sites delivering U.S. diplomacy information

Clement Chau, research manager for the Developmental Technologies Research Group at Tufts University

Edward Cherlin, activist and founder of Earth Treasury—a program following on One Laptop Per Child to link schools around the world; formerly a board member at WIRE AFRICA

James W. Chesebro, director of the digital storytelling master's program at Ball State University, former editor of *Communication Quarterly*, *Review of Communication*, and *Critical Studies in Media*

Lloyd Chesney, chief technology officer, MDVIP, a health, wellness, and fitness company

Barry K. Chudakov, principal, Chudakov Company, marketing and advertising creative consultant

M. Ford Cochran, director of education and mission programs online, *National Geographic*, and an environmental scientist, writer, and Web producer

Janet D. Cohen, blogger, futurist, and trend analyst

Rollie Cole, director of technology policy, Sagamore Institute for Policy Research, a nonprofit, nonpartisan think tank headquartered in Indianapolis, IN

Jonathon D. Colman, associate director for digital marketing at The Nature Conservancy

Joan Connell, online editor, *The Nation* magazine, formerly an executive producer for MSNBC.com, senior editor for MSN, and a Pulitzer finalist for her reporting

Luca Conti, conversational-media consultant, BuzzParadise, and a professor at Urbino University and the University of Malta in Rome; author of books on Skype, search engines

Tamara Coombs, senior Web and strategy specialist, American Chemical Society

Ted M. Coopman, lecturer, San Jose State University

Dave Coustan, author of the Extraface blog and an independent consultant who formerly wrote EarthLink's official blog, Earthling, and Stuffo, a subbrand of HowStuffWorks.com

Susan Crawford, founder of OneWebDay, celebrated each September 22; ICANN board member; associate professor, Cardozo Law School; visiting professor, Yale Law School

Trisha Creekmore, interactive executive producer, Discovery Channel Interactive

Jutta Croll, managing director, Stiftung Digitale Chancen—Digital Opportunities Foundation, Berlin, Germany, promoting and supporting access and equal opportunities for all online

Basil Crozier, senior strategist for cyber security, Public Safety Canada; participated in UN ICT Task Force, preparatory phase for the World Summits on the Information Society

Martin Curley, senior principal engineer and global director of IT People, intellectual capital and solutions at Intel; formerly of General Electric in Ireland and Phillips in The Netherlands

Michael Dahan, professor, Sapir Academic College, Israel; Digital Jerusalem; his works include the paper "National Security and Democracy on the Internet in Israel"

Hussein Dajani, knowledge director and strategic planner, JWT, a global communication and advertising company, based in Kuwait City, Kuwait

Rafik Dammak, software engineer, STMicroelectronics, Tunisia; DiploFoundation participant in the study of Internet

Dick Davies, partner, Project Management and Control, Inc.; past president of the Association of Information Technology Professionals

Hank Dearden, director of business development, Digital Industry, Inc., a provider of technology services in the Washington, DC, area

Nick Dearden, campaigns manager, Amnesty International, the human rights organization

Duane Degler, user-centered designer and strategist for Design for Context, writer and editor for IPGems, focused on semantic integration and performance improvement

Susan Dellinger, Web developer, CSC, U.S. Environmental Protection Agency

Walt Dickie, executive vice president and chief technology officer, C&R Research, one of the principal developers of CheckMetrix in the early '90s.

Ed Dieterle, Harvard Graduate School of Education; research tied to handheld devices for ubiquitous learning

Fred Dillon, director of product development, HopeLab, working to improve the quality of life of young people with chronic illness

Drew Diskin, director of e-strategy, Johns Hopkins Medicine

Alex Don, linguist and educator

Xin Dong, researcher concentrating on effective search techniques, data integration, record linkage, XML query optimization, AT&T Research; originally from China

Patricia Dowling, trend manager, Commission for Communications Regulation, Ireland

Diane M. Drew, health education and communication specialist, Centers for Disease Control and Prevention

Jonathan Dube, president of Online News Association, director of digital media at CBC News, the Canadian Broadcasting Corporation, publisher of CyberJournalist.net

Emma Duke-Williams, lecturer in the School of Computing and researcher, University of Portsmouth, U.K.

Brian Dunbar, Internet services manager, NASA office of public affairs

Katie Dunneback, novelist and blogger; founder of the Young Librarian Web site and weblog

Jeska Dzwigalski, director of community and product development, Linden Labs, the company behind Second Life

Peter Eckart, director of health information technology, Illinois Public Health Institute

John Eckman, practice director, Next Generation Internet, Optaros, Inc., a professional services firm offering strategy, design, development, consulting tied to open-source software

Michael Edson, director for Web and new-media strategy, Smithsonian Institution

Bruce Edwards, associate dean for distance education, Bowling Green State University, OH

Robert Eller, Concept Omega, a media marketing and communication company

Tyler Emerson, executive director, Singularity Institute for Artificial Intelligence

Mary Emery, associate director, North Central Regional Center for Rural Development, Iowa State University

Olivier Ertzscheid, lecturer in information science, Nantes University, France

Charles Ess, a professor of philosophy and religion and researcher on online ethics, Drury University, Springfield, MO, and a leader of Association of Internet Researchers

Glen Eustace, director at GodZone Internet Services, part of AGRE Enterprises Ltd.; infrastructure development engineer, Massey University, New Zealand

David Farrar, public policy chair for InternetNZ, the nonprofit organization for Internet service providers, public information groups, and Internet users that represents New Zealand

Stan Felder, president and CEO, Felder Communications, a marketing and advertising firm in Grand Rapids, MI

Cliff Figallo, social innovator and original member of the first online community—The WELL, now of AdaptLocal.org; expert in fitting and implementing social Web applications to groups

João Miguel Rocha Filho, director, DataOne, a provider of software for connecting to Linux; based in Brazil

Kate Finberg, retired research scientist, Digital Futurist Consultancy

Seth Finkelstein, anticensorship activist and programmer, author of Infothought blog and an EFF Pioneer Award winner

Catherine A. Fitzpatrick, Open Society Institute, Physicians for Human Rights; lecturer on humanitarian issues; formerly UN representative for International League of Human Rights

Fabrice Florin, executive director, NewsTrust.net, a nonprofit social news network that allows people to rate the news on quality

Micheál Ó Foghlú, research director, Telecommunications Software & Systems Group, Waterford Institute of Technology, Ireland advisory committee; member of W3C; blogger

Laura Forlano, NYCwireless board member with research interest in mobile and wireless technology; visiting fellow at the Information Society Project at Yale University

Marcus Foth, senior research fellow, Queensland University of Technology, Brisbane, Australia

Vassilys Fourkas, professor, Aristotle University of Thessaloniki, Greece, department of spatial planning and development

Mark Fowler, director and specialist on technology trends, research and strategy, Global Foresight Network, an international consultancy

Susan White Frazier, research manager, National Geographic

Amy Friedlander, director of programs for the Council on Library and Information Resources, a nonprofit that services research and higher education

Yves Froidevaux, Swiss Federal Statistical Office

Gerhard Fuchs, senior lecturer, Munich, Germany

Roberto Gaetano, ICANN board member; also responsible for software development for International Atomic Energy Agency; an active participant in the ICANN policy-making process

Beth Gallaway, Information Goddess Consulting

Matt Gallivan, senior research analyst, National Public Radio

Oscar Gandy, author, activist, retired emeritus professor of communication, University of Pennsylvania

Iddo Genuth, founder and chief editor of The Future of Things, a science and technology e-magazine

Mitko Gerensky-Greene, manager of Web and data services for Universal Service Administrative Company, providing e-rate telecommunications to schools and libraries

Stephanie Geyer, associate vice president e-communications and Web strategy, Noel-Levitz, a division of Sallie Mae

Heath Gibson, manager of research and market analysis, Big Pond, a competitive intelligence company and provider of broadband customer Web sites in Australia

Kathryn K. Goldfarb, president, KG Communications, an independent consultancy

Steve Goldstein, ICANN board member, retired from National Science Foundation, where his job in the 1990s was to help diffuse the Internet globally

Seth Gordon, director, user experience, Infospace and Saltmine; popular speaker and lecturer at CNet Conferences, ASIS, Harvard Cyberposium

James Gorman, principal, Working Technology Partners, a company offering technology solutions to businesses

Patricia Graham, chief marketing officer and executive vice president, Knowledge Networks, a consumer-information and trends-research company

Jeremy Swinfen Green, Telecom Express, an interactive marketing company

Hinda Feige Greenberg, PhD, director of the information center for the Robert Wood Johnson Foundation, dedicated to improving health care for Americans

Paul Greenberg, president, The 56 Group LLC; BPT Partners LLC; MyCRMCareer.com, social media companies

Kathryn Greenhill, emerging technologies specialist, Murdoch University

Shane Greenstein, professor of management and strategy, Kellogg School of Management, Northwestern University

Robert Grenier, partner, Third Wave Systems, a consultancy on management information systems

Bruno Guglielminetti, producer for Radio-Canada, multimedia producer and blogger

Robin Gunston, consulting futurist for Mariri Consulting, a strategic and business planning company

David Hakken, Indiana University School of Informatics and a professor of anthropology who studies social change and the use of automated information and communication technologies

Alexander Halavais, professor and social informatics researcher, Quinnipiac University; explores the ways in which social computing influences society

Richard Hall, professor and codirector of the Laboratory for Information Technology Evaluation, Missouri University of Science and Technology

Richard Hammond, knowledge-management team leader, United States Environmental Protection Agency; currently examining Semantic Web and RFID

June Hansen, Web data analyst and developer, Washington State Department of Information Services; formerly Internet consultant at state board for community and technical colleges

Fred Hapgood, technology author and consultant; freelance writer in technology and science; in the 1990s, he took on the role of moderator of the Nanosystems Interest Group at MIT

Maz Hardey, social analyst, blogger, "defender of new media" completing a doctorate funded by the Economic Social Research Council in the U.K., based at the University of York

Paul Harwood, assistant professor and director of the Public Opinion Research Lab at the University of North Florida; his research has examined the Internet's impact on civil society

Bertil Hatt, researcher of Internet and social services, innovation valuation; employed by France Telecom and Orange

Steven Hausman, president, HausmanTech Consulting

Don Heath, Internet pioneer; former president and CEO of the Internet Society; on U.S. State Department Advisory Committee on International Communication and Information Policy

Ellen Helsper, research fellow, Oxford Internet Institute; responsible for the design, analysis and coordination of the Oxford Internet Surveys and World Internet Project Surveys

Havi Hoffman, senior editor, product development, Yahoo; coauthor of "The Tech Buzz Game," and blogger

Yuri E. Hohlov, PhD, chairman of the board of the Institute of the Information Society, based in Moscow, Russia; also works with the World Bank's Development Gateway

Megan Holbrook, partner, Kapow, Inc., a site-design and development company

Henry Holcombe, vice president IS, Universal Service Administrative Co.

Sarah Holyfield, editor and journalist for Joint Information Systems Committee and the Centre for Educational Technology Interoperability Standards to support education and research

James Jay Horning, chief scientist, information systems security, SPARTA, Inc.; a former fellow at Xerox's Palo Alto Research Center; fellow of the ACM

Sarah Houghton-Jan, consultant for the Infopeople Project; digital futures manager, San Jose Public Library; author of the Librarian in Black technology blog

Louis Houle, president of the Internet Society's Quebec, Canada, chapter

Bernardo Huberman, senior fellow and director of the Social Computing Lab, HP Laboratories; consulting professor in the Department of Applied Physics at Stanford University

Tom Hughes, chief operating officer at The Connors Group, a financial markets information company; formerly senior vice president for technology strategy at Moody's Investors Service

Christian Huitema, distinguished engineer, Microsoft Corporation; pioneering Internet engineer (on the Internet Architecture Board 1991–1996; Internet Society leader 1995–2001)

J. W. Huston, president of Huston Consultancy and futurist

Paul Hyland, executive producer, edweek.org; formerly a member of the board of directors for Computer Professionals for Social Responsibility and director at Media Matters for America

Christopher Jacobs, COO, Solutions for Progress, Inc.; formerly with KnowledgeFlow Inc. and Unisys Corp.

Eliza Jacobs, associate director, research, U.S. Public Broadcasting System

Ken Jarboe, president, Athena Alliance, a nonprofit exploring the potential of a global information economy; author of the research study, *Inclusion in the Information Age*

Jennifer Jarratt, principal, Leading Futurists LLC; works with methodologies to assess and interpret potential futures

Jeff Jarvis, top blogger at Buzzmachine.com; professor at City University of New York Graduate School of Journalism; contributor and consultant to the *Guardian*; advisor to start-ups

Claudia Jellett, Web managing editor for Booz Allen Hamilton, a global strategy and technology consulting firm

Tom Jennings, University of California–Irvine, creator of FidoNet, the first message and file-networking system online; builder of *Wired* magazine's first online presence

Bob Johnson, president, Bob Johnson Consulting, LLC; author of the *Higher Education Marketing Newsletter* and an expert on building Web sites for brand engagement

Rick Johnston, senior Web strategist, Ironworks Consulting

Paul Jones, director of ibiblio.org at University of North Carolina–Chapel Hill; the original manager of SunSITE, one of the first Web sites in North America

Steve Jones, professor of communication and associate dean of liberal arts and sciences, University of Illinois–Chicago, cofounder of Association of Internet Researchers

Rachel Kachur, behavioral researcher, U.S. Centers for Disease Control & Prevention

Yoram Kalman, a researcher and author of *Silence in Text-Based Computer-Mediated Communication,* based at the University of Haifa's Center for the Study of the Information Society

Ilima Kane, program manager, Klein Buendel, Inc., a Colorado-based health education firm

Don Kasprzak, CEO of Panaround.com, a Web-solutions design company; former system engineer at Apple Computer

Helen Keegan, founder of Beep Marketing, a self-employed consultant and a judge for the Webby Awards

Amit Kelkar, consultant and sociology researcher, Postmodern

Tom Keller, business development, 1&1 Internet AG (one of the largest ICANN registrars in Europe); member of the Registrar Constituency and the GNSO Council

Charles Kenny, senior economist for the World Bank; author of many reports, including *ICT: Promises, Opportunities and Dangers for the Rural Future*

Peter Kim, senior analyst, Forrester Research, Boston; specializes in e-strategy and management, social marketing, blogs; recent reports include *Microblogging for Marketers*

Kent Kirschner, media specialist, Neighborhood America, inviting companies to understand the power of community-building online

Helen Klaebe, senior research fellow, Queensland University of Technology, Australia

Norbert Klein, member of ICANN's GNSO Council and Internet Society leader who works with Open Institute Cambodia, a company whose primary focus is on information

Jim Kohlenberger, director of Voice on the Net Coalition; a senior fellow at the Benton Foundation; former White House senior domestic policy adviser and specialist on telecom policy

Gary Kreps, chair of the department of communication, George Mason University; founding chief of health informatics branch of the National Cancer Institute

Gerard LaFond, founder and chief strategy officer, red TANGENT, and cofounder of Persuasive Games; works with leading brand companies such as Sony, Ubisoft, and Kraft

Maurice Lagasse, founder and chief executive officer, netXmedia, a company that specializes in vertical-search portals that connect people and businesses

Cheryl Langdon-Orr, a leader in the Internet Society, ICANN and ALAC (At-Large); a board member of AUDA, the group in charge of Australian Domain Name registration

Larry Lannom, a director for the Corporation for National Research Initiatives, a nonprofit organization that undertakes, fosters, and promotes network research in the public interest

Dan Larson, president and CEO of PKD Foundation, a nonprofit organization working for patient advocacy and education

Greg Laudeman, utilization catalyst and facilitator, Georgia Tech Enterprise Innovation Institute

Fred Ledley, founder and chairman of Mygenome; professor of natural and applied sciences, Bentley College, Waltham, MA

Alan Levin, a director and cofounder of the Future Perfect Corporation; former chair of the Internet Society of South Africa and worked with AfriNIC and the .za Domain Name Authority

John Levine, founder of Taughannock Networks; a leader of the Internet Research Task Force's Anti-Spam Research Group and the Coalition Against Unsolicited Commercial Email

Jeffrey Levy, director of Web communications, U.S. Environmental Protection Agency; an original member of the Web Content Managers Group, now the Web Managers Advisory Council

Andy Linton, longtime network engineer, CityLink, Wellington, New Zealand

Fredric M. Litto, consultant for Pearson Education Global e-Learning, president, Brazil Distance Learning Association

Shuan Lo, interaction designer, Yahoo Taiwan; formerly of Damai Sciences, Kuala Lumpur

Dorothea A. Lorber, president, NetHead, LLC

Geert Lovink, professor, expert on culture and the Internet; based in Amsterdam; author of *Dark Fiber* and *Uncanny Networks*; responsible for the Institute of Network Cultures

Philip Lu, vice president and manager of research analysis, Wells Fargo Bank Internet Services Group; formerly a senior analyst at Gartner and a senior manager at Schwab

Bernie Lubran, sales manager, ForeSee Results, a provider of Web services, and DeVille Consulting

Jim Lucas, Web manager, CACI, a provider of U.S. national security, defense, and intelligence-related solutions

Edward Lundquist, senior science advisor, Alion Science and Technology

Ed Lyell, professor of business and economics, Adams State College, Regis University, San Luis Valley Board of Educational Services; pioneer in issues regarding Internet and education

Clifford Lynch, executive director, Coalition for Networked Information; past president of the American Society for Information Science; member, Internet2 Applications Council

Dan Lynch, founder of CyberCash, Inc. and Interop Company and an Internet pioneer; board member of Santa Fe Institute; director of computing for SRI International in the late 1970s

Hamish MacEwan, consultant, Open ICT, New Zealand

David W. Maher, senior VP for law and policy, the Public Interest Registry, the Internet top-level domain registry; formerly vice president of public policy for the Internet Society

Katherine Mancuso, independent scholar and freelance Web designer, LiveJournal writer, technology coordinator, Alternate Roots; now studying creative collaboration in online worlds

Ruth Martin, National Chengchi University, Taiwan

Craig J. Mathias, principal of the Farpoint Group, an advisory firm specializing in wireless networking and mobile computing based in Massachusetts

Sam McAfee, CTO and owner, RadicalFusion, a company that offers consulting on Web design and development, server-side programming and database development

Jerry McCann, vice president and director of the social marketing group at Carton Donofrio Parners Inc., marketing and advertising firm in the Washington, DC, area

Joe McCarthy, principal instigator, MyStrands; formerly principal scientist at Nokia Research Center in Palo Alto, CA; he also has worked at Intel, Accenture Technology Labs, and Nokia

Jim McConnaughey, public policy-maker, National Telecommunications and Information Administration; a senior economic adviser in U.S. policy on access and the digital divide

Garland T. McCoy, founder, Technology Policy Institute, a think tank focused on the economics of innovation; formerly senior vice president at the Progress and Freedom Foundation

Sean McGinnis, manager of search marketing, Findlaw, Thomson-Findlaw, Thomson-West; earlier built and developed DVD Verdict

Andy McGregor, program manager for the Joint Information Systems Committee, a U.K. provider of innovative ICT solutions for education and research

Neil McIntosh, director of editorial development for guardian.co.uk; one of the first to report on the commercial possibilities of weblogs; one of the first blogging journalists

Timothy J. McManus, vice president of health care product management for Nuance Communications, Inc., a software technology company known for speech-recognition work

Michel Menou, professor and ICT consultant; information policy and development researcher since 1966 and prolific author of reports and journal articles

Susan Mernit, independent consultant and former senior director for product development, Yahoo; blogger; previously a vice president at America Online

Jerry Michalski, founder and president, Sociate, a technology consulting firm; formerly managing editor of the *Release 1.0* newsletter and cohost for the annual PC Forum

Paul Miller, technology evangelist on the senior management team at Talis, a company delivering human-centric Web applications, based in Birmingham, U.K.; a blogger for ZDNet

Rich Miller, chief executive officer, Replicate Technologies, Inc.; an Internet pioneer with ARPAnet who implemented some of the first computer messaging and conferencing systems

Mario Morino, author and chairman of Morino Ventures LLC; a trustee of The Brookings Institution

David Moschella, global research director for the Computer Sciences Corporation's Leading Edge Forum; *Computerworld* columnist

Milton Mueller, professor, Syracuse University and Technology University of Delft; expert on Internet governance; active in ICANN's Non-Commercial User's Constituency

Kirk Munsell, webmaster and science writer for the Jet Propulsion Laboratory

John Murphy, director, USA.gov technologies, General Services Administration of the U.S. government

Mary Murrell, anthropologist specializing in participatory media, University of California–Berkeley

Brian T. Nakamoto, cofounder of MrJoy Inc. and product-line manager for Everyone.net (a leading provider of outsourced e-mail solutions for individuals and companies around the world)

Victoria Nash, PhD, Oxford Internet Institute, University of Oxford, U.K.; formerly a research fellow at the Institute of Public Policy Research

Louis Naugès, president, Revevol, an enterprise 2.0 company with offices in France, Spain, the U.K., and the U.S.; a founder of Microcost, a French IT services and hardware company

Jay Neely, social strategist for News Armada, a Boston-based company working to advance Internet-based news and commentary and community online

Scott D. Nelson, senior engineer, Honda Research & Development Americas; president of mObridge Inc.

Chris Nodder, user-experience specialist, Nielsen Norman Group; formerly a senior user researcher at Microsoft Corp.

Cameron Norman, assistant professor in the department of public health sciences, the University of Toronto; actively engaged in use of the Internet to help tackle tough health issues

Jill O'Neill, director of planning and communication, National Federation of Abstracting and Information Services; author of the Infotoday blog; based in the Philadelphia area

Sooyoung Oh, PhD candidate, department of computer science and engineering, Penn State University; researches cognitive-inspired intelligent agents

Gilda Olinto, researcher, Brazilian Institute for Information in Science and Technology

Francis J. L. Osborn, futurist and activist, philosophy department, University of Wales Lampeter (formerly St. David's University College)

Richard Osborne, Web manager for the School of Education & Lifelong Learning, University of Exeter

Olav Anders Övrebö, media blogger and former *Netzeitung* journalist, University of Bergen

Sam Ozay, e-learning and e-communication specialist and solutions architect at Postmodern (Asia/Pacific)

Han Woo Park, assistant professor, YeungNam University, South Korea; research interests include social networks

Rich Parker, managing partner, the BigMind Group LLC, an Internet communications and software company specializing in online collaborative workspaces based in Langley, WA

Craig Partridge, chief scientist, BBN Technologies; Internet pioneer and early leader in the IETF; active member of the ACM's SIGCOMM and the IEEE Communications Society

Joel Wiramu Pauling, cross-cultural technologist, owner of Ænertia.net Consulting in New Zealand; research focus on intellectual property, society, and technology

Adam Peake, executive research fellow and telecommunications policy analyst, the Center for Global Communications; studies public policy and the Internet

Michele Perras, artist, consultant, researcher, and futurist with Interactive Ontario

Ian Peter, Ian Peter and Associates, Internet Mark 2 Project; Internet pioneer, helped develop the Internet in Australia and the Asia-Pacific region in the 1980s; maintains www.nethistory.info

Brian Peters, director of government relations, Information Technology Industry Council, in charge of telecom, trade, and privacy issues for IT companies

Johanna Pimentel, an expert on communications law and information law and director of privacy for Europe, Middle East, Africa, and South Asia at Western Union

Alejandro Pisanty, director of computer services at the Universidad Nacional Autónoma de México; active in leadership with Internet Society, ICANN, and Internet Governance Forum

Giulio Prisco, chief executive officer of Metafuturing Second Life; former department head at European Satellite Centre, analyst at European Space Agency, and IT specialist for CERN

Sam Punnett, president, FAD Research Inc.; has worked in the field of interactive digital media since the 1980s, for the last 9 years on strategy, marketing, and e-business development

Aaron J. Quigley, PhD, directs a team of 21 faculty and students in research at University College Dublin, Ireland; research interests are pervasive computing and visualization

Thomas Quilty, president, BD Consulting and Investigations, Inc., a firm that investigates and provides services tied to software piracy, trade-secret theft, and other high-tech concerns

Oliver Quiring, Institute for Communication Science and Media Research, LMU, Munich, Germany

Josh Quittner, executive editor, *Fortune* magazine; formerly editor of Business 2.0, Time.com, technology editor of *Time* magazine; freelancer, *Wired* magazine

Faraz Qureshi, Diamond Consulting, a company that brings together companies and IT professionals

Ross Rader, director of retail services, Tucows Inc.; works with ICANN in the Registrars Constituency, part of the Generic Names Supporting Organization (GNSO)

Hakikur Rahman, chairman, SchoolNet Foundation; coordinator of Sustainable Development Networking Programme in Bangladesh; active in Internet Society Board

Lutfor Rahman, chairman of the department of computer science, Stamford University, Bangladesh; leader of the Association for Advancement of Information Technology

Don Ranly, PhD, professor emeritus, University of Missouri School of Journalism

Cambria Ravenhill, manager of national channel planning at TELUS Communications, IT sales executive with expertise in IT hardware, telecom, and wireless space

Marion Rawson, associate at ICF International, management consultancy; formerly energy analyst at Energetics and senior energy analyst at McNeil Technologies

Elena Razlogova, assistant professor of history, Concordia University Digital History Lab, Montréal, Quebec, Canada; research interests include digital cultures and history of media

Howard Rheingold, Internet sociologist and author; one of the first to illuminate virtual communities; author of *Virtual Reality*, *Smart Mobs*, and *Virtual Community*

Sebastian Ricciardi, associate with Jauregui & Associates, a law firm in Buenos Aires; leader in the Internet Society–Argentina, formerly of ICANN's At-Large Advisory Committee

Ronald E. Rice, professor, University of California–Santa Barbara; past president of the International Communication Association; author, *The Internet and Health Communication*

Robert H. Rich, PhD, strategic planning and evaluation, American Chemical Society

Marco Rivera, Internet specialist, Vistronix, Inc., an information management, technological solutions, and managed services firm serving federal, state, and local agencies

Sarah Robbins, director of emerging technologies, Media Sauce; Web 2.0 expert and coauthor of *Second Life for Dummies*

C. R. Roberts, Internet journalist based in Vancouver, Canada

Dennis Rodrigues, chief of the online information branch, U.S. National Institutes of Health

Alex Rodriguez, senior electrical engineer, Telefonica Colombia; formerly interconnect engineer at Avantel

Paul Rodriguez, director of online content, National Cable & Telecommunications Association; crafts online strategies; expert on broadband technology and its influence on society

Paolo Rossi, electrical engineer and computer scientist, University of Pavia, Italy

Anni Rowland-Campbell, working in innovation and research at Fuji Xerox Australia and as a director at Intersticia Pty Ltd; formerly executive director at GAMAA

Anthony M. Rutkowski, cofounder of the Internet Society and founding trustee; longtime leader in International Telecommunication Union; VP for regulatory affairs, VeriSign

D. K. Sachdev, founder and president of SpaceTel Consultancy LLC; early developer of XM Radio

Erkan Saka, lecturer in media and communications systems at Istanbul Bilgi University

Fadi Salem, research associate, Dubai School of Government; research focuses on e-government and development

David F. Salisbury, science and research communications, Vanderbilt University; formerly science and technology reporter for *Christian Science Monitor*

Scott Salyer, Internet services and strategy manager, Resources for the Future, a think tank in Washington, DC

Syamant Sandhir, director, Futurescape Netcom Pvt. Ltd., an experience-design and implementation company

Luis Santos, Universidade do Minho–Braga, Portugal

Christine Satchell, PhD, Institute for Creative Industries and Innovation, Queensland University of Technology

Michael Satterwhite, managing consultant, IBM

Steve Sawyer, associate professor, college of information sciences and technology, Penn State University; research interests include computing and knowledge workers

Buddy Scalera, vice president for interactive content and market research for CommonHealth Qi, in charge of interactive online strategies, including social and viral marketing

Jan Schaffer, executive director of J-Lab, the Institute for Interactive Journalism; former business editor and a Pulitzer Prize winner for the *Philadelphia Inquirer*

Dian Schaffhauser, writer and editor for CampusTechnology.com, THEJournal.com, *Redmond Magazine*, *Computerworld*, and Web Worker Daily

Aaron Schmidt, Walking Paper Consulting, a blogger who writes about libraries, technology, and usability

Karen G. Schneider, research and development, College Center for Library Automation, Tallahassee, FL; expert and thought leader in the libraries and technology community

Adrian Schofield, manager of the applied research unit, Johannesburg Centre for Software Engineering, South Africa; World Information Technology and Services Alliance

Robert Schrott, market and industry intelligence, LexisNexis

Mihai Seceleanu, chief executive officer, Internet Corp SRL, Romania, a managing partner of the online publishing group that developed Wall-Street.ro, 9AM.ro, Kudika.ro, Yuppy.ro

Denise Senecal, research manager, Callahan and Associates, an Internet strategy consortium in the Washington, DC, area; expert on Internet and mobile banking

Gbenga Sesan, Internet-for-development consultant, Paradigm Initiative, Nigeria; works with ICTs in socioeconomic transformation, focusing on underserved groups

Tiffany Shackelford, consultant who works with clients such as Phase 2 Technology, Stateline.org, Foneshow, WebbMedia, and Daily Me

Peter Shane, Jacob E. Davis and Jacob E. Davis II Chair in Law, Ohio State University; author of *Democracy Online: The Prospects for Political Renewal Through the Internet*

Jeremy J. Shapiro, a professor of critical social theory at Fielding Graduate University with a research emphasis on the social and cultural effects of information technology and systems

Joanna Sharpe, senior marketing manager, Microsoft

Lea Shaver, A2K (Access to Knowledge) Program Coordinator, The Information Society Project at Yale Law School; developing a cross-national access-to-knowledge index

Heather Shelsta, manager, ICF International, a global professional services firm that offers technology solutions for government and commercial clients; based in Fairfax, VA

Glenna Shields, information-management specialist for the U.S. Environmental Protection Agency

Clay Shirky, consultant and professor in the interactive program at New York University; author of *Here Comes Everybody: The Power of Organizing Without Organizations*

Tiffany Shlain, founder of the Webby Awards, cofounder of the International Academy of Digital Arts and Sciences; named one of *Newsweek*'s "Women Shaping the 21st Century"

Judith Siess, president of Information Bridges International, Inc. and publisher and editor of the *One-Person Library* newsletter, author, and blogger

Kelly Sims, group research manager, consumer business unit, Symantec Corporation

Michael Skarzynski, managing director, Red Lion Technologies Inc.

Scott Smith, principal, Changeist LLC; consultant, writer, and futurist; formerly with Yankee Group and Jupiter

Jonne Soininen, manager of Internet affairs for Nokia Siemens Networks; formerly system engineering manager for Nokia; Internet Society and Internet Engineering Task Force

Juan B. Solana, director of new media for HispanicBusiness.com

Dan Sonnett, president, Sonnett Media Group, LLC, a company that explores ways to engage audiences through video and Web projects

Ulrich Spalthoff, director of advanced technology, Alcatel-Lucent, Paris; expert in solid-state science, semiconductor materials, optical fiber communications

Stuart Spector, senior vice president, U.S. National Council on Aging; career professional technology and the insurance industry

Anthony Spina, PhD, founder and president of Knowledge Resources, a consultancy focused on assisting organizations meet challenges of complex environments

Todd Spraggins, strategic architect, Nortel Carrier Networks; president and chairman of the board of directors of the Communications Platforms Trade Association

Bill St. Arnaud, chief research officer, CANARIE Inc., an industry-government consortium that promotes and develops information highway technologies in Canada; active in Internet2

Sean Steele, CEO and senior security consultant, infoLock Technologies; leads corporate business strategy for insider-threat management consulting services and solutions company

Sherwin Steffin, self-employed, formerly a senior data analyst at Earthlink and CEO at Brainpower

Ed Steinmueller, professor, science and technology policy research, University of Sussex; researches industrial structure of high-technology industries, coevolution of technology

Jason Stoddard, managing partner/strategy at Centric/Agency of Change, an interactive strategies company; he is also a popular speaker on social media and virtual worlds

David Stott, director, Azimuth Limited, software developer and engineering firm for the Department of Defense and other U.S. government and commercial entities

Joseph Straubhaar, professor, University of Texas–Austin; research interests include international communication and cultural theory and information sciences and the digital divide

Roger Strouse, vice president and lead analyst, Outsell Inc.; researches and writes about information professionals and corporate information best practices

Colby Stuart, creative director, Quantum Brands BV, and the Kids 2020 Foundation; member of the Club of Amsterdam, a futurists organization

Kevin Taglang, independent communications policy researcher and writer based in Evanston, IL; previously editor and senior policy analyst, the Benton Foundation

Elizabeth Talerman, chief operating officer, Talerman+Partners LLC, a marketing integration company; previously senior VP at Martha Stewart Living Omnimedia

Tze-Meng Tan, Multimedia Development Corporation (MDeC), Malaysia, and director at OpenSOS; Internet pioneer as a member of the team that started Malaysia's largest ISP, TMnet

Humphrey Taylor, chairman, the Harris Poll; founded Opinion Research Centre in England in 1966; was a close advisor to prime ministers Edward Heath and Margaret Thatcher

Brad Templeton, chairman of the Electronic Frontier Foundation, also of the Foresight Institute; an Internet pioneer active in the computer-networking community since 1979

Roy Tennant, senior program officer, Online Computer Library Center (OCLC); developer of the eScholarship Repository and eScholarship Editions publishing services

Sharon Tettegah, professor of human perception and performance at University of Illinois–Urbana Champaign; uses Second Life in her research and courses

Ivor Tossell, technology columnist/journalist for the *Toronto Globe and Mail*, known as "the blogging journalist" and a social observer

Anthony Townsend, research director for the Technology Horizons Program of the Institute for the Future; also a cofounder of NYCwireless

Mike Treder, director, Center for Responsible Nanotechnology; research fellow with the Institute for Ethics and Emerging Technologies

Bryan Trogdon, president, First Semantic, working on providing a repository for news, research, ideas, and tools for implementing the Semantic Web

Michael Turk, vice president for National Cable & Telecommunications Association, an advocacy group for U.S. telecommunications providers

Brough Turner, chief technology officer and cofounder of NMS Communications; oversees evolution of technology and product architectures

S. Derek Turner, research director, Free Press, a public interest group that studies and reports on telecommunications issues; author of reports such as *Broadband Reality Check*

Joseph Turow, professor of communication, University of Pennsylvania; author of more than 60 articles and nine books, and an expert on the online marketing industry

Wim van de Donk, professor of public administration and chairman of the Scientific Council for Government Policy in The Netherlands

Peter W. Van Ness, president, Van Ness Group, a Web-development company; founded Personal Computer Solutions in 1983; cofounded StockPlan, Inc. and MyStockOptions.com

Thomas Vander Wal, principal and senior consultant, InfoCloud Solutions Inc.; coined the term *folksonomy*; expert on tagging, the social Web, and social information use and reuse

Hal Varian, chief economist, Google, and on the faculty at the University of California–Berkeley; expert on the economics of information technology

Tapio Varis, expert in global learning environments, University of Tampere, Finland; UNESCO chair in global e-learning; active leader of the Global University System

Cristos Velasco, director general of North American Consumer Project on E-Commerce; Mexico-based attorney and active in the Internet Governance Forum and Internet Society

Tom Vest, IP network architect, RIPE NCC Science Group, Office of the Chief Scientist; consultant for the Internet Society, and for OECD Economics and Statistics Division

Cynthia Villareal, public affairs director of Foley & Lardner LLP, a Washington, DC–area law firm known for work with intellectual property and technology issues

Robert Visser, owner, PageRank-SEO, a search engine optimization and Web marketing company in the Washington, DC, area

Todd Wagner, health economist, Health Economics Resource Center, Palo Alto, CA, part of the U.S. Veterans Administration; also involved with the Center for Health Care Evaluation

Jesse Ward, industry analyst, National Telecommunications Cooperative Association, U.S.

Bill Warren, vice president of government relations, Walt Disney World; founding editor of the *Orlando Business Journal*

Rainer Wasserfuhr, chief coder at MindBroker, a virtual, transparent enterprise based in Germany aimed at creating new concepts for virtual economies and social networks

Nikki Waters, product manager, Internet Services Group, Kaiser Permanente (medical HMO)

J. M. Watts, director, MTM London, works on strategic projects with broadcasters, mobile and fixed-line operators, U.S. studios, and international media companies

Alan Webber, senior analyst, Forrester Research, focused on business-to-business, e-government, and public sector linkages

Philip Webre, senior analyst, microeconomic and financial studies division, Congressional Budget Office, U.S. government

Dan Weingrod, vice president for digital operations, Cronin and Co.; oversees creative online initiatives for integrated marketing communications company

Roy Wetherbee, director of operations for interactive technology, PARTNERS+simons Inc.

Roderick White, editor, *Admap* magazine, World Advertising Research Center

Hal Widsten, general manager, KWED/Seguin Daily News, Guadalupe Media Ltd.

Michael Williams, associate professor of interactive media, University of Maryland–College Park; former director of Internet development for Kalmbach Publishing Company

Alick Wilson, ICT consultant, Xendra Limited, New Zealand; former member of At-Large Advisory Committee review team and councilor for GNSO at ICANN

Paul Wilson, director, Asia Pacific Network Information Centre, the IP address registry for the Asia Pacific region; active in World Summit on the Information Society, IGF, ICANN

Alexandre Winter, cofounder and chief executive officer, LTU Technologies, a global leader in image search and recognition technologies

Leonard Witt, associate professor in communication, Kennesaw State University, GA; researches citizen journalism and user-generated content; author of weblog PJNet.org

Jim Witte, professor of sociology, Clemson University; research is focused on differences between online and offline society, including a special interest in activity in Second Life

Mark Youman, principal, ICF International, a Washington, DC, consulting-services company that works with government and commercial clients

Joshua Young, IMLP systems analyst for General Electric; previously director of quality teams at College of Information Sciences and Technology

Jeremy Yuille, RMIT Communication Design and ACID in Melbourne, Australia; previously director at IXDA, the Interaction Design Association

Jacqueline Zalewski, assistant professor of sociology, West Chester University; research includes distributed teams, change management, and the standardization of work

Xiang Zhou, PhD, Cheung Kong School of Journalism and Communication, Shantou University, China; research includes cultural dimensions and framing the Internet in China

Michael Zimmer, PhD and resident fellow at the Information Society Project at Yale Law School; research includes social, cultural dimensions of information technologies

References

Aboba, B. (2006, November). Reflections on Internet transparency. *IETF Journal, 2*(3). Retrieved January 11, 2009, from http://www.isoc.org/tools/blogs/ietfjournal/?p=114#more-114

Akamai. (2008, April). *Report: The state of the Internet.* Retrieved January 11, 2009, from http://www.akamai.com/stateoftheinternet/

America Online, Opinion Research Corp. (2007, July 26). *E-mail addiction.* Retrieved January 11, 2009, from http://corp.aol.com/press_releases/2007/07/think-you-might-be-addicted-email-youre-not-alone

Anderson, J., & Rainie, L. (2006). *The future of the Internet II.* Pew Internet & American Life Project. Retrieved January 11, 2009, from http://www.pewinternet.org/pdfs/PIP_Future_of_Internet_2006.pdf

Anthes, G. (2007, February 12). A clean slate for the Internet: Is it time to start anew with the Internet? Is it possible? *Computerworld.* Retrieved January 11, 2009, from http://www.computerworld.com/action/article.do?command=viewArticleBasic&articleId=279934&source=NLT_PM&nlid=8

Anthes, G. (2008, February 1). Give your computer the finger: Touch-screen tech comes of age. *Computerworld.* Retrieved January 11, 2009, from http://www.computerworld.com.au/index.php/id;1847998544;fp;;fpid;

Anton, P., Siberglitt, R., & Schneider, J. (2001). *The global technology revolution: Bio/nano/materials trends and their synergies with information technology by 2015.* RAND National Defense Research Institute. Retrieved January 11, 2009, from http://www.rand.org/pubs/monograph_reports/MR1307/MR1307.pdf

Ard, S. (2008, January 7). Intel CEO sees Net boosting consumer electronics. *CNET News*. Retrieved January 11, 2009, from http://news.cnet.com/Intel-CEO-sees-Net-boosting-consumer-electronics/2100-7353_3-6225031.html?tag=nw.1

Bailey, B. (2000, September). *Human interaction speeds*. Retrieved January 11, 2009, from http://webusability.com/article_human_interaction_speeds_9_2000.htm

Barlow, J. P. (1994, March). The economy of ideas: A framework for patents and copyrights in the digital age (Everything you know about intellectual property is wrong). *Wired*. Retrieved January 11, 2009, from http://www.wired.com/wired/archive/2.03/economy.ideas_pr.html

BBC Online. (2008, April 30). Luminaries look to the future Web. *BBC News*. Retrieved January 11, 2009, from http://news.bbc.co.uk/2/hi/technology/7373717.stm

Block, J. J. (2008, March). Issues for DSM-V: Internet addiction. *American Journal of Psychiatry*. Retrieved January 11, 2009, from http://ajp.psychiatryonline.org/cgi/content/full/165/3/306

Bowles, S. (2008, June 20). Policies designed for self-interested citizens may undermine "the moral sentiments": Evidence from economic experiments. *Science*. Retrieved January 11, 2009, from http://www.sciencemag.org/cgi/content/abstract/320/5883/160

Briggs, C. F., & Maverick, A. (1858). *The story of the telegraph and a history of the great Atlantic cable*. New York: Rudd and Carleton.

Carlson, T. (2007, December). The unique political soul of the IETF. *IETF Journal*, *3*(3). Retrieved January 11, 2009, from http://www.isoc.org/tools/blogs/ietfjournal/?p=173#more-173

Carnoy, M., & Castells, M. (2001). Globalization, the knowledge society and the network state: Poulantzas at the millennium. *Global Networks*, *1*(1), 1–18.

Carty, J. J. (1891). "The prophet's column." *Electrical Review*. Retrieved January 11, 2009, from http://www.archive.org/stream/

nationalacademyo009315mbp/nationalacademyo009315mbp_djvu.txt

Cellan-Jones, R. (2008a, March 17). Questions and answers: Tim Berners-Lee. *BBC Online*. Retrieved January 11, 2009, from http://news.bbc.co.uk/2/hi/technology/7300434.stm

Cellan-Jones, R. (2008b, March 17). Web creator rejects Net tracking. *BBC Online*. Retrieved January 11, 2009, from http://news.bbc.co.uk/2/hi/technology/7300434.stm

Center for Democracy and Technology. (2007, August). *Search privacy practices: A work in progress*. Retrieved January 11, 2009, from http://www.cdt.org/privacy/20070808searchprivacy.pdf

Clark, D., Sollins, K., Wroclawski, J., & Braden, R. (2002). *Tussle in cyberspace: Defining tomorrow's Internet*. Special Interest Group on Data Communication (SIGCOMM) of the Association for Computing Machinery (ACM) 2002 conference proceedings. Retrieved January 11, 2009, from http://conferences.sigcomm.org/sigcomm/2002/papers/tussle.pdf

Clark, D., Sollins, K., Wroclawski, J., Katabi, D., Kulik, J., & Yang, X. (2003). *New arch: Future-generation Internet architecture*. Defense Advanced Research Projects Agency Information Technology Office. Retrieved January 11, 2009, from http://www.isi.edu/newarch/iDOCS/final.finalreport.pdf

Computer & Communications Industry Association. (2007, September 12). *Fair use in the U.S. economy*. Retrieved January 11, 2009, from http://www.ccianet.org/artmanager/uploads/1/FairUseStudy-Sep12.pdf

comScore. (2008, January 7). *2007 online holiday shopping season surpasses $29 billion in sales*. Press release. Retrieved January 11, 2009, from http://www.comscore.com/press/release.asp?press=1990

Defense Advanced Research Projects Agency. (2007, August 7–9). *DARPATech 2007 proceedings*. Retrieved January 11, 2009, from http://www.darpa.mil/darpatech2007/proceed.html

Doctorow, C. (2001, August 26). *Metacrap: Putting the torch to seven straw-men of the meta-utopia.* Retrieved January 11, 2009, from http://www.well.com/~doctorow/metacrap.htm

Driver, E., & Jackson, P. (2008, January 7). *Getting work done in virtual worlds*. Forrester. Retrieved January 11, 2009, from http://www.forrester.com/Research/Document/Excerpt/0,7211,43450,00.html

Dux, P., Ivanoff, J., Asplunc, C., & Marois, R. (2006, December 21). Isolation of a central bottleneck of information processing with time-resolved fMRI. *Neuron*. Retrieved January 11, 2009, from http://www.psy.vanderbilt.edu/faculty/marois/Publications/Dux_et_al-2006.pdf

Dworak, M., Schierl, T., Bruns, T., & Struder, H. (2007, November 1). The impact of singular excessive computer game and television exposure on sleep patterns. *Pediatrics*. Retrieved January 11, 2009, from http://pediatrics.aappublications.org/contents-by-date.0.dtl

Economist Intelligence Unit. (2006). *Foresight 2020: Economic, industry, and corporate trends.* Retrieved January 11, 2009, from http://graphics.eiu.com/files/ad_pdfs/eiuForesight2020_WP.pdf

Electronic Frontier Foundation. (n.d.). *Statement on WIPO: World Intellectual Property Organization.* Retrieved January 11, 2009, from http://www.eff.org/issues/wipo

Electronic Privacy Information Center, Privacy International. (2007, December). *Privacy and human rights 2006 executive summary.* Retrieved January 11, 2009, from http://www.privacyinternational.org/article.shtml?cmd%5B347%5D=x-347-559060

Estabrook, L., Witt, E., & Rainie, L. (2007, December 30). "Information searches that solve problems." Pew Internet & American Life Project. Retrieved January 11, 2009, from http://www.pewinternet.org/pdfs/Pew_UI_LibrariesReport.pdf

Fildes, J. (2006, May 23). Web inventor warns of "dark" Net. *BBC Online*. Retrieved January 11, 2009, from http://news.bbc.co.uk/2/hi/technology/5009250.stm

40th anniversary issue interviews, Bill Gates, William Gibson. (2007, November 1). *Rolling Stone.* Retrieved January 11, 2009, from http://www.rollingstone.com/news/story/17168320/the_rolling_stone_40th_anniversary_issue/3

Fox, S. (2005, July 6). *Spyware: The threat of unwanted software programs is changing the way people use the Internet.* Pew Internet & American Life Project. Retrieved January 11, 2009, from http://www.pewinternet.org/ppf/r/160/report_display.asp

Fox, S., Anderson, J., & Rainie, L. (2005, January 9). *The future of the Internet.* Pew Internet & American Life Project. Retrieved January 11, 2009, from http://www.pewinternet.org/pdfs/PIP_Future_of_Internet.pdf

Freedom and its digital discontents: Information overload. (2008, February 26–March 7). *The Economist* [Online]. Retrieved January 11, 2009, from http://www.economist.com/debate/index.cfm?action=summary&debate_id=5

Gartner. (2007a, November 13). *Gartner says "generation virtual" will have a profound influence on culture, society, and business.* Press release. Retrieved January 11, 2009, from http://www.gartner.com/it/page.jsp?id=545108

Gartner. (2007b, April 24). *Symposium/ITxpo 2007: Emerging trends.* Retrieved January 11, 2009, from http://www.gartner.com/it/page.jsp?id=503861

Gilder, G., & Swanson, B. (2008, January 29). *Estimating the Exaflood: The impact of video and rich media on the Internet—A zettabyte by 2015?* Discovery Institute. Retrieved January 11, 2009, from http://www.discovery.org/a/4428

Gillies, J. (2008, April 30). The World Wide Web turns 15 (again). *BBC Online.* Retrieved January 11, 2009, from http://news.bbc.co.uk/2/hi/technology/7375703.stm

Glenn, J., & Gordon, T. (2007). *2007 state of the future.* American Council for the United Nations University Millennium Project.

Retrieved January 11, 2009, from http://www.millennium-project.org/millennium/sof2007.html

Global Environment for Network Innovation (GENI). (n.d.). *Overview: What is GENI?* Retrieved January 11, 2009, from http://www.geni.net/overview.html

Global Environment for Network Innovation (GENI). (n.d.). *GENI design documents.* Retrieved January 11, 2009, from http://www.geni.net/documents.html

Global Environment for Network Innovation (GENI). (2007, May 21). *BBN Technologies and National Science Foundation to build vast experimental network: NSF Global Environment for Network Innovations Initiative to develop nationwide network capable of running 1000s of experiments simultaneously.* Press release. Retrieved January 11, 2009, from http://www.geni.net/news_52107.html

Glover, B. (n.d.). *History of the Atlantic cable and undersea communications: From the first submarine cable of 1850 to the worldwide fiber optic network; Cabot Strait cable and 1857–58 Atlantic cables*. Retrieved January 11, 2009, from http://www.atlantic-cable.com/Cables/1857-58Atlantic/index.htm

Gunn, A. (2007, March 17). FAQ: Web 2.0 basics. *Computerworld.* Retrieved January 11, 2009, from http://www.computerworld.com/action/article.do?command=viewArticleBasic&articleId=9013518

Hamblein, M. (2008, June 24). Cisco chief lauds coming virtual-world technology. *Computerworld.* Retrieved January 11, 2009, from http://www.computerworld.com/action/article.do?command=viewArticleBasic&articleId=9103079&source=NLT_AM&nlid=1

Hammond, A., Kramer, W., Katz, R., Tran, J., & Walker, C. (2007, March). *The next 4 billion.* World Bank International Finance Corporation and the World Resources Institute. Retrieved January 11, 2009, from http://pdf.wri.org/n4b_full_text_lowrez.pdf

Havenstein, H. (2008, April 23). "Revolutionary" collective intelligence of users touted at Web 2.0 Expo. *Computerworld.* Retrieved

January 11, 2009, from http://www.computerworld.com/action/article.do?command=viewArticleBasic&articleId=9080038&source=NLT_AM&nlid=1

Howard, P. N. (2007, April). Testing the leap-frog hypothesis: The impact of existing infrastructure and telecommunications policy on the global digital divide. *Information, Communication & Society, 10*(2), 133–157.

Horrigan, J. (2007, May). *A typology of information and communication technology users*. Pew Internet & American Life Project. Retrieved January 11, 2009, from http://www.pewinternet.org/pdfs/PIP_ICT_Typology.pdf

Horrigan, J. (2008, March). *Mobile access to data and information*. Pew Internet & American Life Project. Retrieved January 11, 2009, from http://www.pewinternet.org/pdfs/PIP_Mobile.Data.Access.pdf

Huesmann, R., & Bushman, B. (2007, December). The impact of electronic media violence: Scientific theory and research. *Journal of Adolescent Health*. Retrieved January 11, 2009, from http://www.jahonline.org/article/S1054-139X(07)00391-6/fulltext

Huston, G. (2006, Spring). The evolution of an Internet standard. *IETF Journal, 2*(1). Retrieved January 11, 2009, from http://www.isoc.org/tools/blogs/ietfjournal/?p=66#more-66

IDC. (2008a, May). *The hyperconnected: Here they come!* White paper. Retrieved January 11, 2009, from http://www.nortel.com/promotions/idc_paper/index.html

IDC. (2008b, June 25). *Worldwide spending on Internet advertising will soar past $106 billion in 2011.* Press release. Retrieved January 11, 2009, from http://www.idc.com/getdoc.jsp?containerId=prUS21304208

International Telecommunication Union. (2006a). *ITU Internet report 2006: Digital.life*. Retrieved January 11, 2009, from http://www.itu.int/osg/spu/publications/digitalife/docs/digital-life-web.pdf

International Telecommunication Union. (2006b). *World Summit on the Information Society report 2006*. Retrieved January 11, 2009, from http://www.itu.int/osg/spu/publications/worldinformationsociety/2006/WISR-low-web.pdf

International Telecommunication Union. (2006c, February). *WSIS golden book 2006*. Retrieved January 11, 2009, from http://www.itu.int/wsis/goldenbook/Publication/GB-final.pdf

International Telecommunication Union. (2007a). *Measuring the information society 2007: ICT opportunity index and world telecommunication/ICT indicators*. Retrieved January 11, 2009, from http://www.itu.int/ITU-D/ict/publications/ict-oi/2007/index.html

International Telecommunication Union. (2007b, May). *World Information Society Report 2007: Beyond WSIS.* Published by Author and the United Nations Conference on Trade and Development. Retrieved January 11, 2009, from http://www.itu.int/osg/spu/publications/worldinformationsociety/2007/report.html

Internet Corporation for Assigned Names and Numbers. (2006). *Joint Project Agreement between the U.S. Department of Commerce and the Internet Corporation for Assigned Names and Numbers.* Retrieved January 11, 2009, from http://www.icann.org/en/general/JPA-29sep06.pdf

Internet Engineering Task Force. (2007, October). Paving the way for IPv6: Community meets to discuss the challenges. *IETF Journal*, *3*(2). Retrieved January 11, 2009, from http://www.isoc.org/tools/blogs/ietfjournal/?p=151#more-151

Internet Engineering Task Force. *Archives of conference proceedings*. Retrieved January 11, 2009, from http://www.ietf.org/proceedings_directory.html

Internet World Stats. (2008, June). *Usage and population statistics, world Internet users.* Miniwatts Marketing Group. Retrieved January 11, 2009, from http://www.internetworldstats.com/stats.htm

Ipsos and Lexmark. (2007, February 20). *Knowledge workers putting in longer workweeks* (Report). Retrieved January 11, 2009, from http://www.lexmark.com/lexmark/pressrelease/home/0,6930,204816596_653271419_1002010230_en,00.html

Iqbal, S., & Horvitz, E. (2007). *Conversations amidst computing: A study of interruptions and recovery of task activity*. Department of Computer Science, University of Illinois. Microsoft. Retrieved January 11, 2009, from http://research.microsoft.com/~horvitz/conversational_interruptions.pdf

Jipguep, J. (1995). *The global telecommunication infrastructure and the information society*. Statement presented at the Internet Society INET '95 conference. Retrieved January 11, 2009, from http://www.isoc.org/HMP/PLENARY/L1-6/html/paper.html

Johnson, B. (2008, June 23). Top 100's ad-spend growth grinds to halt. *Advertising Age*. Retrieved January 11, 2009, from http://adage.com/article?article_id=127793

Kanaracus, C. (2008, March 4). Is the "4.0" era upon us? *InfoWorld*. Retrieved January 11, 2009, from http://www.infoworld.com/article/08/03/04/Is-the-4.0-era-upon-us_1.html

Karrenberg, D. (2007, September). *DNS root servers frequently asked questions*. Internet Society. Retrieved January 11, 2009, from http://www.isoc.org/briefings/020/

Kenny, C. (2006). *Overselling the Web, chapter 1, will the Internet change the world?* Retrieved January 11, 2009, from http://charleskenny.blogs.com/weblog/files/chapterone.pdf

Klensin, J. (2004, September). *Internationalizing top-level domain names: Another look*. Internet Society. Retrieved January 11, 2009, from http://www.isoc.org/briefings/018/

Kohlenberger, J. (2007, September). *Universal, affordable broadband for all Americans*. Benton Foundation. Retrieved January 11, 2009, from http://www.benton.org/benton_files/BroadbandForAll.pdf

Kohler, C. (2008, January 17). NPD December: $18 billion year for game industry. *Wired*. Retrieved January 11, 2009, from http://blog.wired.com/games/2008/01/npd-december-18.html

Kowal, J. (2007, 9 August). *The never-ending rush hour: Internet growth requires continual investment in capacity and innovation in network management.* Analysys Research Ltd. Retrieved January 11, 2009, from http://newmillenniumresearch.org/archive/Rush_Hour_August2007.pdf

Lenhart, A., Madden, M., Rankin Macgill, A., & Smith, A. (2007, December 19). *Teens and Social Media*. Pew Internet & American Life Project. Retrieved January 11, 2009, from http://www.pewinternet.org/pdfs/PIP_Teens_Social_Media_Final.pdf

Madden, M., Fox, S., Smith, A., & Vitak, J. (2007, December 16). "Digital footprints." Pew Internet & American Life Project. Retrieved January 11, 2009, from http://www.pewinternet.org/pdfs/PIP_Digital_Footprints.pdf

Massachusetts Institute of Technology. (2007, October 11). *11th international symposium on wearable computers.* Retrieved January 11, 2009, from http://www.iswc.net/iswc07/

Max Planck Institute for Software Systems. (2008, May). *Glasnost: Results from tests for BitTorrent traffic blocking*. Retrieved January 11, 2009, from http://broadband.mpi-sws.mpg.de/transparency/results/

Meeker, M., Joseph, D., & Thaker, A. (2008, March 18). *Internet trends*. Retrieved January 11, 2009, from http://www.slideshare.net/techcrunch/internet-trends031808meeker?src=embed

Minnesota Internet Traffic Studies. (2007). *Year-end 2007 monthly Internet traffic estimate.* Retrieved January 11, 2009, from http://www.dtc.umn.edu/mints/home.html

National Science Foundation, Department of Commerce. (2002, June). *Converging technologies for improving human performance*. Retrieved January 11, 2009, from http://www.amsoftwareservices.

net/KnowledgeBase/Converging%20Technologies%20for%20Improving%20Human%20Performance.pdf

Needle, D. (2008, March 31). Multitasking: Getting things half-done? *Internetnews.com*. Retrieved January 11, 2009, from http://www.internetnews.com/bus-news/article.php/3737601

Nemertes. (2007, November). *The Internet singularity delayed: Why limits in Internet capacity will stifle innovation on the Web*. Retrieved January 11, 2009, from http://www.nemertes.com/internet_singularity_delayed_why_limits_internet_capacity_will_stifle_innovation_web

Netcraft. (2008, June). *Web server survey*. Retrieved January 11, 2009, from http://news.netcraft.com/archives/web_server_survey.html

Nikander, P. (2005–2006, Winter). Reflections on architecture. *IETF Journal*, *1*(2). Retrieved January 11, 2009, from http://www.isoc.org/tools/blogs/ietfjournal/?p=39#more-39

Nordland, R. (2008, February 29). How grim was my valley. *Newsweek*. Retrieved January 11, 2009, from http://www.newsweek.com/id/117749/page/1

Organisation for Economic Cooperation and Development. (n.d.). *Broadband portal*. Retrieved January 11, 2009, from http://www.oecd.org/document/54/0,3343,en_2649_33703_38690102_1_1_1_1,00.html

Organisation for Economic Cooperation and Development Committee for Information, Computer, and Communications Policy. (2007a, April 6). *Internet traffic prioritization: An overview*. Retrieved January 11, 2009, from http://www.oecd.org/dataoecd/43/63/38405781.pdf

Organisation for Economic Cooperation and Development Committee for Information, Computer, and Communications Policy. (2007b, April 12). *Participative Web: User-created content*. Retrieved January 11, 2009, from http://www.oecd.org/dataoecd/57/14/38393115.pdf

Organisation for Economic Cooperation and Development Committee for Information, Computer, and Communications Policy. (2008, March 28). *Shaping policies for the future of the Internet economy*. Retrieved January 11, 2009, from http://www.oecd.org/dataoecd/1/29/40821707.pdf

Pew Global Attitudes Project. (2007, October 4). *World publics welcome global trade, but not immigration.* Pew Research Center. Retrieved January 11, 2009, from http://pewglobal.org/reports/pdf/258.pdf

Pool, I. d. S. (1983). *Forecasting the telephone: A retrospective technology assessment*. Norwood, NJ: Ablex Publishing.

Privacy International. (2007, September). *A race to the bottom: Privacy ranking of Internet service companies*. Retrieved January 11, 2009, from http://www.privacyinternational.org/article.shtml?cmd%5B347%5D=x-347-553961

Pudding Media. (2008, June). *Pudding Media empowers advertising on any phone.* Retrieved January 11, 2009, from http://puddingmedia.com/

Reed, D. P. (2001, February). The law of the pack. *Harvard Business Review*, 23–24. Retrieved January 11, 2009, from http://harvardbusinessonline.hbsp.harvard.edu/hbsp/hbr/articles/article.jsp?value=BR0102&ml_subscriber=true&ml_action=get-article&ml_issueid=BR0102&articleID=F0102C&pageNumber=1

Reed, D. P. (2007). "Group-forming networks resource page." *Reed's Locus.* Retrieved May 5, 2009, from http://www.reed.com/dpr/

Reeves, B., Malone, T., & O'Driscoll, T. (2008, May). Leadership's online labs. *Harvard Business Review*. Retrieved January 11, 2009, from http://harvardbusinessonline.hbsp.harvard.edu/hbsp/hbr/articles/article.jsp?value=BR0805&ml_subscriber=true&ml_action=get-article&ml_issueid=BR0805&articleID=R0805C&pageNumber=1

Rensselaer Polytechnic Institute. (2008, June 11). *Tetherless world research constellation.* Retrieved January 11, 2009, from http://tw.rpi.edu/launch/

Roberts, K. (2007, June 22). U.S. general laments Google Earth capability. *Reuters*. Retrieved January 11, 2009, from http://www.reuters.com/article/technologyNews/idUSN2140694720070622

Sarasohn-Kahn, J. (2008, April). *The wisdom of patients*. California Health Care Foundation. Retrieved January 11, 2009, from http://www.chcf.org/topics/chronicdisease/index.cfm?itemID=133631

Shirky, C. (2003, November 7). *The Semantic Web, syllogism, and worldview.* Networks, Economics, and Culture mailing list. Retrieved January 11, 2009, from http://www.shirky.com/writings/semantic_syllogism.html

Smart, J., Cascio, J., & Paffendorf, J. (2007, August 17). *Metaverse roadmap overview: Pathways to the 3-D Web*. Acceleration Studies Foundation. Retrieved January 11, 2009, from http://www.metaverseroadmap.org/MetaverseRoadmapOverview.pdf

Solove, D. J. (2007, December). *The future of reputation: Gossip, rumor, and privacy on the Internet*. New Haven, CT: Yale University Press. Retrieved January 11, 2009, from http://books.google.com/books?id=42nJLWB0ztYC&pg=PA11&dq=Solove+%22Future+of+Reputation%22+%22scarlet+letter+in+digital+form

Solutions Research Group. (2007, February 15). *BlackBerry backlash? Americans split on "always-on" culture: Digital life America survey.* Retrieved January 11, 2009, from http://www.srgnet.com/pdf/BlackBerry%20Backlash%20(SRG%20%20Report%20Feb%2007).pdf

Stone, L. (2008, June 24). *Thoughts on attention and, specifically, continuous partial attention.* Retrieved January 11, 2009, from http://continuouspartialattention.jot.com/WikiHome

UNESCO. (2008). *ICT-enhanced public service broadcasting.* Retrieved January 11, 2009, from http://www.unesco-ci.org/cgi-bin/media/page.cgi?g=Media_Development%2FICT-enhanced_Public_Service_Broadcasting%2Findex.html;d=1

United Nations Partnership on Measuring ICT for Development. (2005, November). *Core ICT Indicators*. Retrieved January 11, 2009, from http://www.escwa.un.org/divisions/div_editor/Download.asp?table_name=other&field_name=ID&FileID=846

U.S. Copyright Office Summary. (1998, December). *The Digital Millennium Copyright Act of 1998.* Retrieved January 11, 2009, from http://www.copyright.gov/legislation/dmca.pdf

U.S. Government Accountability Office. (2008, May). *Privacy: Agencies should ensure that designated senior officials have oversight of key functions* (Report). Retrieved January 11, 2009, from http://www.gao.gov/new.items/d08603.pdf

U.S. Homeland Security. (2008, June 18). *Testimony of chief privacy officer Hugo Teufel III before the Senate Committee on Homeland Security and Governmental Affairs: Protecting personal information: Is the federal government doing enough?* Retrieved January 11, 2009, from http://www.dhs.gov/xnews/testimony/testimony_1215461682919.shtm

U.S. House of Representatives. (2007, December 5). *Resolution 4279.* Retrieved January 11, 2009, from http://thomas.loc.gov/home/gpoxmlc110/h4279_ih.xml

U.S. National Intelligence Council. (2004, December). *Mapping the global future: A report of the 2020 project.* Retrieved January 11, 2009, from http://www.foia.cia.gov/2020/2020.pdf

U.S. National Science Foundation. (2005, December). *National Science Board 2020 vision*. Retrieved January 11, 2009, from http://www.itu.int/wsis/goldenbook/Publication/GB-final.pdf

Video games blamed for reading problems. (2007, November 28). *Metro.* Retrieved January 11, 2009, from http://www.metro.co.uk/news/article.html?in_article_id=77466&in_page_id=34

Waters, D. (2008, April 30). Web in infancy, says Berners-Lee. *BBC Online*. Retrieved January 11, 2009, from http://news.bbc.co.uk/2/hi/technology/7371660.stm

Waverman, L., Meschi, M., & Fuss, M. (2005, March 12). The impact of telecoms on economic growth in developing countries. *The Economist*. Retrieved January 11, 2009, from http://www.london.edu/assets/documents/PDF/L_Waverman_Telecoms_Growth_in_Dev_Countries.pdf

World Bank ICT Policy Division Global Information and Communications Department. (2008, June 15). *The role of mobile phones in sustainable rural poverty reduction*. Retrieved January 11, 2009, from http://siteresources.worldbank.org/EXTINFORMATIONAND COMMUNICATIONANDTECHNOLOGIES/Resources/The_Role_of_Mobile_Phones_in_Sustainable_Rural_Poverty_Reduction_June_2008.pdf

World Wide Web Consortium. *Semantic Web site*. Retrieved January 11, 2009, from http://www.w3.org/2001/sw/

Wulfhorst, E. (2008, June 25). BlackBerries, blogs create overtime work disputes. *Reuters*. Retrieved January 11, 2009, from http://www.usatoday.com/tech/products/2008-06-25-blackberry-blogs-overtime-pay_N.htm?csp=tech

Yahoo HotJobs. (2008, April). *Wireless working: Rudeness made easy?* Retrieved January 11, 2009, from http://hotjobs.yahoo.com/career-articles-wireless_working_rudeness_made_easy-378

Zeldes, N. (2008, June 14). "Quiet time" and "No e-mail day" pilot data. *IT@Intel*. Retrieved January 11, 2009, from http://communities.intel.com/openport/blogs/it/2008/06/14/-quiet-time-and-no-email-day-pilot-data-is-in

Zittrain, J. (2008, April 14). *The future of the Internet—And how to stop it*. New Haven, CT: Yale University Press. Retrieved January 11, 2009, from http://futureoftheinternet.org/

Zogby. (2007, October 24). *Hello, Internet, I'm home: Zogby/463 Internet attitudes poll*. Retrieved January 11, 2009, from http://www.zogby.com/news/ReadNews.dbm?ID=1378

Contributor Index

Author Index

SUBJECT INDEX

Printed in the United States
151948LV00003BA/11/P

9 781604 976151